Biology of Foraminifera

BIOLOGY OF FORAMINIFERA

edited by

Professor John J. Lee
City University of New York, USA

and

Professor O. Roger Anderson
Columbia University, New York, USA

ACADEMIC PRESS
Harcourt Brace Jovanovich
London San Diego New York Boston
Sydney Tokyo Toronto

ACADEMIC PRESS LIMITED
24–28 Oval Road
London NW1 7DX

United States Edition published by
ACADEMIC PRESS INC.
San Diego, CA 92101

Copyright © 1991 by
ACADEMIC PRESS LIMITED

All Rights Reserved

No part of this book may be reproduced in any form by photostat, microfilm, or any other means, without written permission from the publishers.

ISBN 0–12–440670–X

Typeset by P&R Typesetters Ltd, Salisbury, Wilts, UK
and printed in Great Britain by The University Press, Cambridge

Contents

CONTRIBUTORS vii

PREFACE ix

1. INTRODUCTION 1
   John J. Lee and O. Roger Anderson

2. CYTOLOGY AND FINE STRUCTURE 7
   O. Roger Anderson and John J. Lee

3. HYPOTHESES ON FORM AND FUNCTION IN
   FORAMINIFERA 41
   Pamela Hallock, Rudolf Röttger and Karen Wetmore

4. MECHANISMS FOR CALCIFICATION AND CARBON
   CYCLING IN ALGAL SYMBIONT-BEARING
   FORAMINIFERA 73
   Benno ter Kuile

5. THE MOTILITY OF FORAMINIFERA 91
   Jeffery L. Travis and Samuel S. Bowser

6. SYMBIOSIS IN FORAMINIFERA 157
   John J. Lee and O. Roger Anderson

7. ECOLOGY AND DISTRIBUTION OF BENTHIC
   FORAMINIFERA 221
   John W. Murray

8. ECOLOGY AND DISTRIBUTION OF PLANKTONIC
   FORAMINIFERA 255
   John W. Murray

9. LIFE CYCLES OF FORAMINIFERA 285
    John J. Lee, Walter W. Faber, Jr., O. Roger Anderson
    and Jan Pawlowski

10. COLLECTION, MAINTENANCE AND CULTURE
    METHODS FOR THE STUDY OF LIVING
    FORAMINIFERA 335
    O. Roger Anderson, John J. Lee and Walter W. Faber Jr.

    INDEX 359

List of contributors

O. Roger Anderson, Biological Oceanography, Lamont-Doherty Geological Observatory, Columbia University, Palisades, NY 10964 and Natural Sciences, Columbia University Teachers College, New York, NY 10027, USA.

Samuel S. Bowser, Wadsworth Center for Laboratories and Research, Empire State Plaza, PO Box 509, Albany, New York 12202-0509, USA, and School of Public Health, State University of New York, Albany, New York 12222, USA.

Walter W. Faber Jr., Department of Biology, City College, City University of New York, NY 10031 and Department of Invertebrates, American Museum of Natural History, New York, NY 10024, USA.

Pamela Hallock, Department of Marine Science, University of South Florida, St. Petersburg, FL 33701, USA.

Benno ter Kuile, The Interuniversity Institute of Eilat, H. Steinitz Marine Biology Laboratory, PO Box 469, Eilat 88103, Israel. *Present Address*: International Institute for Cellular and Molecular Pathology, Research Unit for Tropical Diseases, Avenue Hippocrate 74.39, B-1200 Brussels, Belgium.

John J. Lee, Department of Biology, The City College, The City University of New York, New York, NY 10031 and Department of Invertebrates, American Museum of Natural History, New York, NY 10024, USA.

John W. Murray, Department of Geology, University of Southampton, Highfield, Southampton, Hants., SO9 5NH, UK.

Jan Pawlowski, Departement de Zoologie et de Biologie Animale, Universite de Geneva, 13 rue des Maraichers, 1211 Geneva, Switzerland.

Rudolf Röttger, Institut für Allgemeine Mikrobiologie, Universität Kiel, Biologiezentrum, Olhausenstr. 40, 23 Kiel, Federal Republic of Germany.

Jeffrey L. Travis, Department of Biological Sciences, State University of New York, Albany, New York 12222, USA.

Karen Wetmore, Department of Paleobiology, U.S. National Museum of Natural History, Washington, D.C. 20560, USA.

Preface

This is a continuation of a series initially edited by Dr S.R.H. Hedley and C.G. Adams (British Museum of Natural History) embracing a multidisciplinary perspective on all aspects of the science of living and fossil foraminifera. The foraminifera, among other microfossil-producing protista, are of increasing importance to a wide audience of scholars in the biological and earth sciences as we come to appreciate ever more clearly the importance of assessing the history of the earth, monitoring ecosystems to better understand their natural order, and predicting the impact of human activity on the dynamics of our natural environment. The shells of benthic and planktonic foraminifera contribute to marine sedimentary records and are a rich source of information for scholars investigating a wide spectrum of topics in biological evolution, historical geology, biostratigraphy, ancient and modern marine ecosystems, marine microbial ecology, and origins of host–algal symbioses. The complexity of the architecture of their shells, elaborate activity of cytoplasmic flow in their reticulopodia, and complex symbiotic associations with diverse algal groups such as diatoms, red algae, chrysophytes and dinoflagellates suggest that foraminifera are a potentially rich source of information for cell and molecular biological research in addition to classical biological and geological investigations.

In this volume, we have compiled a series of contributions on some of the latest developments in biological research on these remarkable "primitive organisms" with very sophisticated life habits. The contributors are all recognized scholars in their fields of research, and we are very pleased that they have agreed to share their perspectives in this unique forum, and trust that their contributions will provide new insights into possible future paths and new directions in foraminiferal research.

J. J. Lee
O. Roger Anderson

1
Introduction

JOHN J. LEE and O. ROGER ANDERSON

Over a decade has passed since the last comprehensive review of foraminifera in this series edited previously by Drs R. H. Hedley and C. G. Adams (1978). It has been over a quarter of a century since the last critical comprehensive review of the biology of the foraminifera (Hedley, 1964). Sufficient progress has been made in nearly every aspect of the biology of these remarkable marine protists to warrant an updated sequel. It is becoming increasingly clear that our knowledge of the fundamental biological processes of these complex single-celled organisms is only beginning to unfold and much remains to be done. The basic fine structural features of benthic and planktonic species have been substantially investigated in recent decades, and we have a fairly comprehensive understanding of the organization of the cytoplasm, nuclei, and ontogenetic origin and architecture of the test. Some current knowledge is presented in Chapters 1 and 2. Both fine structural and physiological principles of the host–symbiont association have been studied extensively, more particularly in benthic species, and the diversity of algal symbionts, their compartmentalization within the host cytoplasm, and the dynamics of interaction between the two have begun to be elucidated as reported in Chapter 6.

Major research questions of cellular physiology have only recently been addressed, including more detailed investigations on the biophysical and fine structural basis of cytoplasmic streaming, origin and regulatory mechanisms of reproductive swarmer production including proliferation of swarmer nuclei in the parent cell, principles regulating host–symbiont recognition and specificity of association, isotopic tracer studies of shell biomineralizing processes, and the biology of isotopic fractionation by host and symbionts during uptake of inorganics. The major competing hypotheses about rhizopodial streaming include (1) the fountain flow model (Mast, 1925; Jahn and Rinaldi, 1959) based on the assumption that pressure from behind drives cytoplasm forward, and (2) the frontal contraction model (Allen, 1961) based on the concept that gelation of cytoplasm at leading edges of the pseudopod draws the cytoplasm forward. These models have been intensely debated on theoretical grounds and more recently have been critically tested based on high-resolution microscopic evidence. Increasing evidence points towards the role of pressure flow (Grebecki, 1979, 1986) as a major explanatory principle for cytoplasmic streaming in amoebae. In foraminifera, bidirectional streaming of rhizopodial cytoplasm parallel to the long axis of the extended rhizopodia appears to be mediated by microtubules, delicate, tubular, cytoskeletal elements providing mechanical support and cytoplasmic flow along their surfaces (Travis et al., 1983; Bowser et al., 1984; Bowser and Rieder, 1985; Weiss et al., 1987). Some of the current fine structural evidence for rhizopodial organization and activity in foraminifera is presented in Chapter 5 based on high-resolution transmission electron microscopy.

The fundamental ecology of foraminifera (e.g. Chapters 6 and 7) including categories of prey, habitat, tolerance of environmental variables (temperature, salinity, light, etc.) has been substantially improved through combined observational studies in the natural environment and experimental studies using laboratory cultures (Bé et al., 1977; Anderson, 1984). Benthic species have been examined more extensively, possibly due to their relative ease of culture in the laboratory (Lee et al., 1970; Lee, 1974, 1980). However, current progress towards maintaining planktonic foraminifera in short-term cultures (growing small individuals collected from open ocean in glass vials immersed in environmentally controlled, illuminated water baths) has provided increasing evidence about the dynamics of organismic–environmental interactions (e.g. Hemleben et al., 1989). Transmission electron microscopic investigation of food vacuolar contents, and cytochemical localization of digestive enzymes in benthic and planktonic species, has begun to elucidate the kind of prey consumed and the location of major digestive activity (Anderson and Bé, 1976; Bé et al., 1977; Anderson et al., 1979; Lee et al., 1991).

The goal of these volumes is to bring together perspectives on the current biological knowledge of benthic and planktonic foraminifera, useful particularly to workers in the biological and earth sciences, contributed by a broad range of expert scholars. Our intent, where possible, has been to create a synthesis of knowledge encompassing benthic and planktonic species. This is intended to complement specialized treatises (e.g. planktonic foraminifera: Hemleben et al., 1989). Although there are clear differences in form and function among benthic and planktonic species, we believe that a concerted effort to understand the common places as well as differentiating characteristics of the two groups can provide new insights into the phylogenetic histories of the two, and perhaps to inspire additional comparative physiological and biomolecular research using modern cell biological techniques. Foraminifera provide excellent material for investigations into fundamental cellular processes in single-celled organisms of relatively large size and remarkable diversity. Their widespread occurrence in marine ecosystems, significant role as palaeoecological markers owing to the abundance of fossil tests deposited in marine sediments, complex cellular motility, life-cycles, and architecture of the calcite test indicate the potential significance of these species for continued basic and applied biological research.

As the various experts in our field review our current knowledge on the biology of foraminifera, it has become clearer to us, from our perspective as editors, that we really know very little about most aspects of foraminiferan biology. For example, we know almost nothing about the molecular biology of the group. Within the reasonable future, questions of evolution and systematic relationships within foraminifera and the relationships of foraminifera among the protista will be addressed by molecular genetic techniques. How is cell wall morphology reproduced so predictably? What life functions do various cell structures (e.g. septal canals) serve? Can we make more links between physiology and morphology? If, as K strategists, foraminifera have many synecological aspects to their niches, then there is much to discover about their community contextual relationships. The discovery of many more species of "nanno foraminifera" in the last few years (Pawleski and Lee, in press; Pawleski, 1991), with their unusually brief life-cycles, is certainly a reminder of the fragmentary knowledge of foraminiferal life-cycles in general. With the exception of a few experiments by Grell's students (Weber, 1965), almost nothing is known about sexual differentiation or the genetics of the group. Except for some simple probing by Zech (1964), the molecular genetics of heterokaryosis in the foraminifera has not yet been examined. Much further research is required on the life-cycles of foraminifera to provide background information essential for classical and molecular genetic research. We summarize current knowledge of foraminiferal life-cycles in Chapter 9 and present practical information on their culture and maintenance in laboratory cultures in Chapter 9.

Because our own most recent research has focused on foraminifera with endosymbionts, we can raise many more questions on the topic as reflected in our chapter on endosymbiosis (Chapter 6). We hope that the contributions by the experts assembled here will serve as a benchmark to measure our progress in this exciting field and set a foundation for further modern research to be reported in future volumes of this series.

REFERENCES

Allen, R.D. (1961). A new theory of amoeboid movement and protoplasmic streaming. *Exp. Cell Res.* **8**:17–31 (suppl.).

Anderson, O.R. (1984). Cellular specialization and reproduction in planktonic foraminifera and radiolaria. In *Marine Plankton Life Cycle Strategies* (eds K.A. Steidinger and L.M. Walker), pp. 35–66. Chemical Rubber Co. Press, Boca Raton, Florida.

Anderson, O.R. and Bé, A.W.H. (1976). A cytochemical fine structure study of phagotrophy in a planktonic foraminifer *Hastigerina pelagica* (d'Orbigny). *Biol. Bull.* **151**:437–449.

Anderson, O.R., Spindler, M., Bé, A. and Hemleben, Ch. (1979). Trophic activity of planktonic foraminifera. *J. Mar. Biol. Assoc., U.K.* **59**:791–799.

Bé, A.W.H., Hemleben, Ch., Anderson, O.R., Spindler, M., Hacunda, J. and Tuntivate-Choy, S. (1977). Laboratory and field observations of living planktonic foraminifera. *Micropaleontol.* **23**:155–179.

Bowser, S.S. and Rieder, C.L. (1985). Evidence that cell surface motility *Allogromia* is mediated by cytoplasmic microtubules. *Can. J. Biochem. Cell Biol.* **63**:608–620.

Bowser, S.S., Israel, H.A., McGee-Russell, S.M. and Rieder, C.L. (1984). Surface transport properties of reticulopodia: Do intracellular and extracellular motility share a common mechanism? *Cell Biol. Int. Rep.* **8**:1051–1063.

Grebecki, A. (1979). Organization of motory functions in amoebae and in slime moulds plasmodia. *Acta Protozoologica* **18**:43–58.

Grebecki, A. (1986). Relationship between cytoskeleton and motility. *Insect Sci. Applic.* **7**:379–386 (special issue on Progress in Protozoology).

Hedley, R.H. (1964). The biology of foraminifera. *Int. Rev. Gen. Exp. Zool.* **1**:1–84.

Hedley, R.H. and Adams, C.G. (eds) (1978). *Foraminifera*, Vol. 3. Academic Press, London.

Hemleben, Ch., Spindler, M. and Anderson, O.R. (1989). *Modern Planktonic Foraminifera*. Springer-Verlag, New York.

Jahn, T.L. and Rinaldi, R.A. (1959). Protoplasmic movement in the foraminiferan *Allogromia laticollaris*; and a theory of its mechanism. *Biol. Bull. Mar. Biol. Lab., Woods Hole* **117**:100–118.

Lee. J.J. (1974). Towards understanding the niche of a foraminifera. In *Foraminifera* (eds R.H. Hedley and C.G. Adams), Vol. 1. pp. 201–260. Academic Press, London.

Lee, J.J. (1980). Nutrition and physiology of the foraminifera. In *Biochemistry and Physiology of Protozoa* (eds M. Levandowsky and S.H. Hutner), pp. 43–46. Academic Press, London.

Lee. J.J., Tiejen, J.H., Stone, R.J., Muller, W.A., Rullman, J. and McEnery,

M.E. (1970). The cultivation and physiological ecology of members of a salt marsh epiphytic community. *Helgolander Wiss. Meeres.* **20**:136–156.

Lee. J.J., Faber, W.W., Jr and Lee, R.E. (1991). Granuloreticulopodial digestion – a possible preadaptation to benthic foraminiferal symbiosis? *Symbiosis* **10**:47–61.

Mast. S.O. (1925). Structure, movement, locomotion and stimulation in amoeba. *J. Morph.* **41**:347–425.

Pawleski, J. (in press). Distribution and taxonomy of some benthic tiny foraminifera from the Bermuda rise. *Micropaleontol.* **38**.

Pawleski, J. and Lee, J.J. (in press). Taxonomic notes on some tiny shallow water foraminifera from the north Gulf of Elat (Red Sea). *Micropaleontol.* **38**.

Travis, J.L., Kenealy, J.F.X. and Allen, R.D. (1983). Studies of the foraminifera: 2. The dynamic microtubular cytoskeleton of the reticulopodial network of *Allogromia laticollaris*. *J. Cell Biol.* **97**:1668–1676.

Weber, H. (1965). Über die paarung der Gamonten und den kerndualismus der Foraminifere *Metarotaliella parva* Grell. *Arch. Protistenk.* **108**:217–270.

Weiss, D.G., Langford, G.M. and Allen, R.D. (1987). Implications of microtubules in cytomechanics: Static and motile aspects. In *Cytomechanics: The Mechanical Basis of Cell Form and Structure* (eds J. Bereiter-Hahn, O.R. Anderson and W.-E. Reif), pp. 100–113. Springer-Verlag, Heidelberg.

Zech. L. (1964). Zytochemische Messungen an den Zellkernen der Foraminiferen *Patellina corrugata* und *Rotaliella heterokaryotica*. *Arch. Protistenk.* **107**: 295–330.

2
Cytology and fine structure

O. ROGER ANDERSON and JOHN J. LEE

1.0	The concept of foraminifera	7
2.0	Comparative morphology of benthic and planktonic foraminifera	8
	2.1 Intrashell and extrashell cytoplasm of benthic and planktonic foraminifera	9
	2.2 Major cellular organelles	15
3.0	Algal symbionts	29
4.0	General morphology of mineralized tests	31
5.0	Summary	35
	References	37

1.0 THE CONCEPT OF FORAMINIFERA

Modern five-kingdom systems of biological classification make clear distinctions between the prokaryotes (cyanobacteria, archaebacteria and other bacteria) lacking true nuclei, and eukaryotes (more advanced organisms) with membrane-bound nuclei, microtubules, intracellular membranous organelles and flagella with axonemes (central shaft composed of microtubules). Prokaryotes are assigned to the kingdom Monera. The eukaryotes are placed in four kingdoms: (1) Protista (non-tissue-forming eukaryotes including protozoa and algae), (2) Fungi, (3) Plantae and (4) Animalia.

Foraminifera are included among the Protista and exhibit cytoplasmic organization and pseudopodial streaming characteristic broadly of amoeboid organisms. However, unlike other amoebae, foraminifera protrude a net-like array of pseudopodia with a halo of thin, web-like, granular rhizopodia and pencil-shaped filopodia emerging from the cell body. They are surrounded by an organic test which, in the simplest members, is a single-chambered, imperforated capsule with an aperture through which rhizopodia emerge. Some recent approaches to higher-level taxonomy have included naked forms, within a larger group at the phylum level "Granulareticulopodea" (Lee, 1989). Other species contain mineral grains collected from the environment embedded in the organic test, or calcitic deposits secreted by the cell on the organic substratum of the test. The latter are often perforated by pores and have one or more apertures through which rhizopodia emerge. Test morphology and architectural detail, however, vary extensively among benthic and planktonic species. Moreover, some planktonic species possess spines that increase surface area and provide anchorage for the attached rhizopodia. Non-spinose, planktonic species at maturity often have small peg-like protrusions, called pustules, on the surface of the test.

The complexity of the tests, involving elaborate adaptive structures (e.g. see Chapter 5) to enhance digestive efficiency, compartmentalize cytoplasm and accommodate algal symbionts, suggests these are highly evolved species, even though they are classified as primitive organisms based on a broad phylogenetic perspective. The elaborate organization of the cytoplasm and test, and the diverse habitats of foraminifera, spanning deep-ocean benthos, salt marsh and surface strata of the open ocean, suggest that these unicells have evolved to a level of complexity and adaptive efficiency that is near optimum for a single-celled organism. Without the advantage of tissue organization and specialized organ functions, the foraminifera have none the less produced remarkably advanced cytoplasmic adaptations, mimicking in some cases the specialized functions of organ-forming metazoa. In brief, the foraminifera have probably reached an apex in evolutionary advancement for unicells.

2.0 COMPARATIVE MORPHOLOGY OF BENTHIC AND PLANKTONIC FORAMINIFERA

The major morphological features of living benthic and planktonic foraminifera are shown in Figs 1–6. The test of benthic species varies from simple, single-chambered (monothalamic), non-mineralized forms (Figs 5, 6) to mineralized multi-chambered forms (Figs 1, 2) with a remarkably complex architecture. The rhizopodia emerge from one or more apertures and, though variable in overall organization, are characterized by a flattened halo of web-like, granular pseudopodia attached to the substratum (Fig. 1). Occasional branches project from the surface of the fan-like array into the surrounding water. The organization of the rhizopodial network is dynamic. Cytoplasmic streaming, associated with convergence and divergence of individual strands, produces a fluid and ever-changing pattern of organization

(Figs 7, 8). The large surface area of the rhizopodial array provides an efficient mechanism for gathering prey from the surrounding environment. Rhizopodia are also utilized for locomotion, allowing the foraminifer to move by a gradual creeping movement from one location to another. Many benthic species are so efficient at grazing on algal mats that they leave cleared "feeding trails" as they move across the surfaces of the culture dishes.

Planktonic foraminifera of the non-spinose (Fig. 9) and spinose (Figs 3, 4, 10) forms are floating organisms and exhibit on the whole a body plan that tends towards a spheroidal mass producing a uniform halo of outward radiating rhizopodia (Figs 3, 4). This provides a large surface area to snare prey and to support algal symbionts, when present. Many spinose species contain algal symbionts that are carried by cytoplasmic streaming outwards along the spines where optimal exposure to illumination and perhaps nutrients permit photosynthesis. The tests are multi-chambered and often contain numerous small pores in addition to one or more apertures through which the rhizopodia emerge.

2.1 Intrashell and extrashell cytoplasm of benthic and planktonic foraminifera

Compartmentalization of cytoplasm within well-defined, non-living structures varies considerably among protista. Three major categories can be distinguished (Anderson, 1984):

1. *Diffuse*, lacking well-defined compartmentalization structures, organelles occur diffusely throughout the cytoplasm and no well-defined, non-living barrier separates major specialized regions of the cytoplasm (e.g. amoebae).
2. *Transitional*, exhibiting localized regions of cellular compartmentalization, but the pattern is fluid and intergradations are common; non-living barriers provide only partial compartmentalization of the cytoplasm and they often possess large apertures or other openings permitting continuity in the intergradation of cytoplasmic organization (e.g. testate amoebae, and benthic and planktonic foraminifera).
3. *Zonal*, where distinct regions of the cytoplasm, containing characteristic organelles, are segregated by enclosing non-living barriers (e.g. the finely perforated central capsule of radiolaria clearly segregating intracapsulum from extracapsulum).

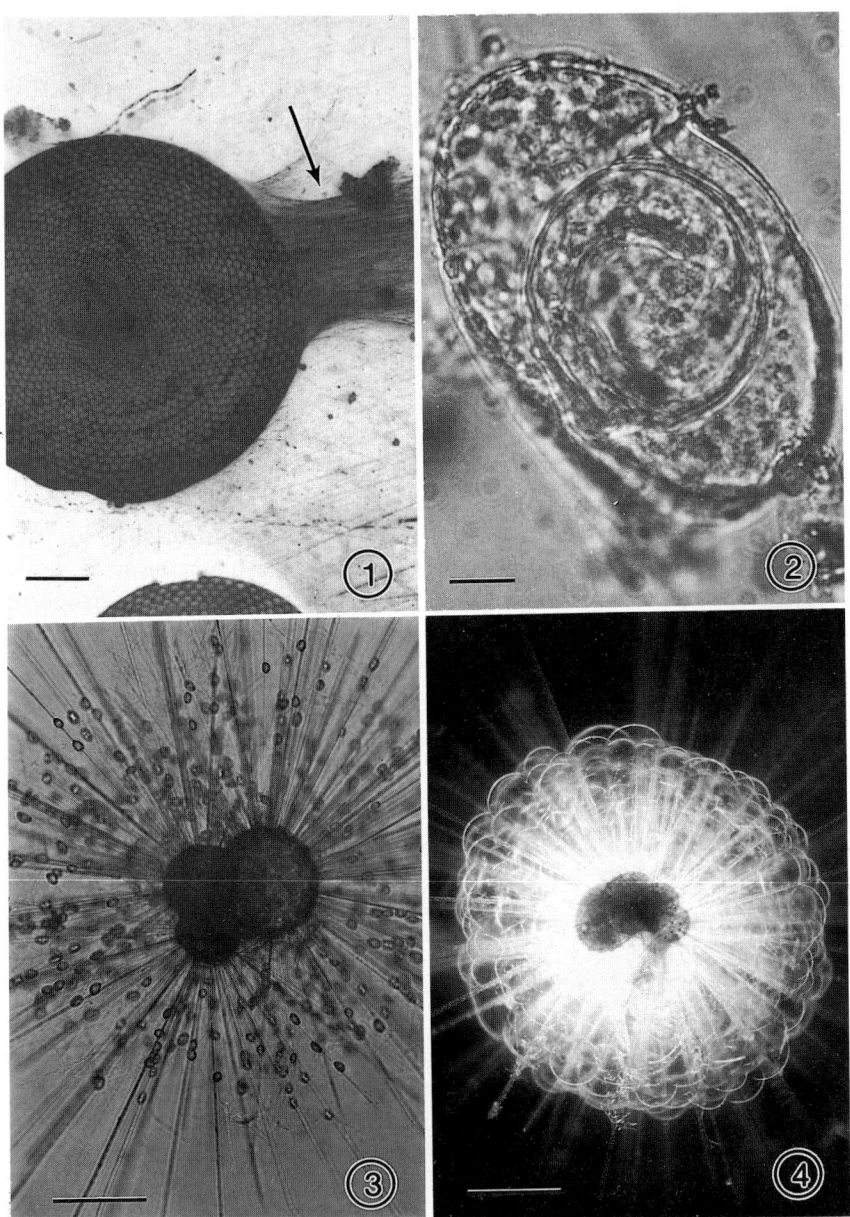

Figs 1–4. Light microscopic views of living benthic foraminifera (Figs 1, 2) and planktonic foraminifera (Figs 3, 4). Fig. 1. *Amphisorus* sp. with a broad mass of rhizopodia (arrow) extending from one side of the shell and attached to the substratum (scale bar = 400 μm). Fig. 2. *Spiroloculina* sp. showing the inner chamber and organization of the cytoplasm within the shell (scale bar = 20 μm). Fig. 3. *Globigerinoides ruber* with numerous dinoflagellate symbionts held in the rhizopodia attached along the spines (scale bar = 200 μm). Fig. 4. *Hastigerina pelagica*, a large species lacking symbionts, and surrounded by a voluminous cytoplasmic bubble capsule that enhances buoyancy (scale bar = 100 μm).

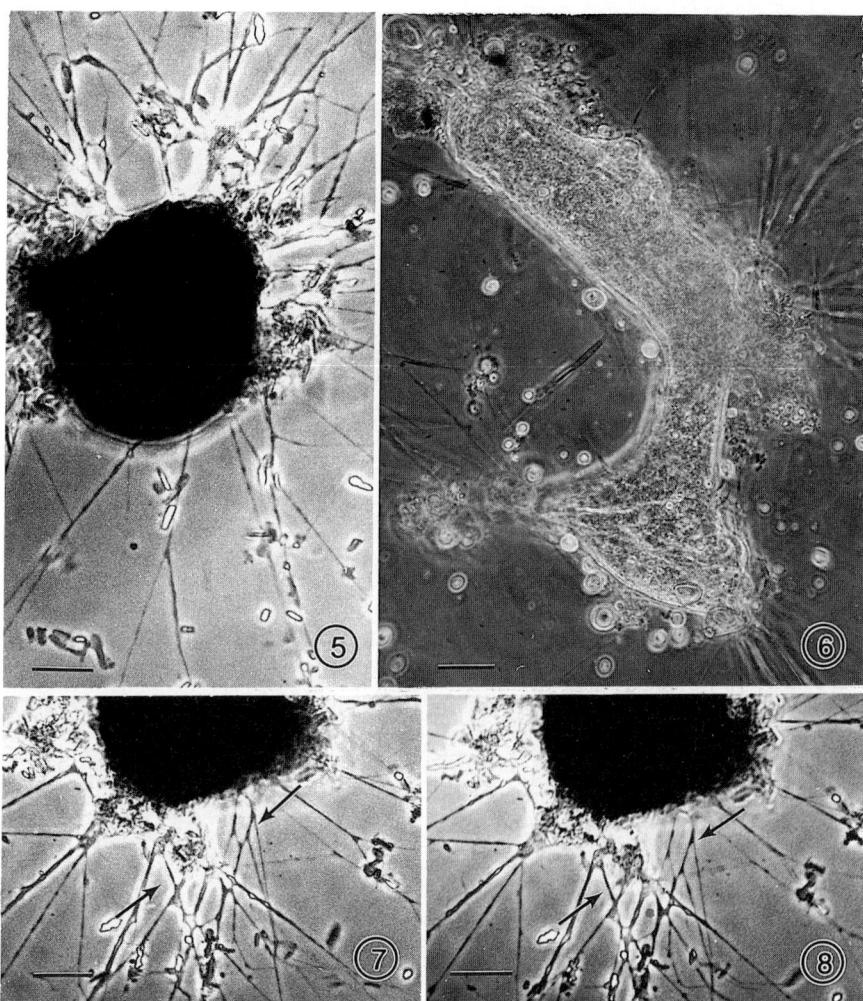

Figs 5, 6. Light microscopic views of living *Allogromia* sp., a benthic foraminifer enclosed by a flexible organic wall. Fig. 5. A globose cell surrounded by a halo of reticulated rhizopodia with attached detrital particles and small prey. Fig. 6. A more mature stage showing the plasmodial form with rhizopodia emerging from several apertures. **Figs 7, 8.** Images of the rhizopodia at successive intervals of time over several minutes showing the subtle changes in rhizopodial configuration (arrows) produced by cytoplasmic streaming. All scale bars = 100 μm.

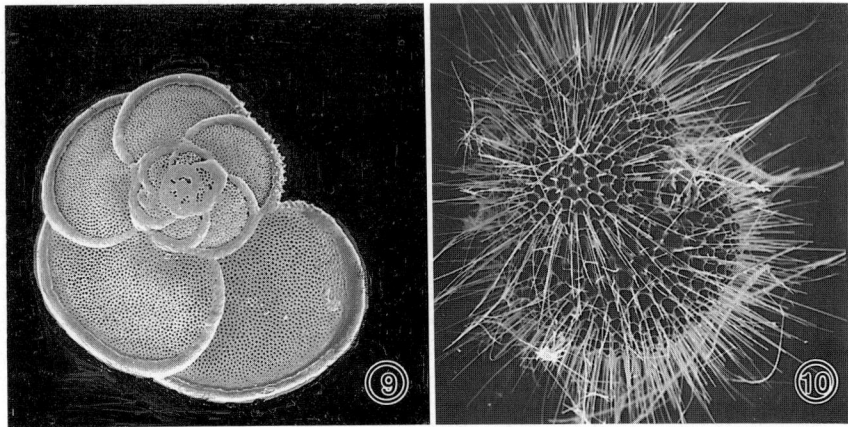

Figs 9, 10. Comparative scanning electron microscopic views of shells from a non-spinose planktonic foraminifer (Fig. 9: *Globorotalia menardii*) and a spinose species (Fig. 10: *Globigerinoides sacculifer*).

The transitional quality of intrashell and extrashell cytoplasm is clearly evident in light and electron microscopic images of sectioned cells of benthic and planktonic foraminifera (Figs 11, 12). The distribution of organelles within intra- and extrashell cytoplasm is continuous through the aperture(s). Cytoplasmic streaming produces a constant exchange of smaller organelles between the internal and external cytoplasm. Some variation in the degree of fluidity of this exchange occurs among species and, on the whole, planktonic foraminifera appear to have less clearly segregated intrashell and extrashell cytoplasmic regions compared to benthic species. For example, the algal symbionts, when present, are moved into and out of the intrashell cytoplasm of planktonic foraminifera and show a marked diel pattern of distribution (symbionts are gathered internally or drawn near to the shell wall at night), whereas some benthic species enclose the symbionts within the intrashell cytoplasm. The variation in the density of intrashell and extrashell cytoplasm, however, is highly variable relative to the physiological state of the organism. Intrashell cytoplasm is usually more compact when the organism is well nourished, but becomes highly vacuolated when poorly nourished. Digestive vacuoles are distributed throughout much of the intrashell and extrashell cytoplasm (Anderson and Bé, 1976a,b; Faber and Lee, in press), but are most commonly observed in the outer chambers of adult

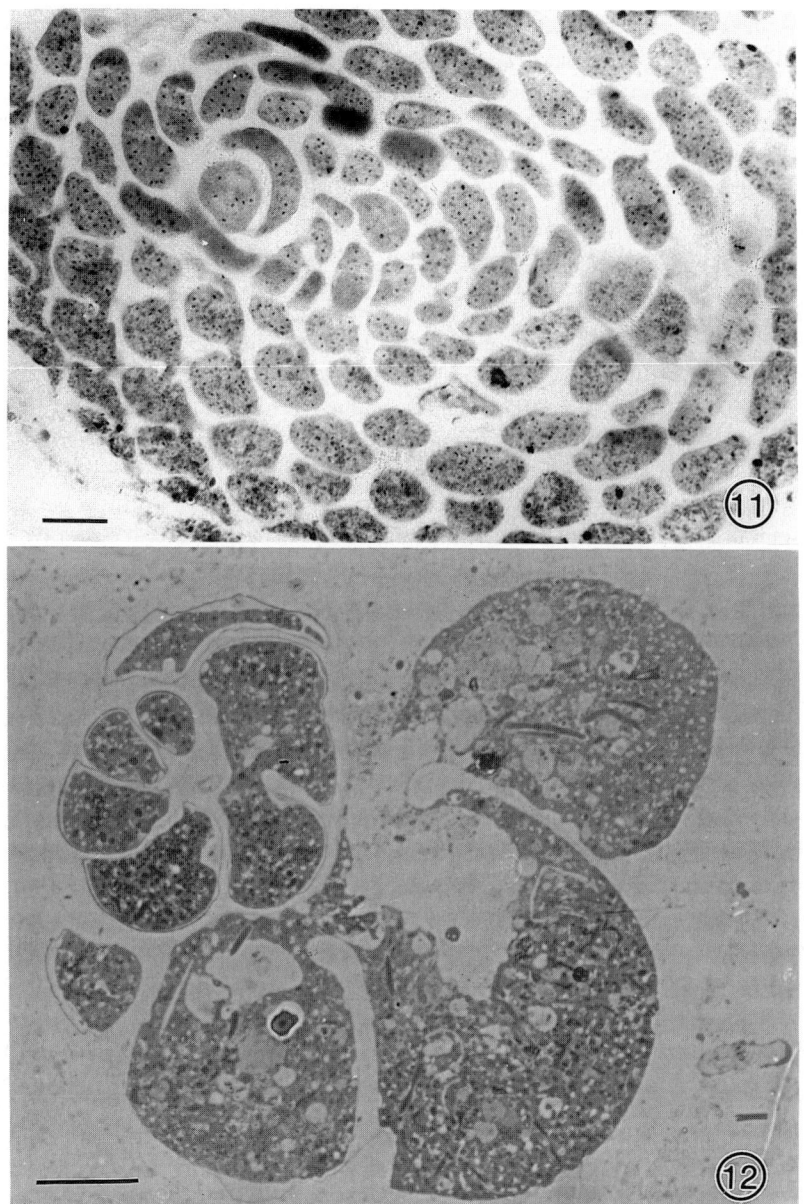

Figs 11, 12. Light micrograph of a thin section exhibiting the cytoplasmic organization within the whorl of chambers in a benthic species (Fig. 11: scale bar = 50 μm) and a planktonic species (Fig. 12: scale bar = 10 μm).

multi-chambered species. Digestion in benthic foraminifera begins in the reticulopodial network, in masses of gathered food, or near the aperture(s) (Lee et al., 1991). Food particles in varying stages of digestion are commonly observed in proximity to the major apertures where cytoplasmic flow permits intake of vacuolar-bound food particles and the expulsion of spent digestive vacuoles (egesta).

Typically, the nucleus is located deep within the intrashell cytoplasm (Fig. 16), and in multi-chambered species is rarely found in the outermost chambers, except in some benthic species with dimorphic (heterokaryotic) nuclei where the larger nuclei occur in the cytoplasm of the outermost whorl of chambers. In some larger foraminifera, the generative nuclei occur in early chambers as in *Sorities marginalis* (Müller-Merz and Lee, 1976), while somatic nuclei are present in the symbiont ring and outer chambers as in *Amphisorus hemprichii* (McEnery and Lee, 1981). Food reserves (e.g. lipid droplets) are usually located throughout the cytoplasm, but are often conspicuous in the innermost chambers of multi-chambered species. With the exception of the ultimate and penultimate chambers, cytoplasm is usually more compact within the test, but varies with the nutritional state of the organism. In well-nourished planktonic foraminifera, the test cytoplasm is uniformly dense and may bulge from the aperture where it becomes more frothy or reticulated and gives rise to the extrashell cytoplasmic coat and numerous rhizopodial strands radiating outward as a halo around the shell. In benthic species, the rhizopodia emerge from the aperture(s) and extend outwards in a fan-like array attached to the substratum. The organization of the rhizopodia changes during cytoplasmic streaming, causing temporal and spatial variations in distribution, and in relation to changes in physiology including nutritional status, reproductive state and general vigour. Some benthic foraminifera establish feeding territories. When the pseudopodia of two organisms contact each other, the organisms retract them and move in other directions. Feeding in many benthic foraminifera is episodic and, after a mass of food is gathered near the aperture, the feeding network is withdrawn to the immediate vicinity of the organism. When planktonic non-spinose species are maintained in culture, individuals become attached to the surface of the dish and exhibit movements similar to benthic species. When, however, two individuals contact one another, their rhizopodia become intricately intertwined and eventually the larger or stronger individual consumes the lesser one. This also occurs when floating spinose or non-spinose species come in contact.

2.2 Major cellular organelles

The many functions of life, typically distributed among tissues and organs in metazoa (multicellular higher organisms), are all present within the single cell of benthic and planktonic foraminifera. Thus, the vital functions of feeding, digestion, metabolism (including respiration and biosynthesis of cellular constituents), secretion, excretion, growth and reproduction are encompassed within the cytoplasm of these single-celled organisms (Figs 13–19). These processes usually occur within membrane-enclosed compartments (organelles) or the surrounding hyaloplasm (typically colourless suspension of organic and inorganic substances composing the intracellular milieu). Some functions such as mineral secretions, more fully explained later, may occur at the external surfaces of membranes and thus are not properly referred to as intracellular processes. Some of the major life processes, and the cytoplasmic structures mediating them, will be described broadly for both benthic and planktonic foraminifera. Distinctive organelles, found only in benthic or planktonic species, will be clearly distinguished by giving their unique attributes, or presented in a comparative perspective between the two groups.

2.2.1 *Nuclei*

As in all eukaryotic cells, the genetic material, composed of long polymers of desoxyribonucleic acid (DNA), is enclosed within the nucleus surrounded by a double membraneous envelope (Figs 13–15). Nuclear shape varies among species and apparently in relation to the physiological state of the cell. Typically, the nucleus is spherical to spheroidal (Figs 13, 16), or lobate (Fig. 14). The DNA within the nucleus is typically in a dispersed state, forming a fine meshwork of DNA fibrils known as euchromatin. However, during cell division, or in some cases during reproduction when gametes are formed, the chromatin becomes condensed (heterochromatin) to form chromosomes. Stained sections of reproducing cells clearly exhibit condensed chromatin in the form of chromosomal bodies (Fig. 17). One or more masses of nucleoprotein (RNA and protein) forming nucleoli (Fig. 14) are commonly observed within the nucleus. Nucleoli may be located near the centre of the nucleus or more peripherally near the nuclear envelope. The nuclear envelope is penetrated by numerous nearly circular pores (Fig. 15). These pores provide continuity between the nucleoplasm and the surrounding cytoplasm. Genetic information is transmitted to the

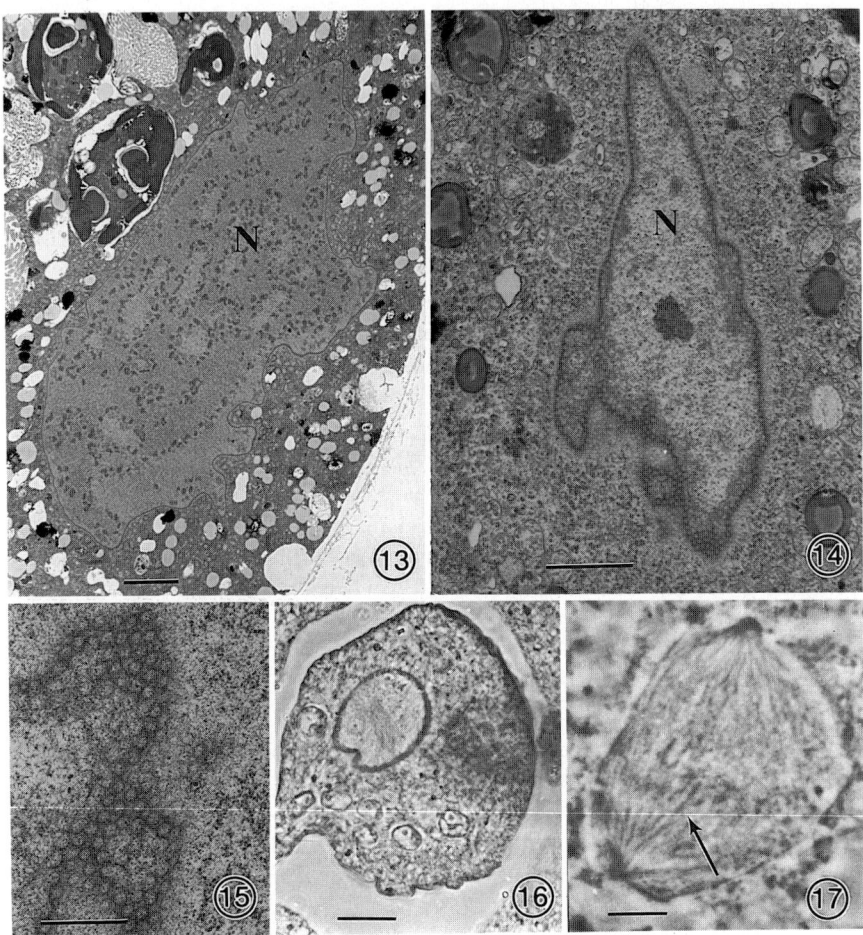

Figs 13–17. Transmission electron microscopic (Figs 13–15) and light microscopic (Figs 16, 17) images of nuclei (N) in planktonic and benthic species respectively. A somewhat globose nucleus (Fig. 13: scale bar = 3 µm) occurs near the periphery of the chamber cytoplasm, while a more lobate form in higher magnification (Fig. 14: scale bar = 1.5 µm) contains an electron-dense nucleolus. A very high magnification of a section tangential to the surface of the nucleus shows the pattern of pores in the nuclear envelope. The pores connect nucleoplasm with the surrounding cytoplasm (Fig. 15: scale bar = 3 µm). Fig. 16. A nucleus within the inner chamber of a benthic species (scale bar = 100 µm). Fig. 17. A dividing nucleus in a benthic foraminifer showing the chromosomes (arrow) and attached spindle fibres radiating from the poles of the dividing nucleus (scale bar = 300 µm).

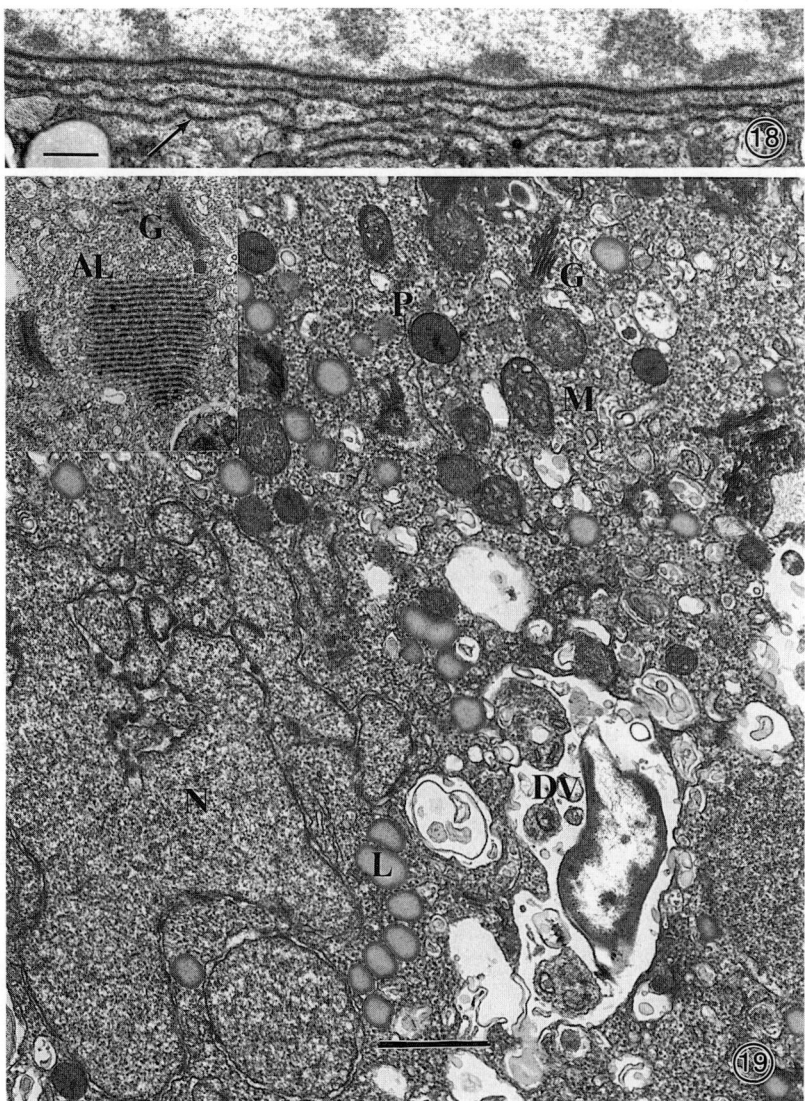

Figs 18, 19. Transmission electron microscopic images of ultrathin sections of the cytoplasm in a planktonic foraminifer showing the organization of the cytoplasm. Fig. 18. A high magnification view of the periphery of a nucleus surrounded by several layers of endoplasmic reticulum (arrow). The endoplasmic reticulum, known as rough endoplasmic reticulum (RER), consists of a network of flattened canals with small nucleoprotein particles (ribosomes) on the cytoplasmic surface (scale bar = 300 nm). Fig. 19. A section showing a lobed nucleus (N), lipid bodies (L) that serve as a food storage product, a nearby digestive vacuole (DV) with enclosed partially digested food, mitochondria (M), peroxisomes (P) with a denser matrix containing an electron opaque inclusion body, Golgi bodies (G) and annulate lamellae (AL). The annulate lamellae develop in the early stages of gametogenesis and may be a source of membranes for the gamete nuclei (scale bar = 2 μm).

cytoplasm by production of messenger RNA molecules (mRNA) that are synthesized on the DNA template and carry a complementary hereditary code in the sequence of ribonucleotides composing the RNA polymer. The code-carrying mRNA molecules diffuse from the nucleus into the cytoplasm through the pores in the nuclear envelope. There is good evidence in other eukaryotic cells that the number of pores and their diameter can vary with physiological state of the cell. During peak metabolic periods or during reproduction when large quantities of information are transmitted to the cytoplasm, the pores are more open and/or more numerous. Clumps of apparently newly synthesized mRNA are often observed on the outer surface of the nuclear envelope.

2.2.2 Ribosomes and endoplasmic reticulum

Messenger RNA (mRNA) in the cytoplasm binds to protein-synthesizing particles known as ribosomes attached to the cytoplasmic surfaces of the rough endoplasmic reticulum (arrow in Fig. 18; RER in Fig. 20). These particles are composed of protein and some intrinsic RNA (ribosomal RNA) and provide the proper physical and chemical environment for enzymatic catalysed synthesis of cellular proteins as directed by the information in the mRNA molecules. The sequence of the ribonucleotide monomers in the mRNA determines the sequence of amino acids polymerized into nascent protein molecules, and thus the hereditary information stored in the nucleus is conserved as protein structure and thence as cytoplasmic form and function. The proteins synthesized by the various mRNA molecules provide the structural and enzymatic components of the cell that directly or indirectly (through control of other chemical processes) determine the structure and function of the cell components. Ribosomes also occur as clusters (polyribosomes) dispersed in the cytoplasm, occasionally associated with mRNA on the outer surface of the nuclear envelope, or more typically bound to the intracellular network of membranous canals known as endoplasmic reticulum (Figs 18, 20). the presence of ribosomes on the cytoplasmic side of the endoplasmic reticular network confers a granular appearance, and hence it is designated rough endoplasmic reticulum (RER). Segments of endoplasmic reticulum lacking ribosomes are designated smooth endoplasmic reticulum (SER). Among other functions, the RER is largely the site of new protein synthesis, whereas the SER is involved in lipid synthesis and other biosynthetic activities. The nearly hyaline aqueous suspension of soluble organic and inorganic substances surrounding the endoplasmic reticulum and other cytoplasmic organelles is known as the hyaloplasm. Although it seldom

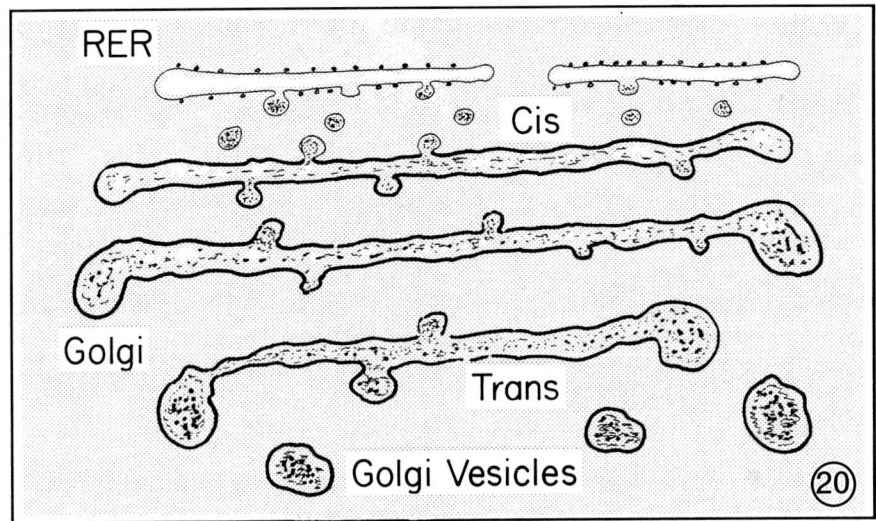

Fig. 20. A line diagram of the Golgi apparatus showing its relationship to the rough endoplasmic reticulum (RER) where protein is synthesized. Ribosomal particles on the cytoplasmic surface of the RER produce protein polymers that are collected in the lumen of the canals of the RER. Membranous vesicles, containing protein products, bud from the surface of the RER and migrate to the flattened saccules of the Golgi that are nearest to the RER known as the *cis* face of the Golgi array. The vesicles fuse with the Golgi saccules and empty the proteins into the cisternae of the Golgi saccule. As the proteins are passed from one Golgi saccule to the next, they are altered, often by addition of carbohydrate chains to form mucopolysaccharides. At the *trans* surface of the Golgi, the various products are packaged into Golgi vesicles that bud off into the cytoplasm. Some of these vesicles contain digestive enzymes and are known as primary lysosomes. Others contain a variety of large organic polymers to be secreted by the cell, including adhesive-like substances used to capture prey.

displays much structure when observed with the conventional stains used for electron microscopy, it is rich in enzymes and small macromolecules that are essential components of the many biochemical reactions sustaining life. For example, the initial stages of glucose metabolism, a significant source of energy for the cell, occur in the hyaloplasm, as do certain biosynthetic processes producing essential organic compounds.

2.2.3 Golgi body

Segments of the endoplasmic reticulum are often in close association spatially and functionally with a major secretory organelle known as the Golgi body (Figs 19, 20). This organelle is composed of flattened membranous cisternae or saccules usually arranged in a stacked configuration with membranous tubular extensions at the periphery. This is a site where membrane-enclosed vesicles are released by budding into the cytoplasm (Fig. 20). The exact organization of the Golgi apparatus and its formation are not fully understood, but current evidence suggests that the Golgi saccules originate by fusion of membraneous vesicles produced by the endoplasmic reticulum. A segment of endoplasmic reticulum that extrudes vesicles is often seen near one side of the Golgi body. The surfaces of the Golgi saccule membranes proximal to this endoplasmic reticulum frequently possess similar vesicles that appear to be fusing with the enclosing Golgi membrane. Hence, this side of the Golgi stack of saccules is known as the forming face or *cis* (same-side) face. The flattened saccules apparently continue to bud-off vesicles that fuse with the successive series of saccules in the Golgi stack to transmit substances throughout the stack of cisternae. As the substances are transmitted across the series of saccules, some become concentrated or altered chemically and eventually are transported to the periphery of the saccules where they are packaged into bud-like protrusions of the peripheral membranes and secreted within cytoplasmic vesicles. The surface of the Golgi body opposite the endoplasmic reticulum is known as the maturing face or *trans* (opposite) face. Among the substances concentrated and/or assembled in the Golgi are:

1. The proteinaceous digestive enzymes secreted in Golgi vesicles known as primary lysosomes.
2. Mucoid substances composed of carbohydrates and proteins.
3. In some cells, but not necessarily in all foraminifera, secretory vesicles containing substances to be deposited on the surface of the plasma membrane surrounding the cell.

In many benthic foraminifera, particularly the heterokaryotic larger foraminifera, numerous Golgi stacks are found in the cytoplasm.

2.2.4 *Lysosomes and digestive vacuoles*

The primary lysosomes containing a variety of digestive enzymes with optimal activity in an acid pH ($c.$ pH $= 5.0$) are dispersed throughout the cytoplasm of outer chambers and reticulopodia, and provide a reserve of digestive enzymes to be utilized when food is engulfed. Prey (including phytoplankton, zooplankton and, in some cases, bacteria) are apprehended in the peripheral rhizopodia of the foraminiferan. The rhizopodia form web-like extensions of cytoplasm that exhibit cytoplasmic streaming and carry captured prey into or near the aperture of the test where digestion occurs (Figs 19, 21, 22). Upon apprehension, individual prey cells (e.g. protozoan cells and algal cells) are first attached to the surface of the feeding rhizopodia by an adhesive substance and subsequently engulfed within vacuoles by invagination of the surface of the rhizopodial membrane. The adhesive substance is produced in Golgi vesicles (Figs 20, 26G) and transported to the peripheral cytoplasm where it is released by expulsion in the vicinity of prey. Larger prey may be invaded by the rhizopodia and segments of prey tissue extracted by the flowing and shearing action of the rhizopodia and subsequently engulfed in vacuoles. These vacuoles, containing prey, are known as food vacuoles or phagosomes (Figs 21, 22). At this stage of metabolism, they do not contain digestive enzymes and are typically alkaline pH. Subsequently, usually as the food vacuole is carried by cytoplasmic streaming towards the main body of the cell,

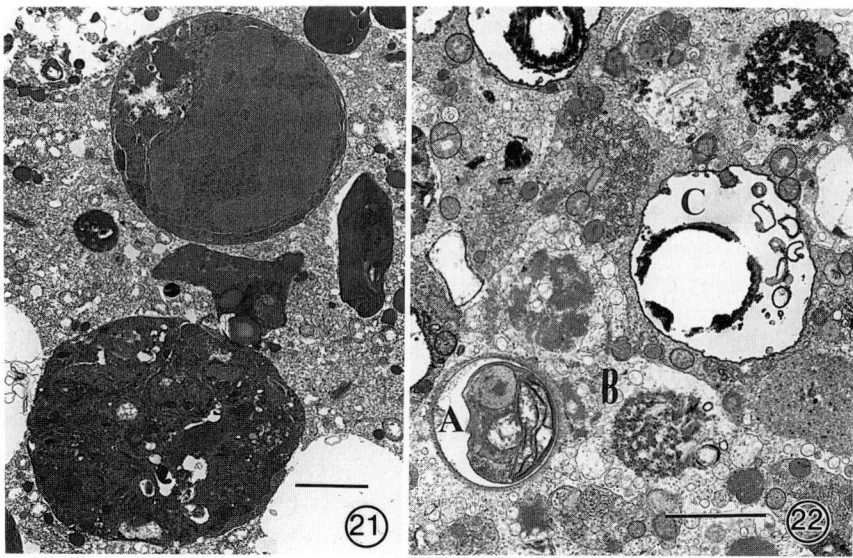

Figs 21, 22. Transmission electron microscopic views of cellular organelles in planktonic foraminifera. Fig. 21. Food vacuoles contain compacted masses of prey tissue from crustacea (scale bar = 2 μm). Fig. 22. A sequence of vacuoles showing a very early food vacuole (A) containing an alga, a later stage when digestion has commenced (B), and a much later stage containing residual non-digested matter (C) (scale bar = 2 μm).

the vacuoles become condensed (smaller and containing less water). The pH is lowered by fusion of the food vacuole with vesicles (acidosomes) containing an acid fluid. Then, the primary lysosome ($c.$ 0.5 μm in diameter) fuses with the food vacuole converting it to a digestive vacuole (Fig. 22). This stage is also known as a secondary lysosome or phagolysosome. In many species of planktonic foraminifera, the digestive vacuoles are carried into the intrashell cytoplasm where the final stages of digestion are completed. In some benthic species, there is increasing evidence that much, if not all, of the digestion occurs in the extrashell cytoplasm. The lysosomes are categorized among those cellular organelles that are surrounded by a single membrane. Their distinctive features are the presence of digestive enzymes or digested food products and a wide variation in size and shape depending in part on the size and form of the prey ingested and the conformation of the surrounding membrane of the vacuoles. In some cases, primary lysosomes appear as electron-dense vacuoles due to the large amount of proteinaceous enzymes.

2.2.5 Microbodies and peroxisomes

Other single-membrane organelles are the peroxisomes or microbodies (Fig. 23). These organelles are about the same size as the primary lysosomes ($c.$ 0.5–1.0 μm in diameter) and contain a lightly granular matrix (internal plasm). In some species of foraminifera, there may be a crystalline inclusion or small cluster of tubular bodies in the centre of the matrix. The crystalline inclusion as observed in other eukaryotic cells is known to be composed of crystallized enzymes. Catalase (among other peroxidases) is an enzyme that transforms hydrogen peroxide (a cellular waste product and toxin) into water while oxidizing other substrates to yield compounds that are metabolically useful. Hence, peroxisomes, among other functions, are significant organelles in the oxidative conversion of waste products to potentially useful metabolic compounds. The chemical composition of the tubular inclusions observed in many planktonic and benthic foraminiferan peroxisomes is not known, although in some planktonic foraminifera, the peroxisomes have been shown to contain peroxidases (Anderson and Tuntivate-Choy, 1984). In general, the peroxisomes serve a wide range of functions depending on the species and physiological state of the cell. These include degradation of nitrogenous waste molecules, conversion of lactic acid and aldehydes into metabolically useful compounds, and the transformation or neutralization of potentially toxic compounds.

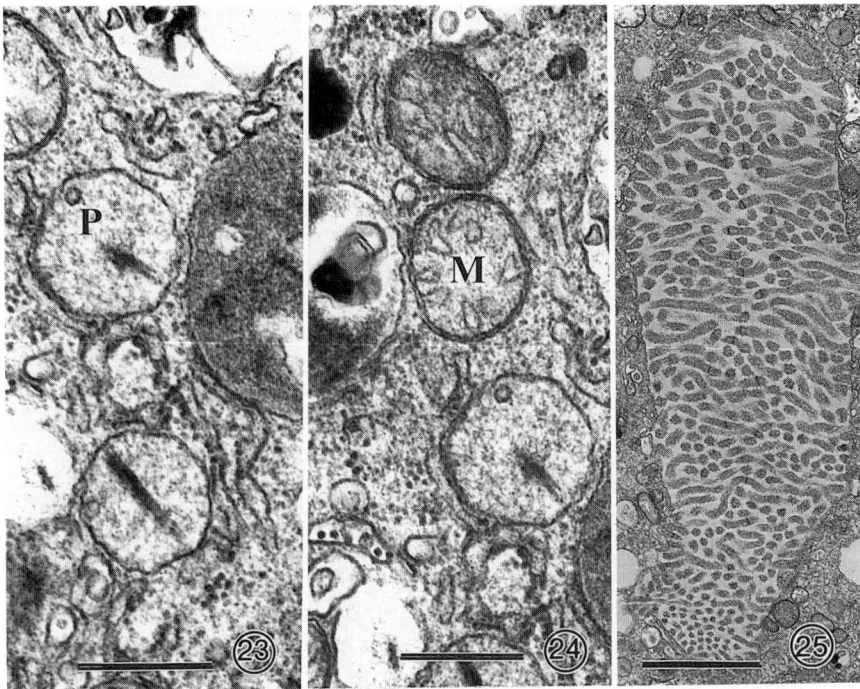

Figs 23–25. High magnification views of some major organelles in planktonic foraminifera. Fig. 23. Peroxisomes (P), surrounded by a single membrane, contain a dense, elongate inclusion body and serve among other functions to degrade toxic compounds (e.g. reducing hydrogen peroxide to water) and convert metabolic products into useful energy-producing molecules. The latter reactions include conversion of alcohol to acetaldehyde and lactic acid to pyruvic acid (scale bar = 1 μm). Fig. 24. Mitochondria (M) are surrounded by a double membrane. The inner membrane forms finger-like invaginations known as cristae and appear frequently as circular bodies in cross-section. Mitochondria are the site of respiration where oxygen is consumed and energy from food molecules is stored in the bonds of ATP molecules (scale bar = 1 μm). Fig. 25. A fibrillar body in *G. sacculifer* is surrounded by a single membrane and contains numerous tubular invaginations (scale bar = 2 μm). See Fig. 26 for a hypothetical explanation of its origin in the cytoplasm based on a reconstruction from transmission electron micrographs.

A specialized group of microbodies, known as glyoxisomes, contain enzymes that mediate the conversion of fats to carbohydrates (e.g. Anderson, 1988, p. 279). The distribution of these organelles among foraminifera needs further investigation. Some of the reactions occurring in microbodies involve oxidation–reduction reactions as also occur in the mitochondria.

2.2.6 Mitchondria

Mitchondria (Figs 19, 24) surrounded by a double membrane are major centres of energy transformation in foraminifera and most eukaryotic cells. A series of oxidation–reduction reactions terminating in the reduction of oxygen to water is accompanied by the generation of ATP molecules. ATP is a fundamental energy storage molecule used by the cell to drive many energy-requiring processes. Mitchondria are absent in prokaryotes and some anaerobic protista, although alternative organelles (e.g. hydrogenosomes) in these species may contain processes involving terminal electron acceptors. The mitochondria of foraminifera exhibit typical protistan fine structure. An outer membrane is smooth and surrounds an inner membrane with finger-like protrusions extending into the central matrix space. These finger-like protrusions, known as cristae, appear elongate and sinusoidal in longitudinal section, and round in cross-section. In some eukaryotic cells, other than foraminifera, the cristae are flattened elongate saccules connected by a narrow neck to the inner membrane, or occur as discoid flattened saccules projecting into the central matrix space. A narrow space separates the outer and inner membranes of mitochondria. The inner membrane is the site of oxidative phosphorylation where ATP is produced. Mitochondria are typically numerous and scattered throughout the intrashell and extrashell cytoplasm. They are relatively small ($c.$ 1.0–2.0 μm) compared to the widely scattered and visually dominant digestive vacuoles that are frequently distributed throughout much of the cytoplasm of a healthy, recently fed foraminiferan. However, most higher magnification images of the foraminiferan cytoplasm contain profiles of the sectioned mitochondria exhibiting their characteristic concentric double-membrane envelope and internal complement of cristae.

Given the constant requirement for energy-yielding compounds, it is not surprising that the mitochondria are widely distributed throughout the cytoplasm. In some planktonic foraminifera, the mitochondria appear to be more densely distributed near the pores of the shell. The distribution of mitochondria in benthic foraminifera has been the focus of at least one fine structural study (Leutenegger and Hansen, 1979). The specimens examined were either collected in box cores from the continental shelf and basin of San Pedro, Southern California, at depths of 29, 44, 750 and 898m, or taken from the aquarium of the Zoological Institute of Zürich University (organisms originally from the Mediterranean Sea). Dissolved oxygen in the various habitats ranged from <0.3 ml O_2/l to 5 ml O_2/l. The density of the mitchondria in the low-oxygen assemblage [*Buliminella tenuata* (small form),

Bolivina argentea and *Loxostomum pseudobeyrichi*] collected at a depth of 898m was generally low. The mitochondrial distribution in the cytoplasm was uneven. They were most abundant close to the perforated walls and formed clusters below the pore terminations. Less pronounced concentrations of mitochondria were found in *Cassidulinoides cornuta* from the same depth and in *Globobulimina pacifica* and *Buliminella tenuata* (large form) from slightly shallower waters (750m). In the species collected from habitats with higher dissolved concentrations of oxygen (*Nonionella stella* and *Bolivina* cf. *subexcavata*), mitochondrial densities were higher and more evenly distributed in the cytoplasm. Some mitochondria were located below the pores (Leutenegger and Hansen, 1979). Although not specifically mentioned, one of the figures in Berthold's (1976b) study of *Patellina corrugata* shows mitochondria grouped beneath a pore. This is probably a physiological adaptation to enhance access to oxygen required for respiration. A similar pattern of a higher density of mitochondria at the periphery of the cell has been observed in ciliates and other protista (e.g. Anderson, 1988, p. 254).

2.2.7 Annulate lamellae and cytoplasmic changes during sexual reproduction

Multilamellar membraneous bodies known as annulate lamellae (AL in Fig. 19) have been observed scattered throughout the cytoplasm of mature individuals of planktonic species (especially in *Hastigerina pelagica*). The number increases immediately before proliferation of nuclei during reproduction, but decreases when the nuclei have fully developed. This suggests that the annulate lamellae provide a source of membranes for the numerous reproductive nuclei (Spindler and Hemleben, 1982).

Cytoplasmic changes during reproduction of benthic and planktonic species have been reviewed in protozoological texts (e.g. Grell, 1973; Anderson, 1988) and extensively in reference works (Anderson and Bé, 1978; Hemleben *et al.*, 1989) and only some recent advances are summarized here. Reproduction in benthic and planktonic foraminifera commences with the proliferation of nuclei, but the time of onset before gamete production can vary substantially among species. Some benthic species produce amoeboid gametes, whereas others release flagellated gametes (Grell, 1973; Goldstein and Barker, 1990). Only flagellated reproductive cells have been observed in planktonic foraminifera. These are produced by the proliferation of smaller nuclei that become segregated into a cytoplasmic network, followed by the development of flagella, and separation of the biflagellated cells into a myriad of swarmers often released by almost explosive force through the aperture.

All of the parent cell cytoplasm is converted into reproductive cells and only the empty test remains after release. In all planktonic species examined thus far, the flagellated reproductive cells of a given species are all of the same size (e.g. Anderson and Bé, 1976a; Hemleben et al., 1989). In *H. pelagica*, however, larger spherical bodies containing a thin layer of cytoplasm surrounding a massive central vacuole are also released. These appear to be residual waste vacuoles that are expelled into the environment during the release of the flagellated reproductive cells. There is no evidence at present that these are macrogametes (Spindler et al., 1978). They are relatively few in number, often contain what appears to be organic debris in the vacuole, and do not disperse very far from the parent test after release. Although gamete fusion has been reported in benthic species and the life-cycles of several species have been carefully documented in laboratory culture, our knowledge of the reproductive cycle in planktonic species is very limited, owing in part to a lack of success in maintaining continuous cultures in the laboratory.

2.2.8 Fibrillar bodies

In all planktonic foraminifera observed thus far, but not in the benthic species, a peculiar ensemble of single-membrane organelles resembling vacuoles or canals filled with a fibrillar substance has been observed in light microscopic and electron microscopic sections (Figs 25, 26). They were described by Rhumbler (1911) in his pioneering studies on foraminifera collected during the Humboldt Expedition. These bodies, designated the fibrillar system or fibrillar bodies (Lee et al., 1965; Zucker, 1973; Anderson and Bé, 1976b), occur widely scattered within the intrashell cytoplasm and sometimes protrude into the aperture and extend into the nearby extrashell cytoplasm. The fibrillar bodies are polymorphic, varying in size and shape, although they tend to be elongate spheroidal bodies. The internal fibrillar masses appear to undergo a predictable developmental sequence commencing as tightly coiled masses of tubular bodies that gradually expand with increasing size of the vacuole (Fig. 26). When fully expanded, the fibrillar substance becomes more puffy in appearance. The organizational pattern of the fibrillar substance, particularly in the early stages, appears to be species-specific. Because fibrillar bodies have never been observed in benthic foraminifera and they are consistently present in planktonic foraminifera, Anderson and Bé (1976b) suggested that they may be flotation organelles. The calcitic shell of planktonic foraminifera is clearly negatively buoyant in sea water and must be

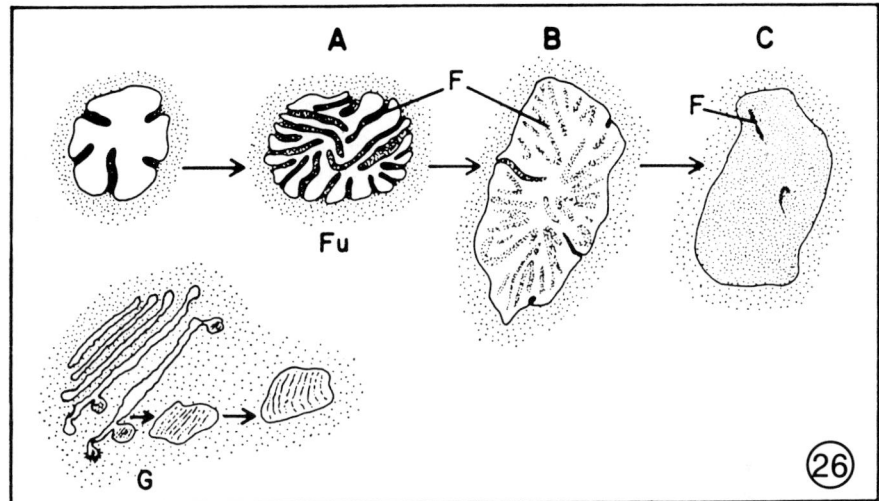

Fig. 26. Two modes of secretory activity in planktonic foraminifera. The first mode, producing fibrillar bodies, is believed to have a buoyancy function. Four phases in the origin of the fibrillar system are shown in the upper half of the diagram. Cytoplasmic invagination into a vacuole produces tubular intrusions, thus giving rise to form A. Form A is transformed into form B (with fibrillar matter), which produces form C (with highly dispersed fibrillar matter). All three phases may occur as a developmental gradient in the same vacuolar space. Each membrane-bound space, containing fibrillar elements (F), is called a "fibrillar unit" (FU). The second secretory mode shown in the lower half of the diagram produces an adhesive substance. The Golgi apparatus (G) secretes small vacuoles containing a finely whorled or lamellated substance commonly seen in the cytoplasm and within the rhizopodia (see also Fig. 20). Reproduced with permission from Anderson and Bé (1976b).

compensated by some mechanism for increasing buoyancy. Hence, large spaces containing substances much lighter than sea water could improve flotation. It has not been determined, however, that the fibrillar bodies contain a substance lighter than sea water. They do not appear to be gas vacuoles. It is possible that in addition to the fibrillar material, an organic solution lighter than sea water such as ammonium chloride could aid flotation. Lighter-than-sea water solutions have been cited as flotation devices in metazoan marine organisms (e.g. Denton and Gilpin-Brown, 1973).

As an alternative, Spero (1988) has suggested that the fibrillar system may be a source of stored calcium ions for calcification. This is based on transmission microscopic observations of ultrathin sections of calcifying stages in *Orbulina universa*, where the fibrillar bodies protruding through the aperture appear to be in close proximity to calcifying surfaces. However, there is no evidence presently that these membraneous spaces are enriched in calcium ions. There is experimental evidence that calcium is stored in the cytoplasm prior to new chamber deposition (Anderson and Faber, 1984). The site of storage has not been determined, though it is likely to be compartmentalized, because Ca^{2+} ions are major signal molecules in many eukaryotic cells and the cytoplasmic concentration is carefully regulated. The absence of fibrillar bodies in benthic species, and their apparently ubiquitous occurrence in planktonic species, suggests that they could at least serve a flotation function. We cannot dismiss the possibility, however, that the fibrillar bodies serve multiple functions; and much more research is needed to clarify the chemical composition of these organelles and their changes during the cell growth cycle.

2.2.9 Cytoskeletal structures

The distribution of organelles within the cytoplasm and the directional flowing of cytoplasm within the shell and among the rhizopodia of the extrashell cytoplasm is mediated by a complex cytoskeletal system consisting of hollow, proteinaceous, rod-like microtubules (c. 30 nm in diameter), composed of helically arranged sub-units. The microtubules (Fig. 27) are part of a cytoskeletal system that provides surfaces for the directional flow of organelles. These scaffold-like structures also provide axial support to cytoplasmic extensions such as pseudopodia. The microtubules are labile structures that are capable of disassembly and reassembly from the constituent tubulin monomers. Hence, the cell shape can be continuously modified, mediated in part by the constant reorganization of the microtubular cytoskeleton. For example, extension of filopodia occurs by elongation of microtubules that provide axial support and directionality during elongation. Occasionally, cytoplasmic particles can be observed attached to a microtubule and apparently moving along its surface. The role of microtubules in cytoplasmic streaming and particle movement has recently been thoroughly reviewed in studies on cytomechanics (e.g. Bereiter-Hahn et al., 1987).

The contractility and motility of cytoplasm is sustained in part by tensile and contractile elements forming fine filamentous networks of microfilaments (c. 5 nm in diameter). Microfilaments are widely distributed throughout the cytoplasm of cells and are particularly abundant as densely interwoven filaments near the plasma membrane of motile surfaces such as pseudopodia in many amoeboid organisms. The contractile quality of microfilaments is attributed to the presence

Cytology and fine structure   29

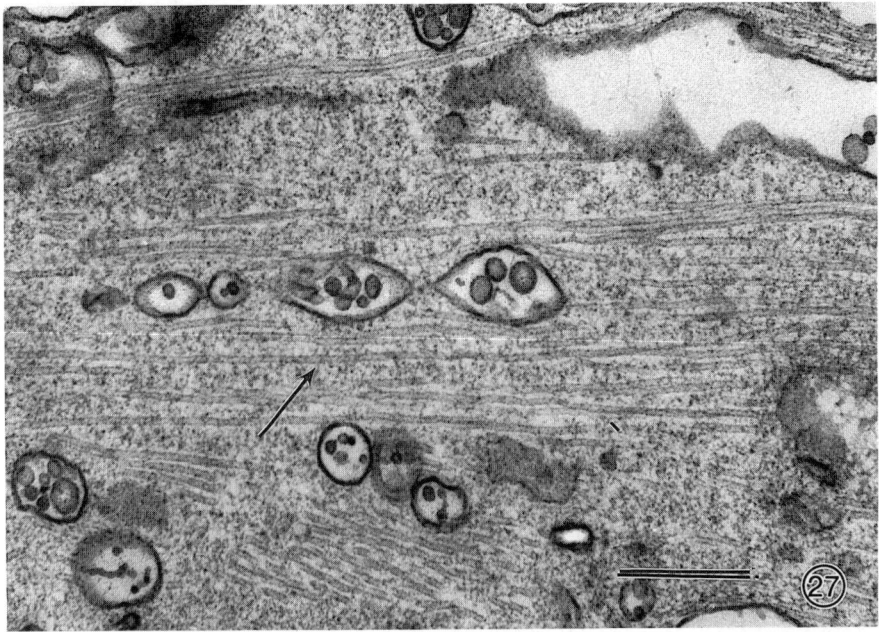

Fig. 27. Transmission electron micrograph of cytoplasmic arrays of microtubules (arrow) that serve as a cytoskeleton and provide surfaces where vesicles attach for transport from one site to another in the cell (scale bar = 400 nm).

of contractile proteins (actin and myosin) chemically similar to the contractile polymers found in the muscle cells of metazoa, but of somewhat different physico-chemical composition. The physiological mechanisms controlling the spatio-temporal patterns of cytoskeletal organizational changes, and organization of contractile and tensile activity, have not been fully elucidated, and therefore we are far from understanding how foraminifera regulate their diverse motile activities (see Chapter 4).

3.0 ALGAL SYMBIONTS

Symbiosis with algal cells is a common feature of many planktonic and some benthic species of foraminifera. The algal symbionts are not present in gametes, and therefore must be selectively gathered from the environment during ontogeny of the foraminifera which have sexually reproduced. During asexual reproduction, symbionts are acquired by individual schizozooites. Algal symbionts are typically

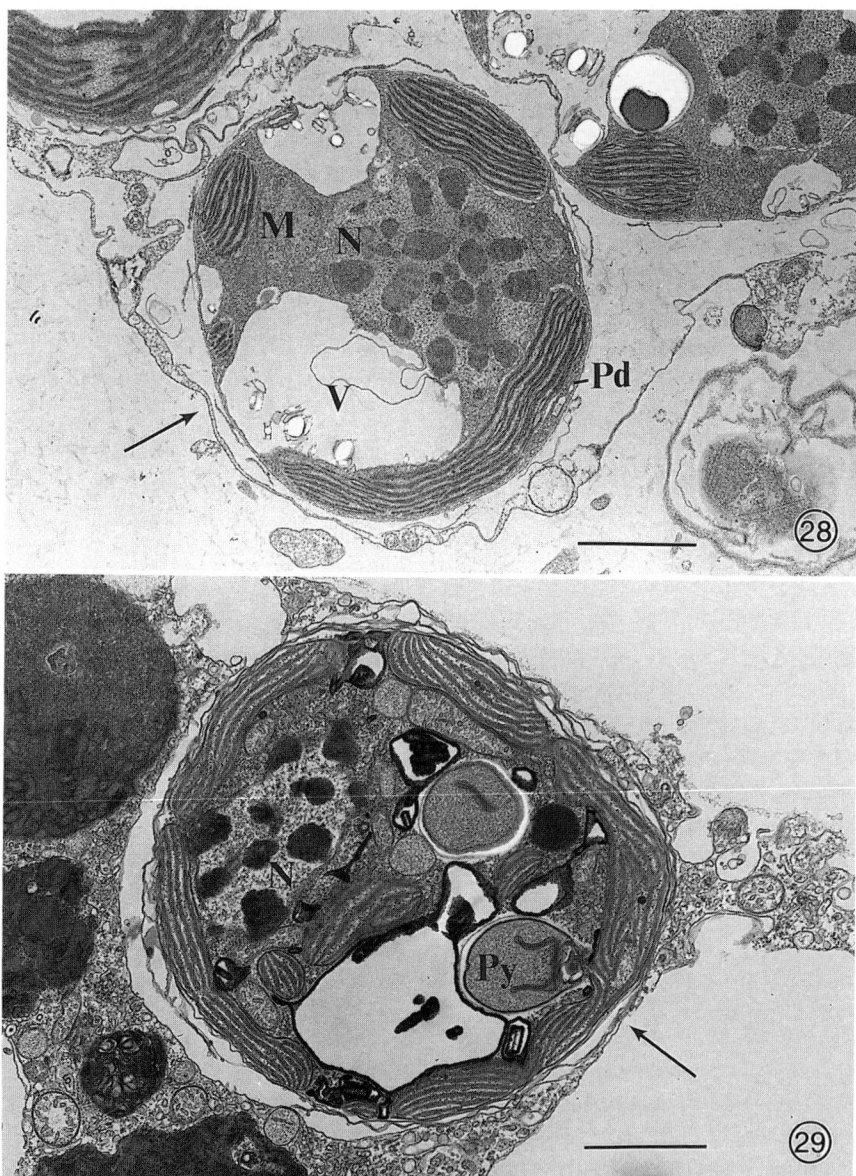

Figs 28, 29. Transmission electron microscopic views of dinoflagellate symbionts contained within rhizopodial sheaths (arrow). The nucleus (N) exhibits typical dinoflagellate puffy chromosomes. Large vacuoles (V) and plastids (Pd) containing chlorophyll, where light energy is trapped during photosynthesis, occur near the periphery. Mitochondria (M) contain tubular cristae. Pyrenoids (Py), where starch is secreted, are attached by a stalk to the surface of the plastid (scale bars = 2 μm).

sequestered within vacuoles bounded by a single membrane in planktonic species (Figs 28, 29), and in benthic species except for the red algal symbionts which are dispersed among the cytoplasmic strands of the intrashell cytoplasm (Lee, 1990; and see Chapter 5). Containment of the algal symbionts within vacuoles provides close control of the exchange of material between symbiont and host and may provide structural separation between the two very diverse genomic species while permitting exchange of beneficial compounds (e.g. Anderson, 1983; Lee and McEnery, 1983). It is not known why the red algae of some benthic species are not always enclosed in vacuoles, but the presence of an organic coat surrounding the alga may provide sufficient separation from the host cytoplasm to eliminate the need for an additional perialgal vacuolar membrane. We have little evidence for the kinds of substances exchanged by symbiont and host, but there is good indirect evidence that the symbionts provide nutrition for the host (Lee, 1980; and see Chapter 5).

4.0 GENERAL MORPHOLOGY OF MINERALIZED TESTS

The chemical composition and ultrastructure of the test varies substantially among foraminiferan species. There are three modes of test construction:

1. Agglutinated tests composed of detrital particles cemented together to form a wall.
2. Tests composed of calcite needles produced within cytoplasmic vesicles and deposited at the surface of the cell.
3. Calcite shells formed by extracellular secretion of calcium carbonate on the surface of an organic anlage or template that determines the shape of the test.

In some benthic species, and in planktonic species, the test is bilamellar, formed by deposition of calcite on the proximal and distal side of an organic membrane (primary organic membrane) that serves as an anlage for the shell shape (Fig. 38). An overview of the fine structural features of shell composition and morphology among some representative types is presented here.

Agglutinated tests of benthic foraminifera consist of accumulated mineral particles cemented together within an organic matrix (Figs 30, 31). In many cases, the particles appear to be random collections of available material in the environment, but some discrimination of

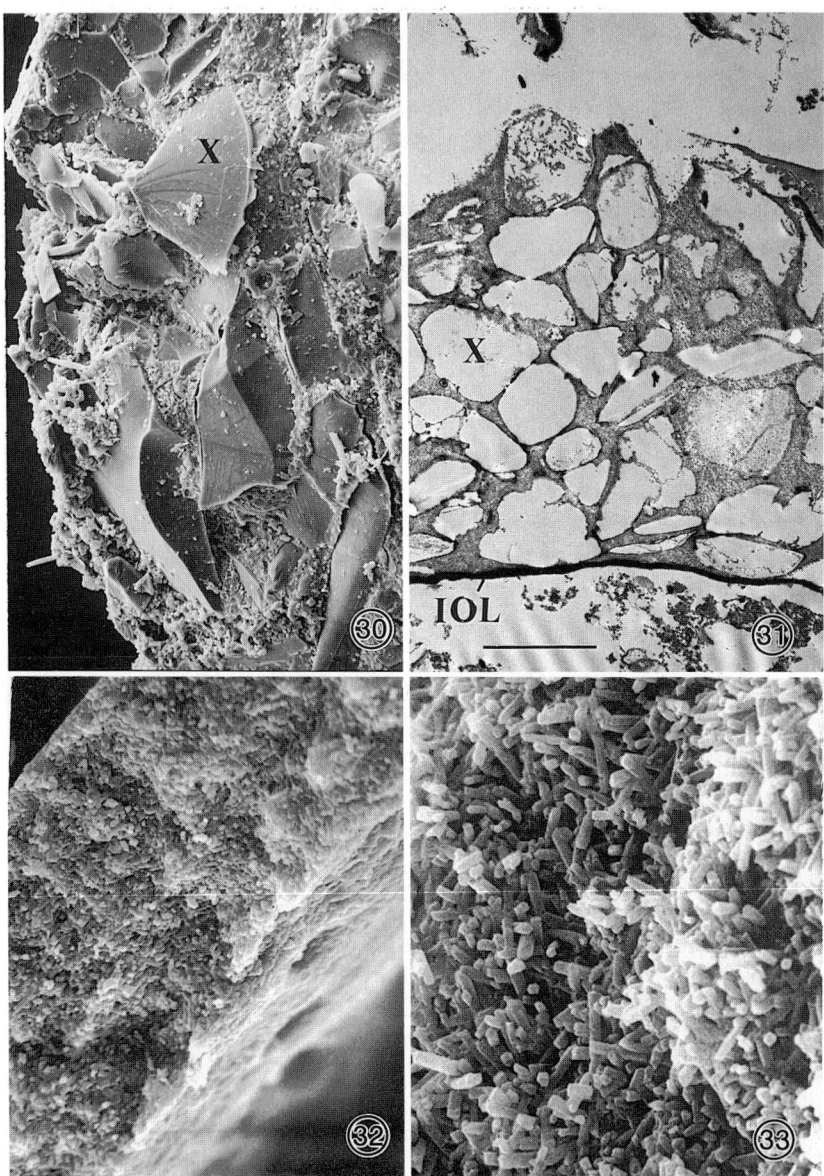

Figs 30–33. Wall structure in benthic species. Figs 30, 31. Obsidian particles (X) gathered by *Abyssotherma pacifica* and cemented into the organic matrix of the shell are shown in SEM view (Fig. 30). TEM views of ultrathin sections exhibit the close packed arrangement of the particles (X) shown in profile after HF dissolution (Fig. 31). A dense organic layer, known as the inner organic lining (IOL), occurs on the inner surface of benthic and planktonic foraminiferal shells (scale bar = 4 μm). Fig. 32. A fractured surface of a calcite shell wall showing the granular texture of the calcium carbonate deposits. Fig. 33. A fractured surface of a shell of a porcellaneous species showing the aggregate of calcite needles that compose the inner portion of the wall.

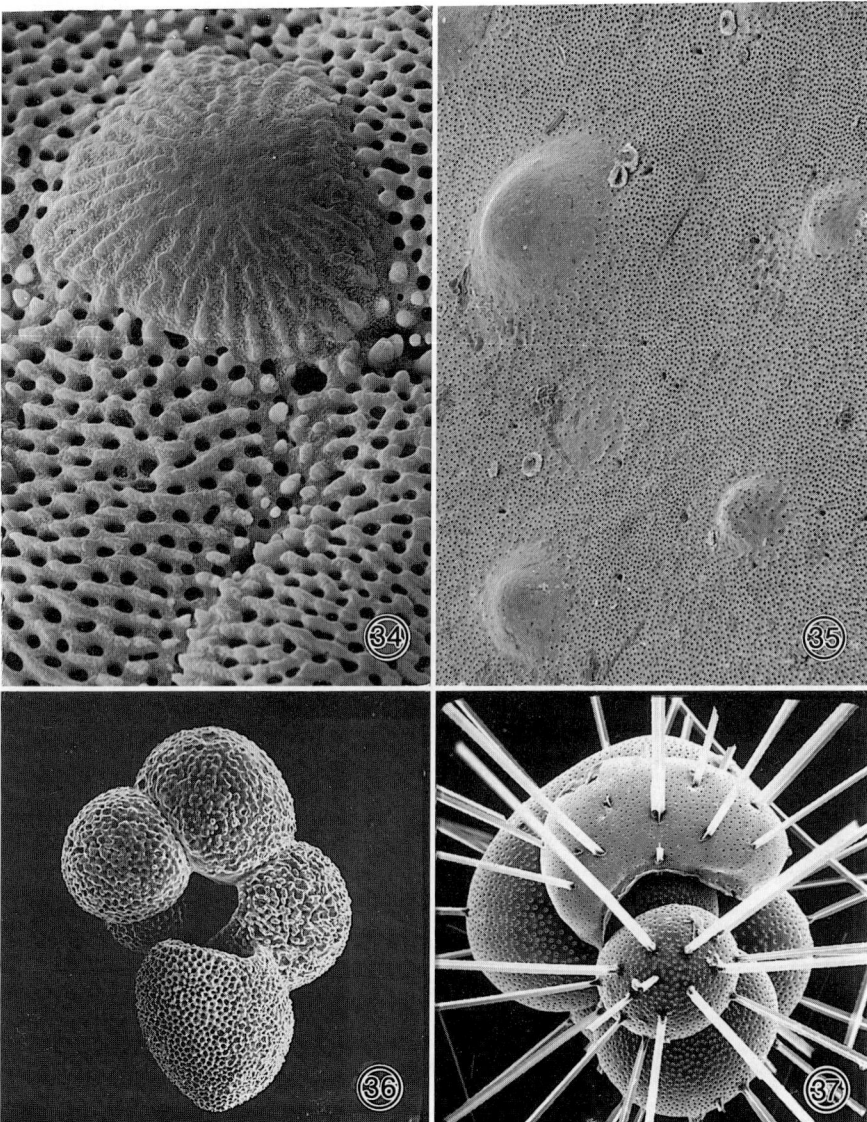

Figs 34–37. Comparative SEM views of the pores in the walls of shells from benthic species in high magnification, *Baculogypsina* sp. (Fig. 34) and *Operculina* (Fig. 35); and planktonic species in lower magnification, *Globigerinoides sacculifer* (Fig. 36) with thickened wall bearing reticulated ridges surrounding the pores, and *Hastigerina pelagica* (Fig. 37) with a thinner and smoother wall.

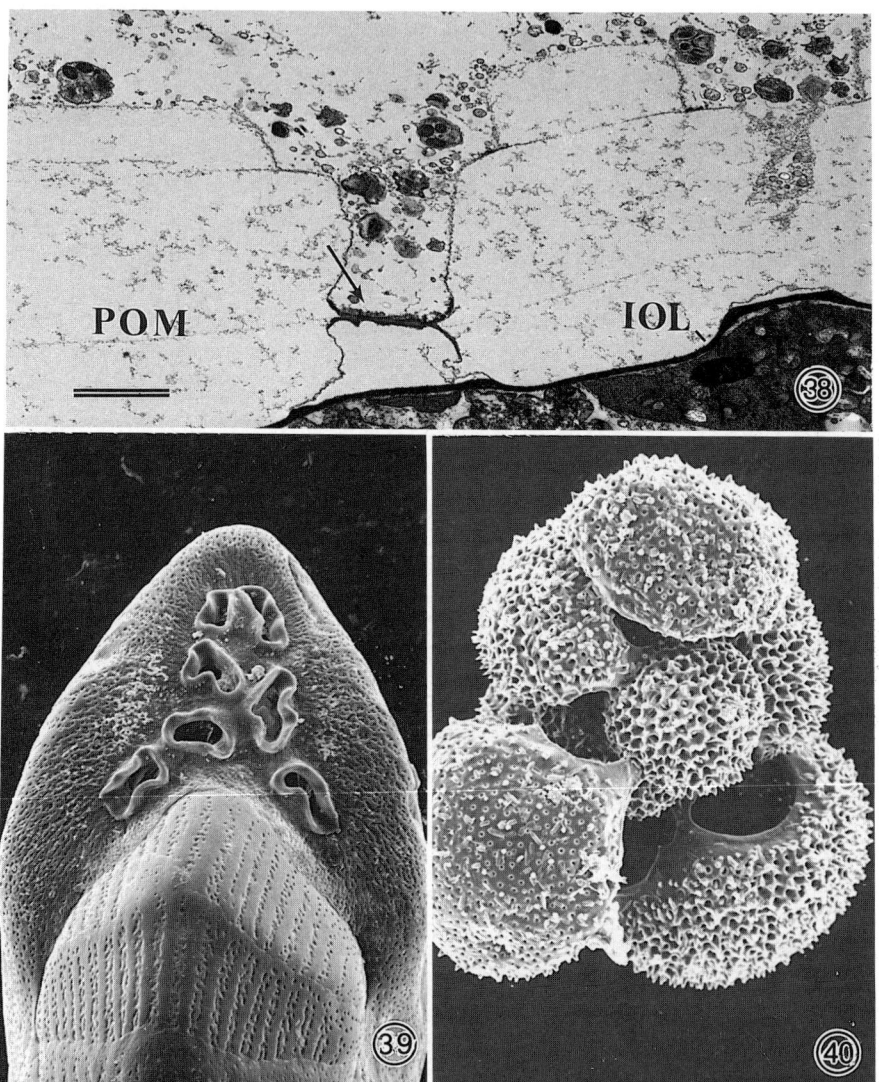

Fig. 38. Transmission electron micrograph of a section through a decalcified wall of a planktonic foraminifer showing the inner organic lining (IOL) covering the inner surface of the wall, the primary organic membrane (POM) that is deposited initially when the wall is being secreted and becomes thickened within the pore forming a perforated pore plate (arrow). The bilamellar construction of the wall is evident forming two distinct calcite layers on either side of the POM (scale bar = 2 μm). Compare with Fig. 31 showing a section through the wall of an arenaceous benthic species. **Figs 39, 40.** Scanning electron microscopic views of multiple apertures in a benthic foraminifer (Fig. 39) and a planktonic foraminifer (Fig. 40).

particle size, composition and shape appears to occur. Monolamellar and bilamellar calcite shells (Fig. 32) are often translucent to light and appear glassy when immersed in water or somewhat opalescent when observed dry. Tests composed of high magnesium calcium carbonate needles consist of an internal layer arranged in an irregular meshwork and a proximal and distal layer of needles arranged approximately parallel to form a smooth veneer (Fig. 33). The tests have an opaque and glossy appearance reminiscent of porcelain when viewed with reflected light; hence they are termed porcellaneous tests. Some genetic variants of the planktonic species *Globigerinoides ruber* have pink-pigmented shells, whereas others are colourless. Pores are present in the tests of planktonic species and many benthic species. Scanning electron microscopic images show the rich diversity of pore configurations and variations in internal organization of the pore walls among species (Figs 34–37). A variety of surface ornamentations also characterize planktonic species, including elevated reticulated ridges, rugose elevations, or smooth final surfaces bearing a thin veneer of highly uniform calcite. For example, in *G. sacculifer*, the pores are situated in depressions within the reticulate pattern of ridges deposited on the surface of the chambers (Fig. 36), while *H. pelagica* has a fine smooth porous surface (Fig. 37). In ultrathin sections, the lamellar organization of the organic layers in the test of planktonic species is clearly visible and the fusion of successively deposited organic layers to form the pore wall and septum is readily observed (Fig. 38). Some benthic species have regularly arranged patterns of pores distributed as zones on the surface of chambers between thickened septal ridges, or grouped in clusters on the surface of the test. The pore rim and internal wall may be smooth in some species or ornamented with tooth-like projections. These projections (also present on the periphery of the aperture) may serve a kind of sieving function during ingestion of prey (particularly diatom prey), permitting the passage of digestible matter and exclusion of the mineralized tests surrounding the prey. Many benthic and planktonic species have simple apertures without ornamentation, organized as circular, oval or slit-like openings through which major strands of cytoplasm communicate with the external environment (Figs 39, 40).

5.0 SUMMARY

The overall cytoplasmic organization of benthic and planktonic foraminifera is remarkably similar. Variations are particularly apparent,

Table 1 Sources of information on cytology and fine structure of benthic and planktonic foraminifera

Category	Benthic	Planktonic
Basic morphology and cytology	Boltovskoy and Wright (1976), Grell (1973), Anderson (1988)	Rhumbler (1911), Adshead (1967, 1980), Bé et al. (1977), Anderson (1984, 1988), Hemleben et al. (1989, pp. 55–65)
Cytoplasmic fine structure	Anderson and Bé (1978), Leutenegger (1977), Lee (1980), Lee and McEnery (1983), Koestler (1985)	Lee et al. (1965), Anderson and Bé (1976a,b, 1978), Anderson and Tuntivate-Choy (1984), Anderson (1984), Spindler and Hemleben (1982), Hemleben et al. (1989, pp. 64–85)
Test morphology and fine structure	Angell (1967a,b, 1971, 1980), Berthold (1976a,b), Spindler (1976, 1978), Spindler and Röttger (1973), Sliter (1974), Hottinger and Dreher (1974), Leutenegger and Hansen (1979), Bender and Hemleben (1988), Hemleben and Auras (1986), Goldstein and Barker (1988)	Anderson and Bé (1978), Hemleben (1969a,b, 1975), Hemleben et al. (1977, 1989), Anderson and Faber (1984), Towe and Cifelli (1967), Towe (1971), Spero (1988)
Reproduction	Grell (1954), Hedley et al. (1968), Hedley and Wakefield (1969), Angell (1971, 1990), Schwab (1976), McEnery and Lee (1976), Goldstein and Barker (1990)	Bé and Anderson (1976), Spindler et al. (1978), Bé et al. (1982, 1983), Anderson (1984), et al. (1989, pp. 140–155)

reflecting adaptations to the very different habitats of the two groups. Planktonic species consistently produce a nearly spheroidal halo of rhizopodia around the test (sometimes with alveoli as in *H. pelagica*), contain intracellular fibrillar bodies – possibly to enhance buoyancy – and enclose the algal symbionts (when present) in perialgal vacuoles to better control interaction including distribution of the symbionts in response to diel light cycles. Although benthic and planktonic foraminifera both exhibit a transitional form of cytoplasmic organization with no clear demarcation between intra- and extrashell cytoplasm, it is not uncommon for the benthic species to exhibit somewhat more clearly defined gradients in cytoplasmic density in the vicinity of the transition from condensed intrashell cytoplasm to the more dispersed and reticulate extrashell cytoplasm. Benthic species, with agglutinated tests, also use rhizopodial streaming to gather particles to be attached to the test during wall construction. Some benthic species contain dimorphic nuclei (large and small), but this has not been demonstrated in planktonic species. Some pertinent sources of information on the cytology and fine structure of the trophic and reproductive stages of benthic and planktonic species are cited in Table 1.

REFERENCES

Adshead, P.C. (1967). Collection and laboratory maintenance of living planktonic foraminifera. *Micropaleontol.* **13**:32–40.

Adshead, P.C. (1980). Pseudopodial variability and behaviour of Globigerinids (Foraminiferida) and other planktonic Sarcodina developing in cultures. In *Studies in Marine Micropaleontology and Paleoecology* (eds W.V. Sliter), pp. 96–126. Special Publication No. 19, Cushman Foundation for Foraminiferal Research, Lawrence, Kansas.

Anderson, O.R. (1983). The radiolarian symbiosis. In *Algal Symbiosis: A Continuum of Interaction Strategies* (ed. L.J. Goff), pp. 69–89. Cambridge University Press, Cambridge.

Anderson, O.R. (1984). Cellular specialization and reproduction in planktonic foraminifera and radiolaria. In *Marine Plankton Life Cycle Strategies* (eds K.A. Steidinger and L.M. Walker), pp. 35–66. Chemical Rubber Co. Press, Boca Raton, Florida.

Anderson, O.R. (1988). *Comparative Protozoology: Ecology, Physiology, Life History.* Springer-Verlag, Heidelberg.

Anderson, O.R. and Bé, A.W.H. (1976a). A cytochemical fine structure study of phagotrophy in a planktonic foraminifer *Hastigerina pelagica* (d'Orbigny). *Biol. Bull.* **151**:437–449.

Anderson, O.R. and Bé, A.W.H. (1976b). The ultrastructure of a planktonic foraminifer *Globigerinoides sacculifer* (Brady), and its symbiotic dinoflagellates. *J. Foram. Res.* **6**:1–21.

Anderson, O.R. and Bé, A.W.H. (1978). Recent advances in foraminiferal fine structural research. In *Foraminifera* (eds R.H. Hedley and C.G. Adams), Vol. 3, pp. 122–202. Academic Press, London.

Anderson, O.R. and Faber, W.W., Jr (1984). An estimation of calcium carbonate deposition rate in the planktonic foraminifera *Globigerinoides sacculifer* using ^{45}Ca as a tracer: A recommended procedure for improved accuracy. *J. Foram. Res.* **14**:303–308.

Anderson, O.R. and Tuntivate-Choy, S. (1984). Cytochemical evidence for peroxisomes in planktonic foraminifera. *J. Foram. Res.* **14**:203–205.

Angell, R.W. (1967a). The process of chamber formation in the foraminifer *Rosalinia floridana* (Cushman). *J. Protozool.* **14**:566–574.

Angell, R.W. (1967b). The test structure and composition of the foraminifer *Rosalina floridana*. *J. Protozool.* **14**:299–307.

Angell, R.W. (1971). Observations on gametogenesis in the foraminifer *Myxotheca*. *J. Foram. Res.* **1**:39–42.

Angell, R.W. (1980). Test morphogenesis (chamber formation) in the foraminifer *Spiroloculina hyalina* Schulze. *J. Foram. Res.* **10**:89–101.

Angell, R.W. (1990). Observations on reproduction and juvenile test building in the foraminifer *Trochammina inflata*. *J. Foram. Res.* **20**:246–247.

Bé, A.W.H. and Anderson, O.R. (1976). Gametogenesis in planktonic foraminifera. *Science* **192**:890–892.

Bé, A.W.H., Hemleben, Ch., Anderson, O.R., Spindler, M., Hacunda, J. and Tuntivate-Choy, S. (1977). Laboratory and field observations of living planktonic foraminifera. *Micropaleontol.* **23**:155–179.

Bé, A.W.H., Spero, H.J. and Anderson, O.R. (1982). Effects of symbiont elimination and reinfection on the lilfe processes of the planktonic foraminifer *Globigerinoides sacculifer* in laboratory culture. *Mar. Biol.* **70**:73–86.

Bé, A.W.H., Anderson, O.R. and Faber, W.W., Jr (1983). Sequence of morphological and cytoplasmic changes during gametogenesis in the planktonic foraminifer *Globigerinoides sacculifer* (Brady). *Micropaleontol.* **29**:310–325.

Bender, H. and Hemleben, Ch. (1988). Calcitic cement secreted by agglutinated foraminifers grown in laboratory culture. *J. Foram. Res.* **18**:42–45.

Bereiter-Hahn, J., Anderson, O.R. and Reif, W.-E. (eds) (1987). *Cytomechanics: The Mechanical Basis of Cell Form and Structure.* Springer-Verlag, Heidelberg.

Berthold, W.-U. (1976a). Test morphology and morphogenesis in *Patellina corrugata* Williamson, Foraminiferida. *J. Foram. Res.* **6**:167–185.

Berthold, W.U. (1976b). Ultrastructure and function of wall perforations in *Patellina corrugata* Williamson, Foraminiferida. *J. Foram. Res.* **6**:22–29.

Boltovsky, E. and Wright, R. (1976). *Recent Foraminifera.* W. Junk, The Hague.

Denton, E.J. and Gilpin-Brown, J.B. (1973). Flotation mechanisms in modern and fossil cephalopods. In *Advances in Marine Biology* (eds F.S. Russell and M. Yonge), Vol. 11, pp. 197–268. Academic Press, London.

Faber, W.W., Jr and Lee, J.J. (1991). Feeding and growth in the foraminifer *Peneropolis planatus* (Fechtel and Moll) Montfort. *Symbiosis* **10**:63–82.

Goldstein, S.T. and Barker, W.W. (1988). Test ultrastructure and taphonomy of the monothalamous agglutinated foraminifer *Cribrothalamina*, N. Gen., *alba* (Heron-Allen and Earland). *J. Foram. Res.* **18**:130–136.

Goldstein, S.T. and Barker, W.W. (1990). Gametogenesis in the monothalamous agglutinated foraminifer *Cribrothalamina alba*. *J. Protozool.* **37**:20–49.

Grell, K.G. (1954). Zur sexualität der Foraminiferen. *Naturwissenschaften* **41**:44–45.

Grell, K.G. (1973). *Protozoology*. Springer-Verlag, Heidelberg.
Hedley, R.H. and Wakefield, J.St. J. (1969). Fine structure of *Gromia oviformis* (Rhizopodea: Protozoa). *Bull. Brit. Mus. (Nat. Hist.), Zool.* 18:5–89.
Hedley, R.H., Parry, D.M. and Wakefield, J.St. J. (1968). Reproduction in *Boderia turneri* (foraminifera). *J. Nat. Hist.* 2:147–151.
Hemleben, Ch. (1969a). Ultramicroscopic shell and spine structure of some spinose planktonic foraminifera. In *Proceedings of the First Conference on Plankton Microfossils 2* (eds P. Brönnimann and H.H. Renz), pp. 254–256. E.J. Brill, Leiden.
Hemleben, Ch. (1969b). Zur Morphogenese planktonischer Foraminiferen. *Zitteliana* 1:91–133.
Hemleben, Ch. (1975). Spine and pustule relationships in some recent planktonic foraminifera. *Micropaleontol.* 21:334–341.
Hemleben, Ch. and Auras, A. (1986). Zooplankton mit Kalkskelett, sedimentierende Partikel und Sediment. *Ber. Polarforsch.* 32/86:34–39.
Hemleben, Ch., Bé, A.W.H., Anderson, O.R. and Tuntivate-Choy, S. (1977). Test morphology, organic layers and chamber formation of the planktonic foraminifer *Globorotalia menardii* (d'Orbigny). *J. Foram. Res.* 7:1–25.
Hemleben, Ch., Spindler, M. and Anderson, O.R. (1989). *Modern Planktonic Foraminifera*. Springer-Verlag, New York.
Hottinger, L. and Dreher, D. (1974). Differentiation of protoplasm in Nummulitidae (Foraminifera) from Elat, Red Sea. *Mar. Biol.* 25:41–61.
Koestler, R. (1985). Cytological investigation of digestion and reestablishment of symbiosis in the larger benthic foraminifer *Amphistegina lessonii*. Ph.D. thesis, City University of New York, New York.
Lee, J.J. (1980). Nutrition and physiology of foraminifera. In *Biochemistry and Physiology of Protozoa* (eds M. Levandowsky and S.H. Hutner), pp. 43–66. Academic Press, London.
Lee. J.J. (1989). Granuloreticulosa. In *Handbook of Protoctista* (eds L. Margulis, J.O. Corliss, M. Melkonian and D.J. Chapman), pp. 524–548. Jones and Bartlett, Boston, Mass.
Lee, J.J. (1990). Fine structure of the rhodophycean *Porphyridium purpureum* in situ in *Peneropolis pertusus* (Forskol) and *P. ascicularis* (Batsch) and in axenic culture. *J. Foram. Res.* 20:162–169.
Lee, J.J. and McEnery, M.E. (1983). Symbiosis in foraminifera. In *Algal Symbiosis: A Continuum of Interaction Strategies* (ed. L.J. Goff), pp. 37–68. Cambridge University Press, Cambridge.
Lee, J.J., Freudenthal, H.D., Kossoy, V. and Bé, A.W.H. (1965). Cytological observations on two planktonic foraminifera, *Globigerina bulloides* d'Orbigny, 1826 and *Globigerinoides ruber* (d'Orbigny, 1839) Cushman, 1927. *J. Protozool.* 12:531–542.
Lee, J.J., Faber, W.W., Jr and Lee, R.E. (1991). Granuloreticulopodial digestion – a possible preadaptation to benthic foraminiferal symbiosis? *Symbiosis* 10:47–61.
Leutenegger, S. (1977). Ultrastructure de Foraminiféres perforés et imperforés ainsi que de leurs symbiontes. *Cahiers de Micropaléontologie* 3:1–52.
Leutenegger, S. and Hansen, H.J. (1979). Ultrastructural and radiotracer studies of pore function in foraminifera. *Mar. Biol.* 54:11–16.
McEnery, M.E. and Lee. J.J. (1976). *Allogromia laticollaris*: A foraminiferan with an unusual apogamic metagenic life cycle. *J. Protozool.* 23:94–108.
McEnery, M.E. and Lee, J.J. (1981). Cytological and fine structural studies of

three species of symbiont-bearing larger foraminifer from the Red Sea. *Micropaleontol.* **27**:71–83.

Müller-Merz, E. and Lee, J.J. (1976). Symbiosis in the larger foraminiferan *Sorites marginalis* (with notes on *Archais* spp.). *J. Protozool.* **23**:390–396.

Rhumbler, L. (1911). Die Foraminiferen (Thalamophoren) der Plankton Expedition; Teil 1 – Die Allgemeinen Organisations-Verhältnisse der Foraminiferen. *Plankton Exped. Humboldt-Stiftung, Ergebn.* **3**,L,c,1–331.

Schwab, D. (1976). Gametogenesis in *Allogromia laticollaris*. *J. Foram. Res.* **6**:251–257.

Sliter, W. (1974). Test ultrastructure of some living benthic foraminifera. *Lethaia* **7**:5–16.

Spero, H. (1988). Ultrastructural examination of chamber morphogenesis and biomineralization in the planktonic foraminifer *Orbulina universa*. *Mar. Biol.* **99**:9–20.

Spindler, M. (1976). Gehäuseanatomie und Ablauf des Kammerbauvorgangs bei *Heterostegina depressa* (Nummulitidae: Foraminifera). Ph.D. thesis, University of Kiel, Kiel, Germany.

Spindler, M. (1978). The development of the organic lining in *Heterostegina depressa* (Nummulitidae: Foraminifera). *J. Foram. Res.* **8**:258–261.

Spindler, M. and Hemleben, Ch. (1982). Formation and possible function of annulate lamellae in a planktic foraminifer. *J. Ultrastruc. Res.* **81**:341–350.

Spindler, M. and Röttger, R. (1973). Der Kammerbauvorgang der Grossforaminifere *Heterostegina depressa (Nummulitidae)*. *Mar. Biol.* **18**:146–159.

Spindler, M., Anderson, O.R., Hemleben, Ch. and Bé, A.W.H. (1978). Light and electron microscopic observations of gametogenesis in *Hastigerina pelagica* (Foraminifera). *J. Protozool.* **25**:427–433.

Towe, K.M. (1971). Lamellar wall construction in planktonic foraminifera. In *Proceedings of the Second Planktonic Conference, 1970*, Vol. 2., pp. 1213–1218, Roma.

Towe, K.M. and Cifelli, R. (1967). Wall ultrastructure in the calcareous foraminifera: Crystallographic aspects and a model for calcification. *J. Paleontol.* **41**:742–762.

Zucker, W. (1973). Fine structure of planktonic foraminifera and their endosymbiotic algae. Ph.D. thesis, City University, New York.

3
Hypotheses on form and function in foraminifera

PAMELA HALLOCK, RUDOLF RÖTTGER and KAREN WETMORE

1.0	Introduction	41
2.0	Test case: Form and function in the larger foraminifera	44
	2.1 Symbiosis	44
	2.2 Size	45
	2.3 Test shape	46
	2.4 Complex internal morphologies	47
	2.5 Nutrient absorption	49
	2.6 Canal systems	52
3.0	Observations and hypotheses on basic foraminiferal morphology	60
4.0	Precautions in hypothesis testing	63
5.0	Concluding remarks	65
	References	66

1.0 INTRODUCTION

For a species to occupy and flourish in a niche, members of that species must be adapted to the range of physical conditions in the environment and they must be able to accumulate the trophic resources necessary to produce new generations at least as fast as predators, disease and physical factors reduce their numbers. However, a species does not have to be perfectly adapted to a niche to occupy it; the species simply must be sufficiently adapted to survive and, if there are competitors for the niche, better adapted to carry out those functions than those competitors. These are important considerations for any ecological study, and are particularly pertinent to understanding form and function in foraminifera. Morphology is a basic adaptive characteristic, as morphological adaptations are typically required to exploit a particular habitat and its resources.

Functional morphology is, of course, studied using the basic scientific methods of observation and hypothesis formulation and testing. During the approximately 200 years that foraminifera have been studied, volumes of observational data have been written on the morphology of foraminifera, most of it descriptive for taxonomic purposes. Periodic summaries of state-of-the-art knowledge of foraminiferal morphology appeared in Cushman (1933, 1940, 1948), Galloway (1933), Loeblich and Tappan (1964), Boltovskoy and Wright (1976), Haynes (1981) and Banner and Lord (1982). In recent years, observation has been greatly expanded with new techniques of measurement and analysis provided by computerized digitization.

Marszalek et al. (1969) discussed the dearth of information on the function of foraminiferal tests and proposed what they called a "general theory of test function". They hypothesized that the function of the most primitive tests is to provide weight to counteract buoyancy; more advanced tubes or series of chambers serve as effective barriers between the cytoplasm and the environment; and that further specialization might represent adaptations to more specialized conditions. Lipps (1975, 1983) proposed a relatively simple model relating coiling mode to feeding strategy: tubular and branched forms, living in quiet waters fixed to the sea-bed or embedded in sediment, are typically suspension feeders; active lenticular forms can be detrital scavengers on soft substrate or macrophytes; elongate forms may also passively scavenge detritus at the sediment/water interface; trochoid or flattened herbivores browse epiphyton on a variety of substrates; and a diversity of forms are found in carnivores, both active and passive. Boltovskoy and Wright (1976), Haynes (1981) and Brasier (1986) provide subsequent discussions of the morphological functions of foraminiferal tests. However, by and large, most of the ideas and hypotheses proposed in these books remain to be rigorously tested and applied to interpreting the fossil record.

Exceptions can be found in studies of the "larger" foraminifera, where research has progressed beyond hypothesis formulation to the testing stage, with the result that their functional morphologies are understood better than in any other group. This progress has occurred during the past 30 years, with the explosion of interest (e.g. Lee et al., 1980) in larger foraminifera that occurred partly in response to Hedley's (1964) challenge to the research community to prove the existence of algal symbiosis in living foraminifera, and coincident with the expansion of research in carbonate sedimentology, as larger foraminifera are prolific producers of carbonate sediment. Smout (1954) suggested that the shapes seen in larger foraminifera tend towards those giving maximum surface-to-volume ratios. Haynes (1965) formulated hypotheses on the interrelationships among symbiosis, wall structure, test shape and habitat in larger foraminifera. These hypotheses helped stimulate research, beginning in the late 1960s, on living taxa from Florida (e.g. Lee and Zucker, 1969; Lee et al., 1974; Lee and Bock, 1976), Hawaii (Röttger, 1972a,b,c, 1974, 1976; Röttger and Berger, 1972; Muller, 1974, 1976, 1978; Hallock, 1979) and the multinational project on the Gulf of Aqaba (e.g. Reiss et al., 1977; Reiss and Hottinger, 1984). These nuclear groups have continued their research throughout the 1980s, often interacting with each other. Field studies and laboratory experiments have directly or indirectly tested Haynes' hypotheses (e.g. Larsen and Drooger, 1977; Hallock, 1979, 1981a; ter Kuile and Erez, 1984; Hallock et al., 1986), with the results not only supporting his models but finding application in palaeoenvironmental research (e.g. Fermont, 1982; Chaproniere, 1984; Hallock and Glenn, 1985, 1986), and in interpreting large-scale events in the geological record (Hallock, 1982, 1987, 1988). Key papers on test structures in larger foraminifera include Hottinger (1978, 1984, 1986).

Other attempts to stimulate research on the functional morphology of foraminifera have not fared as well, partly because the direct economic applications were not as obvious. To our knowledge, no detailed community studies are being conducted to test Lipps' (1975) proposal, although DeLaca et al. (1980) demonstrated the advantages of elongate morphology and elevated apertural openings in suspension feeding. Severin (1983) suggested using test morphology as a discriminator of biofacies; and Wetmore (1987, 1988a,b) has experimentally examined the relationships between morphology, strength, composition and motility.

To review the variety of shapes and coiling modes that occur in foraminifera, one can refer to the summaries in Loeblich and Tappan (1964) or Haynes (1981). Haynes' intense interest in the functional aspects of morphology is reflected in his book; this article will not duplicate his synthesis. Instead, we discuss progress made with larger foraminifera as a possible model for future efforts to understand the functional morphology of a group. We shall identify key hypotheses, their origins and their current status. In looking to other morphological groups, we shall identify and briefly discuss some hypotheses that are currently in need of further testing, and identify some areas where further research may result in hypothesis formulation and testing that will bring understanding of other groups of foraminifera at least up to par with the larger foraminifera.

2.0 TEST CASE: FORM AND FUNCTION IN THE LARGER FORAMINIFERA

Larger foraminifera are not a taxonomically related group, but instead were originally considered together out of methodological necessity: their relatively large size, complex internal structures, often relatively non-descript exteriors and common occurrence in indurated limestones necessitated their identification from thin sections rather than from intact specimens. One of the major factors contributing to the progress in understanding form and function in larger foraminifera was the recognition of the role and implications of algal symbiosis in these protists.

2.1 Symbiosis

Although the possibility of algal symbiosis in foraminifera was widely recognized prior to the surge of interest in larger foraminifera in the mid-1960s (e.g. Cushman, 1922; Myers, 1943; Jepps, 1956), as Hedley (1964) pointed out, it had never been adequately documented. Interestingly, though algal symbiosis in modern larger foraminifera is now well known (see Leutenegger, 1984; or the reviews by Lee and associates, e.g. Lee et al., 1979, 1985; Lee, 1980, 1983; Lee and McEnery, 1983; Lee and Hallock, 1987; ter Kuile et al., 1987), the possible symbioses between algae and a variety of smaller taxa such as *Buliminella* and *Buccella*, which were suggested by Haynes (1965), have still not been verified.

The advantages, and especially the disadvantages, of algal symbiosis are not thoroughly known, even for corals, which have received much more attention that the foraminifera. Nevertheless, it is clear that algal symbiosis provides at least the potential for additional energy supplies and the potential for calcification enhancement, which is useful for the construction of sizeable shells or skeletons. Among the foraminifera, different taxa may derive different benefits from symbiosis. For example, Röttger et al. (1980) found that *Amphistegina* and *Heterostegina*, which harbour diatom symbionts fix CO_2 at significantly higher rates than rhodophyte-bearing peneroplids. Ter Kuile et al. (1987) reported that symbiosis is a more important source of energy for *Amphistegina* than for soritids.

The potential energetic advantages of algal symbiosis, which were explored mathematically by Hallock (1981b), appear to be greatest when abundant light is available and when trophic resources – dissolved inorganic nutrients and particulate organic carbon (POC) – are extremely scarce in the environment. Under these conditions, even though POC is scarce, it represents the only concentration of nutrients available in the water column. Algal symbiosis may enable the host–symbiont system to function as a plant that is equipped to capture those packets of nutrients. Theoretically, the greater the scarcity of nutrients and POC, the greater the advantage of symbiosis. Thus, as Lee and Hallock (1987) suggested, algal symbiosis has the potential to provide a powerful energetic driving force for evolution in nutrient-deficient seas.

2.2 Size

The most basic characteristic of "larger" foraminifera is their relatively large size, commonly in excess of $3\,\text{mm}^3$ in volume (Ross, 1974). Gigantism, whether in corals, bivalves or foraminifera, is often associated with algal symbiosis (e.g. Cowen, 1983); however, as noted by Haynes (1965) and reiterated by Hallock (1985), foraminifera do not have to be large to have algal symbionts, as the symbionts are typically tiny, $4-10\,\mu\text{m}$ in diameter (e.g. Lee et al., 1980; Leutenegger, 1984). Furthermore, all exceptionally large foraminifera do not have algal symbionts: deep-sea (e.g. Brasier, 1984) and high-latitude (e.g. DeLaca et al., 1980) foraminifera may attain large size. Hallock (1985), using mathematical models and basic ecological principles associated with life-history strategies, showed that delayed maturation and growth to large sizes are most advantageous where environmental conditions are relatively stable and food resources are limited. Under either environmental instability or abundant food resources, the most successful life-history strategy should be that of the rapidly maturing opportunist. Conversely, growth to large size and algal symbiosis are both adaptations that are potentially advantageous under conditions where food resources are consistently limited. Because, when food is scarce, algal symbiosis can provide the host–symbiont system with energy that is literally orders of magnitude greater than is available to their asymbiotic competitors (Hallock, 1981b), the occurrence of large, algal symbiont-bearing foraminifera in shallow, oligotrophic, subtropical seas is not surprising. Furthermore, several times in the geological record, particularly during times of major marine transgression and climatic amelioration, larger foraminifera were even more abundant and diverse than in modern seas (e.g. Ross, 1974). With the greater abundance of shallow, oligotrophic habitats afforded by transgressive seas (e.g. Hallock, 1982, 1985, 1987), the proliferation of larger foraminifera associated with shallow-marine carbonate biofacies is not surprising.

2.3 Test shape

Independent lineages of larger foraminifera have arisen and diversified from smaller ancestors numerous times over the past 350 million years (see, e.g. Ross, 1974). One of the first trends to be recognized, besides simply increasing size, was the tendency towards maximizing surface-to-volume ratios (e.g. Smout, 1954). Haynes (1965) proposed that shape is a compromise between hydrodynamic factors and light and metabolic requirements associated with algal symbiosis. This hypothesis held tremendous potential for the use of larger foraminifera as palaeoenvironmental indicators, particularly palaeodepth, for wave motion and light both decline as exponential functions of depth (e.g. Hallock, 1979).

The first major attempt to document the palaeoenvironmental potential of larger foraminifera was launched in the Gulf of Aqaba and published by Reiss et al. (1977). Larsen (1976) and Larsen and Drooger (1977) found that *Amphistegina*, in particular, showed a strong inverse relationship between test thickness and habitat depth. Comparable field documentation of the tendency for thicker-tested forms to live in shallower, more turbulent environments, and thinner-tested forms to live in low-energy and/or deeper environments was reported by Hallock (1979) for two Pacific sites. She found this tendency (1) in algal symbiont-bearing foraminifera as a group, (2) among the members of the genus *Amphistegina* and (3) within individual species of *Amphistegina*. In this and subsequent studies (Hallock, 1981a; Hallock et al., 1986), experimental work (Fig. 1) showed that both light and water motion directly influence test thickness in *Amphistegina* (Hallock, 1979, 1981a; ter Kuile and Erez, 1984) by influencing the thickness of individual lamellae as they are deposited (Hallock and Hansen, 1979; Hallock et al., 1986). Exactly how light and water motion influence individual calcification events is not yet known. Does light enhance calcification by providing additional energy or by removing metabolic CO_2 and phosphates? Does water motion enhance calcification by removing metabolic CO_2 and providing greater access to Ca^{2+}?

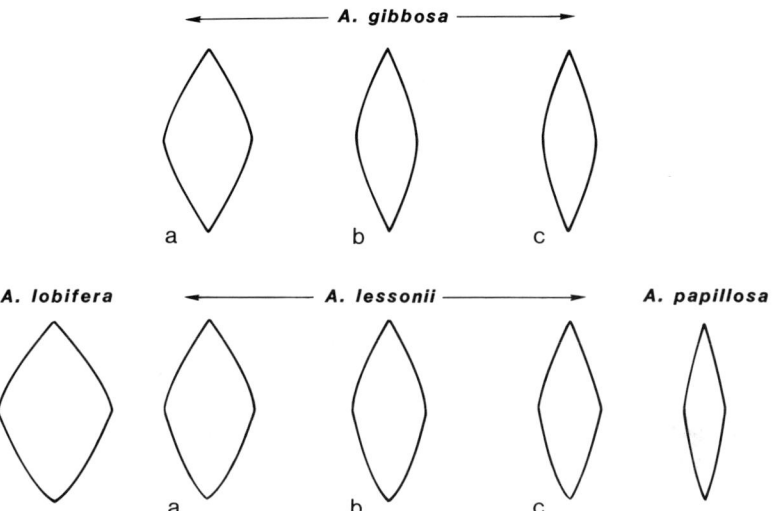

Fig. 1. Silhouettes illustrating the range of shapes in *Amphistegina gibbosa* and *A. lessonii* observed in culture. In each case, "a" was grown under gentle water motion and 23 μE/s-m^2 light intensity, "b" was grown unagitated at 40 μE/s-m^2 and "c" was grown unagitated at 4 μE/s-m^2. *Amphistegina lobifera* and *A. papillosa*, which normally live at depths shallower and deeper than *A. lessonii* respectively in the tropical Indo-Pacific, are also shown for comparison. (Modified from Hallock et al., 1986.)

2.4 Complex internal morphologies

Complex internal morphologies are a characteristic of foraminifera informally considered "larger" forms. Haynes (1965) and many others since have attributed the complexity of the interiors of larger foraminifera as an adaptation to house the algal symbionts efficiently. Hallock (1985) suggested that, if the symbionts serve to remove metabolic wastes from the foraminiferal cytoplasm and to provide the foraminifera with a direct food source, intuitively, it would seem that the most efficient way to serve these functions would be for groups of algae to surround small packets of cytoplasm. Subdivision of the chamber and therefore the cytoplasm into chamberlets would provide these small, efficient packets of cytoplasm and symbionts. However, algal symbiont-bearing foraminifera such as *Amphistegina* and *Operculina* seem to function quite adequately without tiny chamberlets. Furthermore, again, there are asymbiotic foraminifera, such as *Cyclammina*, which are large in size and which have complex interiors. Thus, other functions of complex interiors must be considered.

Brasier (1986) discussed the most obvious advantage of complex interiors – the increased strength that interior partitions impart to the test. In suicide-reproducers like foraminifera, the life-history strategy of delayed maturation and growth over relatively long time periods can only be successful if rates of adult mortality are extremely low (Hallock, 1985). Both test size and test strength may be important deterrents of predation and breakage. To house the same amount of cytoplasm, a test with internal partitions may be both larger and stronger than one without partitions. If the extra energy required to construct the partitions pays off as reduced mortality, then test complexity can be a useful adaptation with or without symbionts. Furthermore, the more complex the internal morphology, the more resistant the foraminifer may be to predators that attack individual chambers by drilling or breaking (e.g. Arnold *et al.*, 1985). The small amount of cytoplasm contained in an individual chamber may not be sufficient to pay the energy costs of getting it. Direct experimental testing of test strength of foraminifera with complex internal morphologies (Wetmore and Plotnick, 1991), as is currently being done on certain smaller taxa (e.g. Wetmore, 1987, 1988a), will provide insight into this question.

However, with algal symbionts, test complexity may be especially important, because the outer walls of the tests must be relatively thin to transmit the light needed by the symbionts. In shallow turbulent water, strengthening of the entire test may be crucial, and abundant light may be able to penetrate even relatively thick test walls, as seen in the Calcarinidae. But in deeper water, where light is more limiting and wave turbulence is negligible, thinner outer walls are needed. A thin outer wall covering a fairly large chamber would have a limited resistance to crushing by predators or the impact of sand grains. Such a test would also have little resistance to bending forces. By reinforcing the test by essentially I-beam construction, not only is the test more difficult to crush, but it is also less apt to break in two from impact or compression.

Strength considerations are particularly applicable when considering algal symbiont-bearing milioline foraminifera, and by analogy, possibly the fusulinids as well. The relative opacity of the milioline-wall structure necessitates extreme thinning of the outer walls to allow sufficient light to enter the test to support symbiotic algae. This is accomplished by pits in the peneroplines, grooves in the alveolinids, pseudopores in archaiasinids and windows in the soritids. In *Archaias*, the pseudopores are covered by thinned portions of the randomly oriented calcite layer; the outer, calcite layer of the test is absent

(Cottey and Hallock, 1988). Windows are recessed below the level of the chamberlet wall in complex soritids like *Marginopora* and *Amphisorus*, and therefore they are protected from direct impact. Such thin outer walls may not be strong enough to resist crushing or bending forces. The real strength of the test is not in the outer walls in this case, but in the lateral supports provided by both the chamber and chamberlet side walls.

2.5 Nutrient absorption

The thinned outer walls in the miliolines probably also allow some degree of gas and possibly ion exchange, serving functions similar to that of the pores in perforate species (e.g. Leutenegger and Hansen, 1979). This indicates another possible function of a high surface-to-volume ratio besides light gathering.

The diets of larger foraminifera present some interesting problems (e.g. Lee and Bock, 1976; Lee et al., 1980; ter Kuile et al., 1987). Röttger (1972b) and Schmaljohann (1980) have observed that *Heterostegina* need not feed, that it can absorb all the nutrients it needs from sea water. *Amphistegina* spp. visibly feed when food is offered, but when kept unfed in nutrient-enriched sea water, they will continue to grow at normal rates (Röttger et al., 1980; Hallock et al., 1986). Those soritids that have been examined are primarily dependent upon active feeding (Lee and Bock, 1976; ter Kuile et al., 1987). Mathematical models of algal symbiosis (Hallock, 1981b) indicate that the relationship provides the greatest energetic benefits when nutrient uptake is via POC capture by the host, particularly in nutrient-deficient environments. Under such conditions, if the foraminiferal symbionts had to rely solely on absorption of dissolved inorganic nutrients from the sea water medium, their position inside the host would be a great disadvantage in competing for those nutrients with free-living algae. However, there are at least two kinds of environments where dissolved nutrients are available at an interface, so that large, flat foraminifera, with their algal symbionts, might compete effectively for those nutrients as they emerge from the substratum, i.e., soft sediment and seagrass blades.

Bacterial activity or organic matter accumulating in the sediments results in nutrient regeneration within the sediments. Thus, there is typically a nutrient gradient at the sediment–water interface between the sediment–pore waters carrying regenerated nutrients and the overlying waters from which phytoplankton and benthic algae have stripped dissolved nutrients. A foraminifer, with a large surface-to-volume ratio, could access regenerating nutrients at the sediment–water interface while capturing light from above for photosynthesis (Fig. 2a). Often bacterial–algal mats develop under these conditions, so if the foraminifer can also feed using pseudopodia emerging from the periphery, the result should be a very efficient system for maximizing limited resources. However, for large, flat foraminifera to survive for several months to several years living on open sediment implies several environmental constraints. First, unless the wave and current motion is relatively limited, discoid foraminifera would be moved around and broken up. Slightly heavier, lenticular or fusiform shapes might be able to tolerate somewhat more water motion. Secondly, the supplies of organic carbon to the sediment must be sufficiently low that surface sediments remain well oxygenated. If too much organic matter is accumulating and breaking down in the sediments, anoxia can develop in the sediments close to the sediment–water interface, particularly at night when only respiration is occurring. Because water motion and therefore exchange rates must be relatively limited, rates of accumulation of organic matter must also be low. Research and hypothesis testing in this area will yield important information on the function and ecology of larger foraminifera, as well as basic sedimentological information on processes at the sediment–water interface.

A variety of soritid foraminifera can utilize shallow environments where both water motion and organic accumulation rates are higher by living vagily, or semi-permanently or permanently attached to seagrass blades. Although seagrasses can tolerate a higher trophic resource availability than coral reefs, if nutrient levels are too high, limited water transparency will shift primary production to the phytoplankton at the expense of benthic plant communities. Thus, dissolved nutrient loads in seagrass beds are often relatively low. However, seagrasses are vascular plants, rooted in the sediment where nutrient regeneration is taking place. Seagrasses absorb ammonia and phosphates from the sediments and transport those nutrients to the leaves where some nutrient is lost into the surrounding environment (McRoy and Barsdate, 1970). If algal symbiont-bearing foraminifera living on the seagrass blades can absorb nutrients being released by the seagrass, again, the discoid shape maximizes absorption of nutrients from below and light from above, while feeding on bacteria or other epiphytes via pseudopodia extruded from multiple apertures along the periphery of the final annular chamber (Fig. 2b), maximizes resource utilization. Obviously, research should be directed towards these foraminifera and their relationships with seagrasses and seagrass epiphytes.

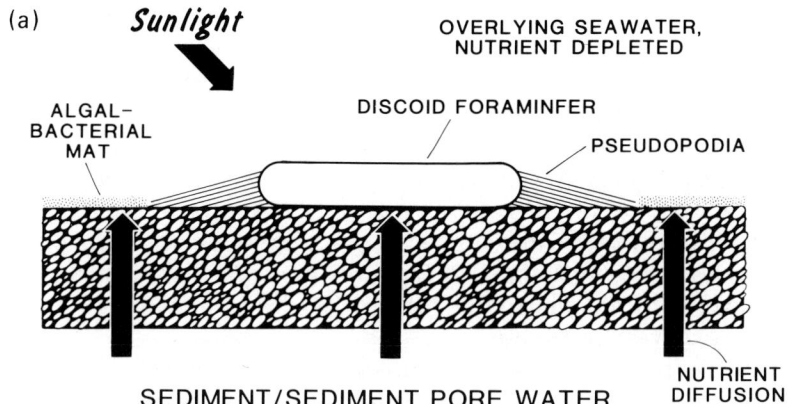

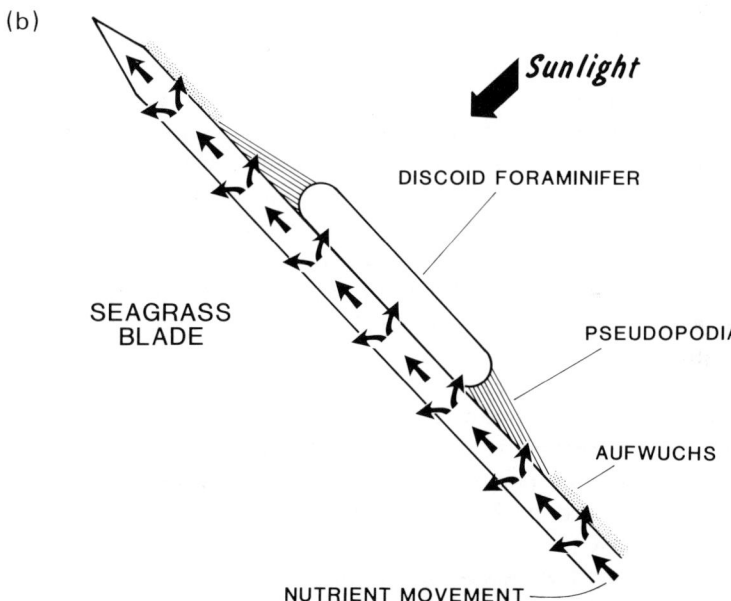

Fig. 2. Cartoon illustrating in cross-section two proposed mechanisms for acquisition of nutrients by sessile and semi-sessile discoid larger foraminifera. (a) Discoid foraminifer living on the sediment: high surface-to-volume ratio provides a large upper surface to capture sunlight for photosynthesis and a large lower surface for uptaking dissolved-inorganic and possibly dissolved-organic nutrients diffusing from sediment pore waters. The foraminifer, with its encircling pseudopodial net, may also be feeding on microalgae and bacteria in the surrounding sediment. (b) Scenario similar to (A) except that the substratum is a seagrass blade: seagrasses absorb ammonia and phosphates from the sediment and transport those nutrients to the leaves where some nutrient is released to the environment.

2.6 Canal systems

Advanced fossil and living rotaliacean foraminifera (Superfamily Rotaliacea) are characterized by canal systems located within their test walls. These canals were described in detail during the nineteenth century (Carpenter, 1850, 1859; Carter, 1852; Carpenter et al., 1862). More recently, artificial casts of test cavities were used in three-dimensional studies of the canal systems, employing scanning electron microscopy (Hottinger, 1977, 1979; Billman et al., 1980; Hottinger and Leutenegger, 1980). Hottinger (1978, 1986) has summarized much of this work on test anatomy.

In most multilocular foraminifera, successive chambers are connected by a single or multiple foramen; the last formed chamber opens into the surrounding medium through the terminal foramen or aperture. Pseudopodia emerge only from the aperture or from multiple apertures, and then only if the cytoplasm has not been retracted from the last chamber. In a foraminifer with a canal system within its chamber walls, these canals allow communication between the chamber cavities and the lateral surfaces of the test. Nummulitids also have a three-dimensional network of canals, which are continuous with the rest of the canal system, within the thickened peripheral keel (marginal cord), which open into ambient sea water. Nummulitids usually lack true primary or secondary apertures (Hottinger, 1977); which is not surprising because the canaliculate marginal cord, together with the canals of the chamber and chamberlet walls, serve as an effective substitute. However, apertures of various shapes and sizes have been observed in at least some individuals of *Heterostegina* (Spindler and Röttger, 1973).

The canal system permits the extrusion of pseudopodia from any point on the marginal cord, even when the cytoplasm has been withdrawn from the peripheral chambers. Thus, motility can be maintained, even when the last chambers of the test are devoid of cytoplasm because of mechanical or chemical irritation. This was, though hypothetical, for some time considered the only function of the canal system (Hottinger, 1977).

Observations and experiments on *Heterostegina depressa* and *Calcarina gaudichaudii* have demonstrated that these species are primarily sessile (Röttger, 1973; Röttger and Richwien, 1977; Röttger and Krüger, 1990). When undisturbed, *H. depressa* only moves a few millimetres every few days, to leave its protective sheath and construct a new one nearby. When subject to unfavourable conditions, individuals do not move at all, but become sickly or move very slowly up to a few centimetres per week. In such protists, the pseudopodia emerging from the canal system may serve functions other than movement.

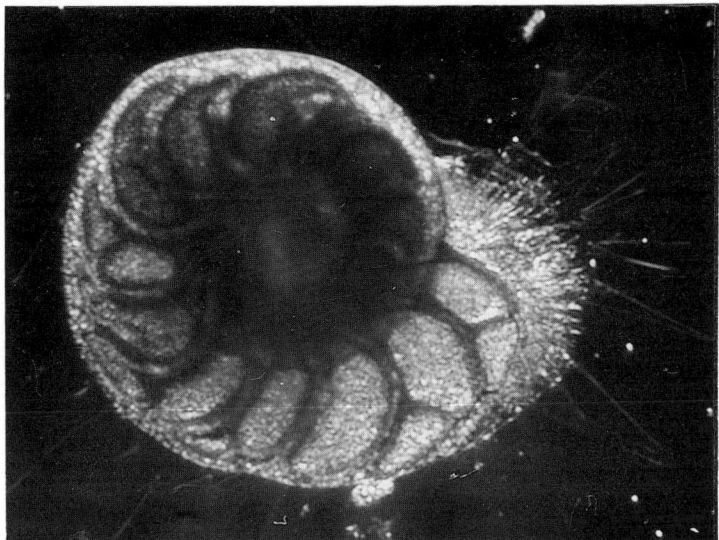

Fig. 3. *Heterostegina depressa.* Process of chamber formation. Cytoplasmic filaments in large numbers emerge from the canals of the marginal cord of the youngest chamber in order to construct the template of a new chamber. Size of specimen = 900 μm. Reproduced from Röttger et al. (1984) with the permission of Macmillan Magazines Ltd.

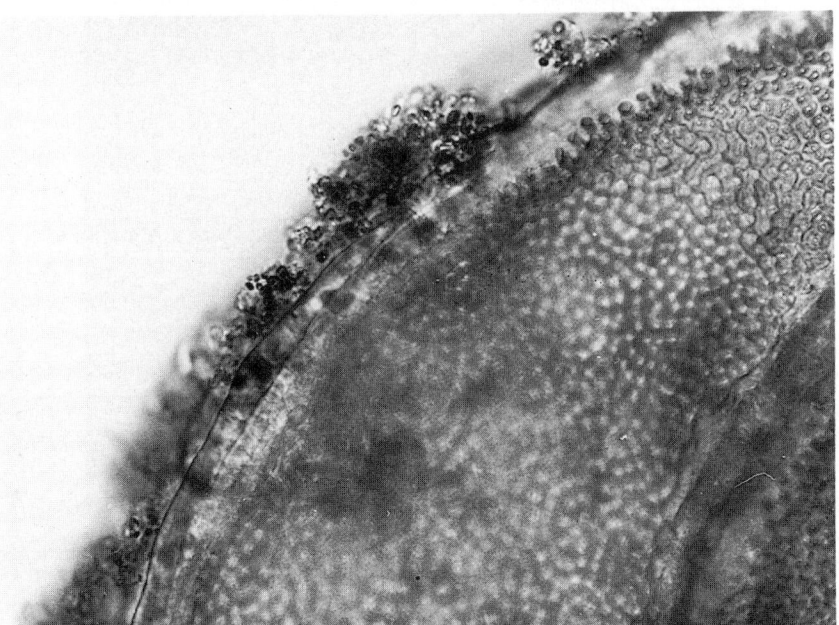

Fig. 4. *Heterostegina depressa.* Part of the lateral surface of the youngest chamber with accumulations of excrement on the marginal cord of the test. Those grains 1–2 μm in size have been excreted through the canals of the marginal cord. Reproduced from Röttger et al. (1984) with the permission of Macmillan Magazines Ltd.

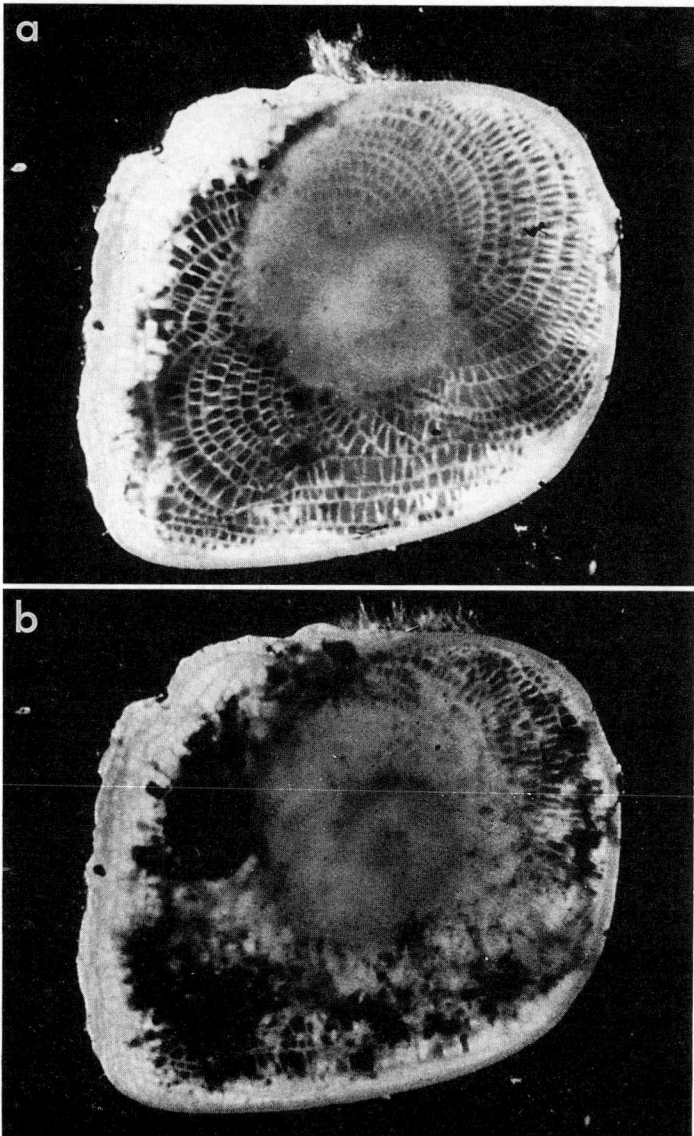

Fig. 5. *Heterostegina depressa*. Microcinematographic shots of an agamont from 45 m water depth, Island of Maui, Hawaii (size 9.8 mm). The cytoplasm emerges from the test and then divides into the daughter cells. (a) The cytoplasm has retracted from the peripheral part of the calcareous test, thus indicating the beginning of multiple fission. (b) The cytoplasm has evacuated most of the test through the canal system, and now covers both lateral surfaces.

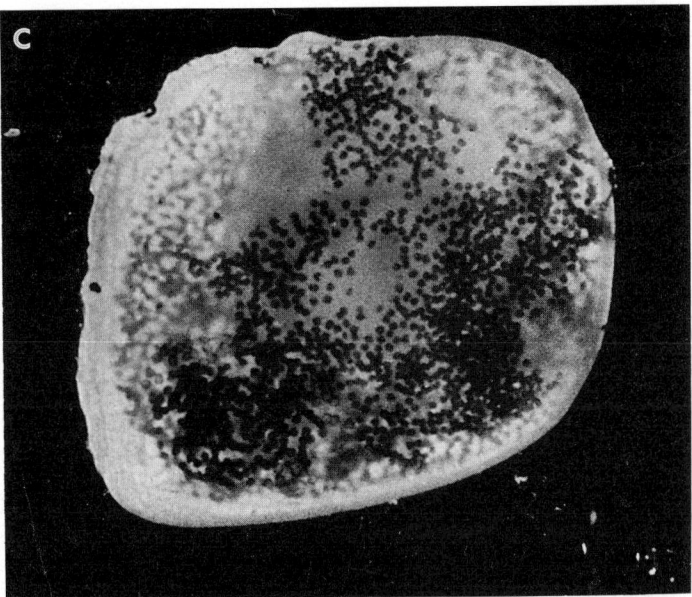

Fig. 5. (c) The cytoplasm has divided into the daughter cells. Juveniles lying below the maternal test are also visible. The duration of the evacuation process is 2 h 17 min. Reproduced from Röttger et al. (1984) with the permission of Macmillan Magazines Ltd.

The primary functions of the pseudopodia, and therefore of the aperture and canal systems that allow the pseudopodia to emerge from the test, are probably motility and material exchange. In the latter role, the canal system of *H. depressa* functions in chamber formation, reproduction, excretion and the formation of the protective sheath (Röttger et al., 1984). The terminal openings of the canals in the marginal cord function as a multitude of small, primary apertures that extrude cytoplasm, which forms the template for a new chamber (Fig. 3). Although this process was described by Spindler and Röttger (1973), its recent recording on film (Röttger and Inst. Wiss. Film, 1984) has vividly demonstrated the significance of the canal system. Nummulitids such as *H. depressa*, *Heterocyclina tuberculata*, and *Operculina ammonoides* would probably be unable to construct large chambers with only a single exit for cytoplasm.

The canal system also serves an excretory function. *H. depressa* is nourished by photosynthates obtained from endosymbiotic diatoms and by digestion of those symbionts throughout the test (Schmaljohann, 1980). Waste products are transported by the cytoplasm from older to younger chambers, and excreted through the openings in the marginal cord on the terminal chamber (Fig. 4), and through the aperture where it exists.

During sexual reproduction, the 2.5-μm gametes are released mainly through the canal system of the marginal cord. During asexual multiple fission, the cytoplasm, along wth the symbionts, evacuates the test and subsequently divides into hundreds or thousands of daughter cells. Microcinematographic time-lapse photographs (Röttger and Inst. Wiss. Film, 1984) have shown that the cytoplasm emerges from the lateral surfaces of the parent test via the canal system (Fig. 5). The symbiotic diatoms, which lack siliceous frustules, undergo severe deformation as they are transported through the 2- to 3-μm wide sutural canals. The diatoms are normally 9–11 μm long, 4–5 μm wide and variable in shape, so that they are well adapted to this mode of transport. Cytoplasmic

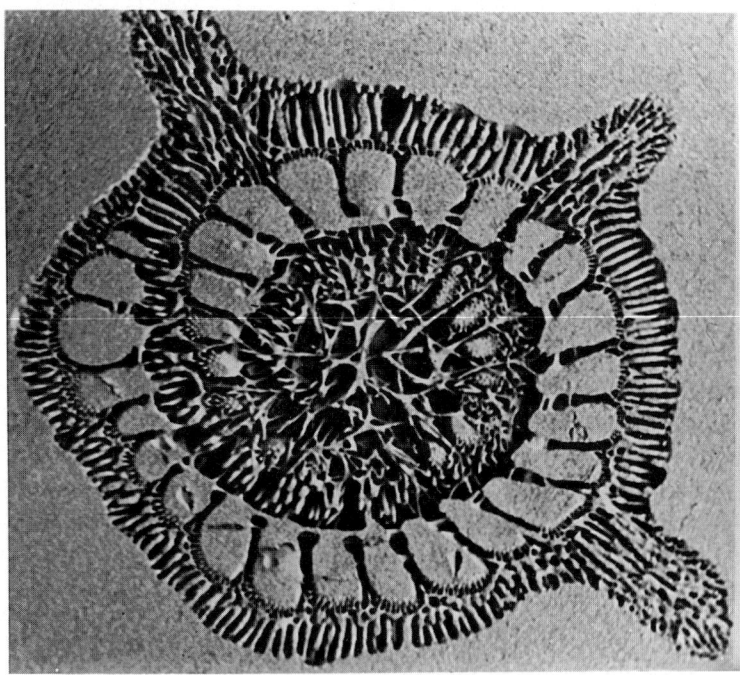

Fig. 6. *Calcarina gaudichaudii*, megalospheric specimen. Section through araldite cast, perpendicular to its axis ('equatorial section'), ventral view. The scanning electron micrograph shows the layer of canals enveloping the spiral of chambers. The spines, representing a bundle of canals each, run from the oldest part of the test to the test surface. In the example shown, the central axis of none of the 3 spines was sectioned. Size about 2.8 mm including spines. (Reproduced from Hottinger and Leutenegger, 1980 with permission from Birkhauser Verlag, Busel.)

evacuation of the test through the long and narrow canal system occurs relatively rapidly (38 min to 4 h 16 min).

During most of its life, *H. depressa* is covered by a transparent sheath attached to the substrate by elastic processes. This prevents displacement in turbulent water and protects the chambers during growth (Röttger, 1973; Spindler and Röttger, 1973; Röttger and Richwien, 1977; Röttger and Inst. Wiss. Film, 1984). Light microscope observations

Fig. 7. *Calcarina hispida* (larger diameter 2500 μm, including spines). Pseudopodia emerge from the openings of the enveloping canal system distributed over the entire test surface and from any part of the spine surface. (After Röttger and Krüger, 1990.)

show that the secretion of the sheath takes a few hours; the material for construction is passed through the openings of the canal system onto the surface of the test. Ultrathin sections of *H. depressa* surrounded by its sheath show a complete cytoplasmic layer situated between the calcareous test and the sheath (Spindler, 1976). Like pseudopodia, this cytoplasmic layer may serve in gas and ionic exchange between the cytoplasm and sea water.

The canal system of the calcarinids differs from that of the nummulitids. Most *Calcarina* spp. possess an enveloping canal system, i.e. a layer formed by innumerable canals, oriented perpendicular to the test surface, that covers each chamber. In addition, bundles of canals form spines that are inserted into the spiral of chambers (Hottinger and Leutenegger, 1980; Fig. 6). These spines run radially from the oldest part of the test to far beyond the periphery of the spiral. Living *C. gaudichaudii* possess the anlagen of spines in the three-chamber stage. The canals open on all exposed surfaces of the test.

Though different in geometry, the canal system of *Calcarina* appears to serve the same functions as that of the nummulitids (Röttger and Krüger, 1990). Pseudopodia protrude from canal openings all over the test surface (Fig. 7). The remains of digested symbiotic diatoms are excreted on the surface of the spines in *C. gaudichaudii* (Fig. 8). This

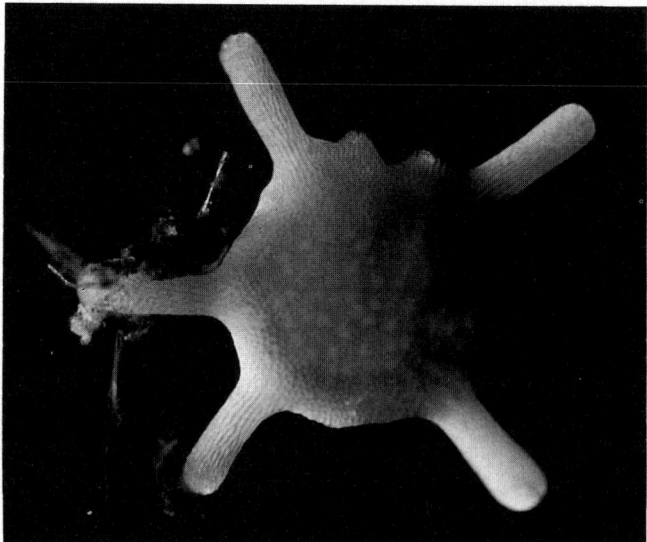

Fig. 8. *Calcarina gaudichaudii* (larger diameter 3250 μm, including spines). Clumps of excrements, the remains of digested symbiotic diatoms, have been excreted through the canals of the spines. (After Röttger and Krüger, 1990.)

species attaches to algal thalli by adhesive elastic plugs secreted at the ends of the spines, a function similar to the protective sheath of *H. depressa* (Fig. 9). During multiple fission of the agamont of *C. gaudichaudii*, the cytoplasm, as in *H. depressa*, emerges from the test lumen through the canal system into the dorsally located brood chamber, where the juveniles are formed.

Fig. 9. *Calcarina gaudichaudii* (size of lower specimen 2800 μm, including spines). Each specimen is firmly attached to an algal stem by an elastic plug secreted through the canals at the end of a spine. Attachment occurs by only one spine in most cases. (After Röttger and Krüger, 1990.)

3.0 OBSERVATIONS AND HYPOTHESES ON BASIC FORAMINIFERAL MORPHOLOGY

Because test structure and morphology are major features used in foraminiferal taxonomy, phylogenetic considerations are basic to understanding functional morphology and genetic constraints on morphology. Thus, functional morphology is interwoven with taxonomy.

Possession of a test indicates that expending energy on test construction is either advantageous, neutral, or at least not so disadvantageous as to allow some other organism to outcompete the foraminifer for its niche. Under the simplest conditions of abundant food and unsophisticated predators, simple binary fission would permit the highest potential for population increase. That is, a successful life-history strategy involving reproduction by multiple fission indicates that mortality declines with size or age, indicating a protective function for the foraminiferal test.

Multiple fission and protective coverings may have evolved in response to reduced food supplies or increased predation pressure. Foraminifera may have been capable of invading niches where binary fission is actually the better reproductive strategy if they were otherwise better adapted to those niches or if there were no binary-fission competitors.

The composition and structure of the test wall are the next most basic characteristics, and are considered of taxonomic importance at the subordinal level. Individual species within suborders are therefore morphologically constrained by the type of test they are genetically capable of constructing. The characteristics of the test do not necessarily prevent a taxon from living in a particular habitat, but, undoubtedly, certain wall structures should be more advantageous than others in certain environments.

For example, an organic test or agglutinate test with organic cements is generally thought to be energetically advantageous in brackish-water and in the deep sea below the calcium carbonate compensation depth, where calcium carbonate is undersaturated. In marsh environments, where food supplies are sufficient enough that rapid growth and short generation times are effective in maintaining population densities and so the major function of the test may be protection from chemical changes, simple organic tests may be advantageous. In the deep ocean, where food supplies are limited, growth rates are slow and generation times are long, the added physical protection of an agglutinated test may be worth the resource investment. In shallower environments, where sea water is supersaturated with respect to calcium carbonate and where food supplies are more limited, a reduction of the organic portion of the test wall in favour of secreted calcium carbonate may be energetically advantageous. Unfortunately, the energetics of calcification versus organic or agglutinated tests have not been investigated.

Within the calcareous taxa, the crystalline arrangement within the test is also of subordinal importance. Haynes (1981) proposed that the porcelaneous structure of the Miliolina, which includes randomly oriented calcite needles (Towe and Cifelli, 1967) along with thick, brown organic membranes, may provide protection in shallow water from short, metagenic wavelengths. Haynes suggested that the agglutinated structure of the Textulariina may provide similar protection. Haynes also suggested that hyaline calcite walls, particularly those with a radial structure, may have an adaptive value in taxa that have algal endosymbionts. A detailed study of the palaeobiology and palaeoenvironments of early hyaline taxa may indicate whether hyaline walls evolved in shallow-dwelling, possibly algal symbiont-bearing taxa, or if hyaline-walled taxa evolved under some other circumstances, but if they acquired algal symbionts, they were better able to house those symbionts than agglutinates or miliolines because their walls were more transparent.

Perhaps the greatest advantage of the hyaline wall structure was that it gave rise to a lamellar structure in some lineages. Lamellarity allowed the evolution of a variety of structures not possible in non-lamellar taxa, including the annular-complexes in larger foraminifera like the miogypsinids and lepidocyclinids, spines and stellate forms found in the calcarinids, and costae and keels in planktonic foraminifera (e.g. Haynes, 1981). In addition, a lamellar structure allows progressive test strengthening throughout the life of the individual; the earlier chambers are continuously re-enforced throughout the life of the individual. Drilling predators or predators that crush the test wall must expend progressively more energy to access cytoplasm in interior chambers (see, e.g. Sliter, 1971), increasing the probability that an individual could survive an encounter with such a predator. Test strength in non-lamellar taxa, on the other hand, is more uniform (Wetmore, 1987) – if a predator can crush or drill one part of the test, they can eventually gain access to all of the chambers.

The superiority of imperforate over perforate tests in environments where rapid chemical changes may occur has been suggested (Boltovskoy and Wright, 1976). This observation, which seems most applicable to littoral environments, certainly requires further testing. In imperforate forms, ionic and gas exchange and feeding may be confined to the apertural area, except in larger, algal symbiont-bearing taxa whose outer walls are thinned by pits, pseudopores and windows. Pores in perforate test have been shown experimentally to facilitate exchange of CO_2 (Leutenegger and Hansen, 1979), as well as some other dissolved substances (Berthold, 1976). Adaptation to environments rich in organic matter, where oxygen levels may be quite low, seems enhanced in perforate taxa like the bolivinids and buliminids. Leutenegger and Hansen (1979) found that the mitochondria were located close to the pores in taxa that are highly adapted to survival on limited oxygen concentrations. Further laboratory experiments are needed to examine the hypothesis that perforate tests provide the pre-adaptation for low oxygen tolerance.

It is curious that dormancy has not been investigated in foraminifera despite the presence of their tests, which could be readily modified to serve as protective cysts. In fact, it is assumed that foraminifera are not capable of achieving dormancy (e.g. Levy 1982, 1984), even though cysts, resting stages and other forms of dormancy are well known among the protozoa (e.g. Grell, 1973). Arnold (1967) reported that the aperture and foramina may be closed off by membranous partitions in unfavourable environments. Muller (1978) observed that *Amphistegina* kept in prolonged darkness became inactive and did not reattach to the substratum when dislodged; she also found indirect evidence of metabolic decline. Most specimens survived several months of darkness. The hypothesis that dormancy does not occur in foraminifera should be rigorously examined using ultrastructural and metabolic studies, because dormancy would provide survival benefit to foraminifera living in a variety of marginal or seasonal environments, and in any soft-bottom setting where burial and subsequent disinterment is likely.

A secreted test rather than an agglutinated test, which requires the acquisition of particles from the environment (Haynes, 1981), was necessary for the planktonic life mode to evolve in the foraminifera. It is also possible that a calcareous perforate test was a necessary prerequisite to planktonic life because pores reduce the density of the test, and provide the adaptive potential for increasing or decreasing the size and numbers of pores, adaptations necessary for planktonic life in water masses of different densities. In addition, lamellarity provided the adaptive potential for test thickenings such as keels.

G.C.H. Chaproniere (personal communication) suggested that algal symbiosis has played a role in the iterative evolution of planktonic foraminiferal morphologies analogous to the role it has played in larger foraminiferal evolutionary patterns. He proposes that certain morphological characteristics such as larger test size, radial symmetry, supplementary spiral apertures, more enveloping final chambers and lower-arched umbilical apertures, may be adaptations to symbiotic existence in warm, oligotrophic surface waters of the oceans.

There are a multitude of other morphological characteristics – most of them of taxonomic importance – that probably have functional significance, including chamber form and arrangement (e.g. Brasier, 1982, 1986), suture characteristics (Marszalek *et al.*, 1969), aperture modifications and ornamentation. Have apertural modifications evolved in response to selective pressure from small predators that can otherwise enter the test through the aperture? What types of test ornamentation strengthen the test wall? Detailed morphometric studies of both planktonic and benthic foraminiferal species and assemblages (e.g. Malmgren and Kennett, 1982; Painter and Spencer, 1984; Gary, 1985; Spencer, 1987; Wei, 1987) provide a major resource for functional morphological hypothesis formulation and testing. Observations and rigorous field and laboratory tests of hypotheses concerning shell function in foraminifera is an area in need of much more work. However, such studies must be carefully formulated and sweeping conclusions should not be made from limited data sets.

4.0 PRECAUTIONS IN HYPOTHESIS TESTING

Because so few species of foraminifera have been studied biologically in any detail, replication of observations is often lacking in descriptions of associations between form and function. Data on a single species in a category is only *one* observation of form and function within that category. Thus, generalizations require observations of many species in the category. Observations of a single species in each of several categories does not establish a pattern for those categories, for with only a single species sampled for each category, nothing is known about the range of variation within the category. For example, the lack of difference in test weight in *Elphidiella hanni* between environments with high and low water turbulence (Wetmore, 1988a) does not rule out the possibility that some species increase in test weight in more turbulent environments in order to reduce buoyancy.

Direct measurements of how well a structure performs an hypothesized function is the most straightforward way to determine relationships between form and function. For example, direct measurements of the transport of molecules through pores (Leutenegger and Hansen, 1979) is more conclusive evidence that gases can be exchanged through the pores than any analysis based on observations of pore morphology alone.

Resistance to crushing is another test function that lends itself to direct measurement (Wetmore, 1987, 1988a). A complete understanding of the relationship between test morphology and test strength is complicated by the vast morphological diversity of the foraminifera. It is difficult to find two species that are identical except for one character hypothesized to affect test strength. Therefore, observations of many species with similar tests are needed to establish correlations between test strength and test morphology (Wetmore, 1988a).

Where direct measurements of functions are impossible, an alternative approach is to test hypothesized relationships using mechanical or computer models. The theoretical validity of interpreting differences between actual morphology and some theoretical optimum or paradigm has been extensively debated (see Reif, 1982; Signor, 1982; Smith, 1982; and references therein). Difficulties arise because there may be more than one form that would function equally well, as has been shown for test strength in benthic foraminifera (Wetmore, 1987, 1988a). In addition, there can be morphological or energetic trade-offs with other functions that result in a sub-optimum form for any one function (Reif, 1982; Signor, 1982).

Any mechanical analysis requires sound theoretical and experimental bases for predictions. When applying mechanical principles, they must be applied correctly to the situation under analysis. Early considerations of relative test strength in benthic foraminifera relied on common sense, which can be misleading when dealing with unfamiliar structures. Two factors are especially important in an analysis of test strength in benthic foraminifera. First, there is a major difference in scale between human-built structures and foraminiferal tests, and therefore in the forces affecting these structures and how they are applied. For example, much of the stress a building has to resist comes from the need to support its own weight, whereas the weight of a foraminiferal test contributes a negligible amount to stresses within the test. Secondly, as engineers discovered through trial and error, some mechanical properties of structures go against "common sense". For example, a small notch or hole in a structure decreases its strength much more than would be predicted simply from the resulting reduction in cross-sectional area (Gordon, 1978; Wainwright et al., 1982).

Because small holes decrease the mechanical strength of a material, fewer, smaller pores would be expected where stresses are likely to be greatest in the test wall of perforate foraminifera. For biconvex-shaped, uniserial and planispirally coiled tests, the outer wall can be modelled as a pair of domed surfaces. When an inward (compressive) load is applied to the apex of such a structure, the greatest stresses are at the periphery of the dome (Lin and Stotesbury, 1981; Telford, 1985). A test subject to crushing forces is likely to experience the greatest stress at the periphery of the test. The test is also subject to puncturing around the periphery when ingested by certain predators (Mageau and Walker, 1976; Hickman and Lipps, 1983). Thus, if test strength is important to survival, fewer pores would be expected at and near the periphery of the test. Indeed, many coiled perforate tests have an imperforate band or thickened keel at the periphery, just where stresses are predicted to be greatest when the test is crushed. One further step in this analysis will be to determine whether species with these characteristics are more abundant in more physically stressful environments or if they are able to better survive ingestion by certain predators than species with a more uniform pore distribution.

The possibility that a structure performs multiple functions that are related to different environmental variables must also be considered. For example, the inner organic lining (IOL) found in tests of many species, has been shown to act as a barrier against unfavourable salinity changes and may also function as a light filter (Banner and Williams, 1973; Banner *et al.*, 1973). Another function may be to strengthen the test wall (Wetmore, 1988a; Wetmore and Plotnick, 1991). The thickness of the IOL might vary in response to one of those environmental variables, but to determine whether the thickness of the IOL in members of a foraminiferal species depends upon a particular environmental variable, the researcher must be able to eliminate the possible effects of other variables. For example, a comparison between a restricted lagoon, in which salinity fluctuations are high and water turbulence is low, and a shallow, open marine locality, in which salinity fluctuations are low and water turbulence high, might show no difference in proportion or thickness of the IOL simply because both environments require a thicker IOL for survival. In an analogous situation mentioned previously, culture experiments on *Amphistegina* spp. demonstrated that lamellar thickness and test shape respond to both light intensity and water motion (Hallock *et al.*, 1986).

5.0   CONCLUDING REMARKS

One, often unappreciated, benefit of hypothesis development and testing is that understanding of the problem is advanced whether or not the results support the hypothesis. In fact, sometimes more progress is made when the results do not appear to support the hypothesis, because those cases often reveal either important exceptions or force the researcher to re-evaluate and refine the hypothesis. A simple example occurred during research on the relationship between test shape and lamellar thickness in *Amphistegina* (Hallock and Hansen, 1979). The hypothesis was that environmental conditions associated with habitat depth influence chamber formation and therefore the thickness of individual lamellae, and thus more spheroidal tests are produced by deposition of thick lamellae and flatter tests by thin lamellae. Upon sectioning and examining thicker and thinner tests from field samples, differences in mean lamellar thicknesses were often only a few micrometres and were statistically insignificant because interlamellar variability was substantial. However, chamber formation and lamellar deposition occur intermittently and are influenced primarily by environmental conditions at the time of chamber formation. Because cloud cover can reduce light and lack of wind can reduce water turbulence, even individuals living in shallow, relatively high-energy environments can occasionally produce thin lamellae. (Experimental work since has confirmed that both light and water motion influence lamellar thickness and test shape: Hallock, 1981a; Hallock et al., 1986). Furthermore, there is an easily visible shape difference between individuals with thickness-to-diameter ratios of 0.45 (typical for *A. lessonii*) and 0.6 (typical for *A. lobifera*). But in foraminifera 1 mm in diameter, that difference is 150 μm. If those individuals have 30 chambers, they have 60 lamellar layers (30 on the spiral side and 30 umbilical), and therefore only a 2.5-μm difference in mean lamellar thickness can result in the observed difference.

Thus, while formulating hypotheses may be relatively simple, the practice of testing them often requires substantial rethinking of the processes and of the magnitude of change necessary to explain the original observations. Experimental laboratory and field studies aimed directly at testing the multitude of hypotheses concerning foraminiferal form and function will continue to advance the scientific understanding of the roles of the test and of the foraminifera themselves within their environments. Such studies will not only advance palaeoecological and palaeoenvironmental research, but also our understanding of evolution and earth history.

REFERENCES

Arnold, A.J., d'Escrivan, F. and Parker, W.C. (1985). Predation and avoidance responses in the foraminifera of the Galapagos hydrothermal mound. *J. Foram. Res.* **15**:38–42.

Arnold, Z.M. (1967). Biological observations on the foraminifer *Calcituba polymorpha*. *Roboz. Arch. Protistenk.* **110**:280–304.

Banner, F.T. and Lord, A.R. (1982). *Aspects of Micropaleontology*. Allen and Unwin, London.

Banner, F.T. and Williams, E. (1973). Test structure, organic skeleton and extrathalamous cytoplasm of *Ammonia* Brünnich. *J. Foram. Res.* **3(2)**:49–69.

Banner, F.T., Sheehan, R. and Williams, E. (1973). The organic skeletons of Rotaline foraminifera. *J. Foram. Res.* **3(1)**:30–42.

Berthold, W.-U. (1976). Ultrastructure and function of wall perforations in *Patellina corrugata* Williamson, Foraminiferida. *J. Foram. Res.* **6**:22–29.

Billman, H., Hottinger, L. and Oesterle, H. (1980). Neogene to Recent rotaliid foraminifera from the Indopacific Ocean; their canal system, their classification and their stratigraphic use. *Schweiz. Paläontol. Abh.* **101**:71–113.

Boltovskoy, E. and Wright, R. (1976). *Recent Foraminifera*. W. Junk, The Hague.

Brasier, M.D. (1982). Architecture and evolution of the foraminiferid test – a theoretical approach. In *Aspects of Micropaleontology* (eds F.T. Banner and A.R. Lord), pp. 1–37. Allen and Unwin, London.

Brasier, M.D. (1984). *Discospirina* and the pattern of evolution in foraminiferid architecture. In *Benthos '83: Second International Symposium on Benthic Foraminifera*, pp. 87–90, Pau and Bordeaux, April.

Brasier, M.D. (1986). Form, function and evolution in benthic and planktic Foraminiferid test structure. In *Biomineralization in Lower Plants and Animals* (eds B.S.C. Leadbeater and R. Riding), pp. 251–268. Clarendon Press, Oxford.

Carpenter, W.B. (1850). On the microscopic structure of *Nummulina, Orbitolites*, and *Orbitoides*. *Quat. J. Geol. Soc. Lond.* **VI**:21–39.

Carpenter, W.B. (1859). Researches on the foraminifera III. *Phil. Trans. R. Soc. Lond.* **14**:1–41.

Carpenter, W.B., Parker, W.K. and Jones, T.R. (1862). *Introduction to the Study of the Foraminifera*. Published for the Ray Society by Robert Hardwicke, London.

Carter, H.J. (1852). On the form and structure of the shell of *Operculina arabica*. *Ann. Mag. Nat. Hist. Ser.* 2 **X**:161–176.

Chaproniere, G.C.H. (1984). Oligocene and Miocene larger Foraminiferida from Australia and New Zealand. *Bureau of Mineral Resources Bull.* **188**:1–98, 26 plates.

Cottey, T.L. and Hallock, P. (1988). Test surface degradation in *Archaias angulatus*. *J. Foram. Res.* **18**:187–202.

Cowen, R. (1983). Algal symbiosis and its recognition in the fossil record. In *Biotic Interactions in Recent and Fossil Benthic Communities* (eds M.J.S. Tevesz and P.L. McCall), pp. 432–478. Plenum, New York.

Cushman, J.A. (1922). Shallow water foraminifera of the Tortugas Region. *Publ. Carnegie Inst. Wash.* **311**:1–85.

Cushman, J.A. (1933). *Foraminifera: Their Classification and Economic Use*, 2nd edn. Harvard University Press, Cambridge, Mass.

Cushman, J.A. (1940). *Foraminifera: Their Classification and Economic Use*, 3rd edn. Harvard University Press, Cambridge, Mass.

Cushman, J.A. (1948). *Foraminifera: Their Classification and Economic Use*, 4th edn. Harvard University Press, Cambridge, Mass.

DeLaca, T.E., Lipps, J.H. and Hessler, R.R. (1980). The morphology and ecology of a new large agglutinated Antarctic foraminifer (Textulariina: Notodendrodidae nov.). *Zool. J. Linn. Soc.* **69**:205–224.

Fermont, W.J.J. (1982). Discocyclinidae from Ein Avedat (Israel). *Utrecht Micropal. Bull.* **27**:1–173.

Galloway, J.J. (1933). *A Manual of Foraminifera.* Principia Press, Bloomington, Ind.

Gary, A.C. (1985). A preliminary study of the relationship between test and morphology and bathymetry in recent *Bolivina albatrossi* Cushman, northwestern Gulf of Mexico. *Trans. Gulf Coast Assoc. Geol. Soc.* **35**:381–386.

Gordon, J.E. (1978). *Structures, or Why Things Don't Fall Down.* Penguin, New York.

Grell, K.G. (1973). *Protozoology.* Springer-Verlag, Berlin.

Hallock, P. (1979). Trends in test shape with depth in large, symbiont-bearing foraminifera. *J. Foram. Res.* **9**:61–69.

Hallock, P. (1981a). Light dependence in *Amphistegina*. *J. Foram. Res.* **11**:42–48.

Hallock, P. (1981b). Algal symbiosis: A mathematical analysis. *Mar. Biol.* **62**:249–255.

Hallock, P. (1982). Evolution and extinction in larger foraminifera. In *Proceedings of the Third North American Paleontology Convention*, Vol. 1, pp. 221–225.

Hallock, P. (1985). Why are larger foraminifera large? *Paleobiology* **11**:195–208.

Hallock, P. (1987). Fluctuations in the trophic resource continuum: A factor in global diversity cycles? *Paleoceanography* **2**:457–471.

Hallock, P. (1988). Diversification in algal symbiont-bearing foraminifera: A response to oligotrophy? *Revue de Paleobiologie* **2**:789–797 (special volume Benthos '86).

Hallock, P. and Glenn, E.C. (1985). Numerical analysis of foraminiferal assemblages: A tool for recognizing depositional facies in lower Miocene reef complexes. *J. Paleont.* **59**:1384–1396.

Hallock, P. and Glenn, E.C. (1986). Larger foraminifera: A tool for paleoenvironmental analysis of Cenozoic carbonate depositional facies. *Palaios* **1**:55–64.

Hallock, P. and Hansen, H.J. (1979). Depth adaptation in *Amphistegina*: Change in lamellar thickness. *Geol. Soc. Denmark Bull.* **27**:99–104.

Hallock, P., Forward, L.B. and Hansen, H.J. (1986). Environmental influence of test shape in *Amphistegina*. *J. Foram. Res.* **16**:224–231.

Haynes, J. (1965). Symbiosis, wall structure and habitat in foraminifera. *Cushman Found. Foraminiferal Res. Contrib.* **16**:40–43.

Haynes, J.R. (1981). *Foraminifera.* Macmillan, London.

Hedley, R.H. (1964). The biology of foraminifera. *Int. Rev. gen exp. Zoo.* **1**:1-45.

Hickman, C.S. and Lipps, J.H. (1983). Foraminiferivory: Selective ingestion of foraminifera and test alterations produced by the neogastropod *Olivella*. *J. Foram. Res.* **12**:108–114.

Hottinger, L. (1977). Foraminifères operculiniformes. *Mém. Muséum natl. Hist. nat. (Paris), ser. C* **40**:3–159.

Hottinger, L. (1978). Comparative anatomy of elementary shell structures in selected larger foraminifera. In *Foraminifera* (eds R.H. Hedley and C.G. Adams), Vol. 3, pp. 203–266. Academic Press, London.

Hottinger, L. (1979). Araldit als Helfer in der Mikropaläontologie. *Ciba-Geigy Aspekte* **3**:1–10.

Hottinger, L. (1984). Larger foraminifera: The significance of complex shell-

structures. In *Benthos '83: Second International Symposium on Benthic Foraminifera*, pp. 309–315, Pau and Bordeaux, April.
Hottinger, L. (1986). Construction, structure, and function of foraminiferal shells. *Symp. Syst. Ass., Spec. Pub.* **30**:219–235.
Hottinger, L. and Leutenegger, S. (1980). The structure of calcarinid foraminifera. *Schweiz. Paläont. Abh.* **101**:115–151.
Jepps, M.W. (1956). *The Protozoa, Sarcodina*. Oliver and Boyd, Edinburgh.
Kuile, B. ter and Erez, J. (1984). In situ growth rate experiments on the symbiont-bearing foraminifera *Amphistegina lobifera* and *Amphisorus hemprichii*. *J. Foram. Res.* **14**:262–276.
Kuile, B. ter, Erez, J. and Lee, J.J. (1987). The role of feeding in the metabolism of larger symbiont-bearing foraminifera. *Symbiosis* **4**:335–350.
Larsen, A.R. (1976). Studies of Recent *Amphistegina*: Taxonomy and some ecological aspects. *Israel J. Earth-Sci.* **25**:1–26.
Larsen, A.R. and Drooger, C.W. (1977). Relative thickness of the test in *Amphistegina* species of the Gulf of Elat. *Utrecht Micropal. Bull.* **15**:225–239.
Lee, J.J. (1980). Nutrition and physiology of the foraminifera. In *Biochemistry and Physiology of Protozoa* (eds M. Levandowsky and S. Hunter), 2nd edn, Vol. 3, pp. 43–66. Academic Press, London.
Lee, J.J. (1983). Perspective on algal endosymbionts in larger foraminifera. *Int. Rev. Cytol.* **11**:49–77 (suppl.).
Lee, J.J. and Bock, W.D. (1976). The importance of feeding in two species of soritid foraminifera with algal symbionts. *Bull. Mar. Sci.* **26**:530–537.
Lee, J.J. and Hallock, P. (1986). Algal symbiosis as the driving force in the evolution of larger foraminifera. *Ann. N.Y. Acad. Sci.* **503**:330–347.
Lee, J.J. and McEnery, M. (1983). Symbiosis in foraminifera. In *Algal Symbiosis* (ed. L.J. Goff), pp. 37–68. Cambridge University Press, Cambridge.
Lee, J.J. and Zucker, W. (1969). Algal flagellate symbiosis in the foraminifer *Archaias*. *J. Protozool.* **16**:71–81.
Lee, J.J., Crockett, L.J., Hagen, J. and Stone, R. (1974). The taxonomic identity and physiological ecology of *Chlamydomonas hedleyi* sp. nov., algal flagellate symbiont from the foraminifer *Archaias angulatus*. *Br. J. Phycol.* **9**:407–422.
Lee, J.J., McEnery, M., Kahn, E. and Schuster, F. (1979). Symbiosis and the evolution of larger foraminifera. *Micropaleontol.* **25**:118–140.
Lee, J.J., McEnery, M.E., Röttger, R. and Reimer, Ch.W. (1980). The isolation, culture and identification of endosymbiotic diatoms from *Heterostegina depressa* d'Orbigny and *Amphistegina lessonii* d'Orbigny (larger foraminifera) from Hawaii. *Bot. Mar.* **23**:297–301.
Lee, J.J., Soldo, A.T., Reisser, W., Lee, M.J., Jeon, K.W. and Görtz, H.-D. (1985). The extent of algal and bacterial endosymbioses in Protozoa. *J. Protozool.* **32**:391–403.
Leutenegger, S. (1984). Symbiosis in benthic foraminifera: Specificity and host adaptations. *J. Foram. Res.* **14**:16–35.
Leutenegger, S. and Hansen, H.J. (1979). Ultrastructural and radiotracer studies of pore-function in foraminifera. *Mar. Biol.* **54**:11–16.
Levy, A. (1982). Sur la survie de certains Foraminifères dan les eaux continentales et sur ses conséquences. *Mém. Soc. Geol. France, N.S.* **144**:161–171.
Levy, A. (1984). Données nouvelles sur la paléogéographie du sud Turisien au Quaternaire Supérieur. In *Benthos '83: Second International Symposium on Benthic Foraminifera*, pp. 369–379, Pau and Bordeaux, April.

Lin, T.Y. and Stotesbury, S.D. (1981). *Structural Concepts and Systems for Architects and Engineers*. John Wiley, New York.

Lipps, J.H. (1975). Feeding strategies and test function in foraminifera. In *Proceedings of Benthonics '75*, p. 26. Dalhousie University Press, Halifax, Nova Scotia.

Lipps, J.H. (1983). Biotic interactions in benthic foraminifera. In *Biotic Interactions in Recent and Fossil Benthic Communities* (eds M.J.S. Tevesz and P.L. McCall), pp. 331–376. Plenum Press, New York.

Loeblich, A.R., Jr and Tappan, H. (1964). Sarcodina, chiefly "Thecamoebians" and Foraminiferida. In *Treatise on Invertebrate Paleontology* (ed. R.C. Moore), Vol. 1, pp. 1–510a, Vol. 2, pp. 511–900. Geological Society of America/University of Kansas Press, New York.

Mageau, N.C. and Walker, D.A. (1976). Effects of ingestion on foraminifera by larger invertebrates. *Maritime Sediments, Spec. Pub.* **1**:89–105.

Malmgren, BA. and Kennett, J.P. (1982). The potential of morphometrically based phylo-zonation: Application of a late Cenozoic foraminiferal lineage. *Mar. Micropaleontol.* **7**:285–296.

Marszalek, D.S., Wright, RC. and Hay, W.W. (1969). Function of the test in foraminifera. *Trans. Gulf-Cst. Ass. Geol. Socs* **19**:341–352.

McRoy, C.P. and Barsdate, R.J. (1970). Phosphate absorption in eel grass. *Limnol. Oceanogr.* **5**:14–20.

Muller, P.H. (1974). Sediment production and population biology of the benthic foraminifer *Amphistegina madagascariensis*. *Limnol. Oceanogr.* **19**:802–809.

Muller, P.H. (1976). Sediment production by shallow-water, benthic foraminifera at selected sites around Oahu, Hawaii. *Maritime Sediments, Spec. Pub.* **1**:263–265.

Muller, P.H. (1978). Carbon-14 fixation and loss in a foraminiferal-algal symbiont system. *J. Foram. Res.* **8**:34–41.

Myers, E.H. (1943). Life activities of foraminifera in relation to marine ecology. *Proc. Am. Phil. Soc.* **86**:439–459.

Painter, P.K. and Spencer, R.S. (1984). A statistical analysis of variants of *Elphidium excavatum* and their ecological control in southern Chesapeake Bay, Virginia. *J. Foram. Res.* **14**:120–128.

Reif, W.-E. (1982). Functional morphology on the procrustean bed of the neutralism–selectionism debate. – Notes on the Constructional Morphology approach. *Neues Jahr. Geol. Pal.* **164**:46–59.

Reiss, Z. and Hottinger, L. (1984). *The Gulf of Aqaba: Ecological Micropaleontology*. (Ecological Studies Vol. 50) Springer-Verlag, New York.

Reiss, Z., Leutenegger, S., Hottinger, L., Fermont, W.J.J., Mevlenkamp, J.E., Thomas, E., Hansen, H.J., Buchardt, B., Larsen, A.R. and Drooger, C.W. (1977). Depth-relations of Recent larger foraminifera in the Gulf of Aqaba-Elat. *Utrecht Micropalontol. Bull.* **15**:1–244.

Ross, C.A. (1974). Evolutionary and ecological significance of large calcareous foraminifera (Protozoa), Great Barrier Reef. In *Proceedings of the Second International Coral Reef Symposium*, Vol. 1, pp. 327–333. Great Barrier Reef Commission, Brisbane, Australia.

Röttger, R. (1972a). Die Kultur von *Heterostegina depressa* (Foraminifera: Nummulitidae). *Mar. Biol.* **15**:50–159.

Röttger, R. (1972b). Die Bedeutung der Symbiose von *Heterostegina depressa* (Foraminifera, Nummulitidae) für hohe Siedlungsdichte und Karbonatproduktion. *Abh. dt. Zool. Ges.* **1971**:42–47.

Röttger, R. (1972c). Analyse von Wachstumskurven von *Heterostegina depressa* (Foraminifera: Nummulitidae). *Mar. Biol.* **17**:228–242.

Röttger, R. (1973). Die Ektoplasmahülle von *Heterostegina depressa* (Foraminifera: Nummulitidae). *Mar. Biol.* **21**:127–138.

Röttger, R. (1974). Larger Foraminifera: Reproduction and early stages of development in *Heterostegina depressa*. *Mar. Biol.* **26**:5–12.

Röttger, R. (1976). Ecological observation of *Heterostegina depressa* (Foraminifera, Nummulitidae) in the laboratory and in its natural habitat. *Maritime Sediments, Spec. Pub.* **1**:75–80.

Röttger, R. and Berger, W.H. (1972). Benthic foraminifera: Morphology and growth in clone culture of *Heterostegina depressa*. *Mar. Biol.* **15**:89–94.

Röttger, R. and Inst. Wiss. Film (1984). Die Großforaminifere *Heterostegina depressa* – Vielteilung der mikrosphärischen und der megalosphärischen Generation. Film C 1506 des IWF, Göttingen 1983. Publikation von R. Röttger, *Publ. Wiss. Film., Sekt. Biol.*, Ser. 16, No. 28/C 1506, 20 pp.

Röttger, R. and Krüger, R. (1990). Observations on the biology of Calcarinidae (Foraminiferida). *Marine Biology* **106**:419–425.

Röttger, R. and Richwien, M. (1977). Sheaths and locomotion in the larger foraminiferan *Heterostegina depressa*. In *Proceedings of the Fifth International Congress on Protozoology*, New York (abstract no. 368).

Röttger, R., Irwan, A., Schmaljohann, R. and Franzisket, L. (1980). Growth of the symbiont-bearing foraminifera *Amphistegina lessonii* d'Orbigny and *Heterostegina depressa* d'Orbigny (Protozoa). In *Endocytobiology, Endosymbiosis and Cell Biology* (eds W. Schwemmler and H.E.A. Schenk), pp. 124–132. Walter de Gruyter and Co., New York.

Röttger, R., Spindler, M., Schmaljohann, R., Richwien, M. and Fladung, M. (1984). Functions of the canal system in the rotaliid foraminifer, *Heterostegina depressa*. *Nature* **309**:789–791. Erratum note: *Nature* **315**:77 (1985).

Schmaljohann, R. (1980). Ernährungsphysiologische Untersuchungen an der Foraminifere *Heterostegina depressa* (Nummulitidae). Ph.D. thesis, University of Kiel.

Severin, K.P. (1983).Test morphology of benthic foraminifera as a discriminator of biofacies. *Mar. Micropaleontol.* **8**:65–76.

Signor, P.W., III (1982). A critical re-evaluation of the paradigm method of functional inference. *Neues Jahr. Geol. Pal.* **164**:59–63.

Sliter, W.V. (1971). Predation on benthic foraminifers. *J. Foram. Res.* **1**:20–29.

Smith, R.J. (1982). On the mechanical reduction of functional morphology. *J. Theor. Biol.* **96**:99–106.

Smout, A.H. (1954). Lower Tertiary foraminifera of the Qatar Peninsula, pp. 1–96. British Museum (Natural History), London.

Spencer, R.S. (1987). Canonical variate analysis of selected benthic foraminifera: A preliminary study of a potential paleobathymetric tool. *Palaios* **2**:91–100.

Spindler, M. (1976). Gehäuseanatomie und Ablauf des Kammerbauvorgangs bei *Heterostegina depressa* (Nummulitidae: Foraminifera). Untersuchungen im Licht- und Elektronenmikroskop. Ph.D. thesis, University of Kiel.

Spindler, M. and Röttger, R. (1973). Der Kammerbauvorgang der Großforaminifere *Heterostegina depressa* (Nummulitidae). *Mar. Biol.* **18**:146–159.

Telford, M. (1985). Domes, arches and urchins: The skeletal architecture of echinoids (Echinodermata). *Zoomorphology* **105**:114–124.

Towe, K.M. and Cifelli, R. (1967). Wall structure in the calcareous foraminifera:

Crystallographic aspects and a model for calcification. *J. Paleontol.* **41**:742–762.

Wainwright, S.A., Biggs, W.D., Currey, J.D. and Gosline, J.M. (1982). *Mechanical Design in Organisms.* Princeton University Press, Princeton, N.J.

Wei, K.-Y. (1987). Multivariate morphometric differentiation of chronospecies in the late Neogene planktonic foraminiferal lineage *Globoconella. Mar. Micropaleontol.* **12**:183–202.

Wetmore, K.L. (1987). Correlations between test strength, morphology and habitat in some benthic foraminifera from the coast of Washington. *J. Foram. Res.* **17**:1–13.

Wetmore, K.L. (1988a). Test strength, mobility and functional morphology of benthic foraminifera. Unpublished Ph.D. dissertation, The Johns Hopkins University.

Wetmore, K.L. (1988b). Burrowing and sediment movement by benthic foraminifera, as shown by time-lapse cinematography. *Revue de Paleobiologie* **2**:921–927 (special volume Benthos '86).

Wetmore, K.L. and Plotnick, R.E. (1991). Correlations between test morphology, crushing strength and habitat in *Amphistegina gibbosa, Archaias angulatus* and *Laevipeneroplis proteus* from Bermuda. *J. Foram. Res.* (in press).

4
Mechanisms for calcification and carbon cycling in algal symbiont-bearing foraminifera

BENNO TER KUILE

1.0	Introduction	73
2.0	Theories of calcification	74
	2.1 The bicarbonate splitting theory	74
	2.2 The organic matrix theory	75
	2.3 Energy-dependent concentration of the reactants	75
	2.4 The poison removal theory	76
3.0	Schemes for calcification in foraminifera	77
	3.1 *Amphistegina lobifera*	78
	3.2 *Amphisorus hemprichii*	81
4.0	Calcification as part of the carbon budget	82
	4.1 Carbon budget for *Amphistegina lobifera*	82
	4.2 Carbon budget for *Amphisorus hemprichii*	83
	4.3 Carbon budgets for other species	83
	4.4 Carbon cycling of planktonic species	86
	References	87

1.0 INTRODUCTION

Carbon is taken up by algal symbiont-bearing larger foraminifera for three reasons: (1) growth and maintenance of the host's organic matter; (2) calcification, i.e. growth of the skeleton; and (3) photosynthesis by the symbionts. The host can derive additional carbon for growth and maintenance from two sources: photosynthates transferred by the symbionts to the host and/or organic carbon obtained from food. Carbonate is the form ultimately used in calcification, although it is not necessarily taken up in this form from sea water. The symbionts, like all algae, fix inorganic carbon as CO_2, but may concentrate bicarbonate (Lucas, 1983), which is consequently converted to CO_2. In this chapter, the cycling of carbon will be discussed in relation to the mechanisms for calcification. Calcification in invertebrate aquatic systems is treated more generally as well, in order to provide a better background for the discussion of calcification in foraminifera.

The available data on calcification in foraminifera suggest that this process cannot be described by one of the existing theories on calcification, though it may be explained by a combination of these. Moreover, it has been demonstrated that important differences have been shown between perforate and imperforate algal symbiont-bearing hosts (Hemleben et al., 1986; ter Kuile and Erez, 1987, 1988; ter Kuile et al., 1989a). Studies by Hemleben et al. (1977) and Erez (1983) suggest that the planktonic foraminifera, which are perforate, behave physiologically like the perforate larger foraminifera.

2.0 THEORIES ON CALCIFICATION

2.1 The bicarbonate splitting theory

This is one of the earliest theories on calcification, having been proposed in the nineteenth century by Cohn (1863) and Pringsheim (1888) for calcification in algae. It has been given several names, e.g. "CO_2 utilization", "bicarbonate usage" (Borowitzka, 1977, 1982a) and "carbonate trapping" (Pentecost, 1985). Basically, these are the same and feature as their driving principle the increase in carbonate concentration as a consequence of the rise in pH following the fixation of CO_2 by photosynthesis. The fixed CO_2 may initially have been taken up in the form of HCO_3^- that is then split into CO_2 and OH^-. Removal of the latter from the site of photosynthesis will cause a rise in pH. The actual splitting of two molecules of bicarbonate into CO_2 and CO_3^{2-}, the only alternative to a pH-driven mechanism, has never been proposed and is contradicted by most available evidence. Therefore, this mechanism is applicable only to situations where little or no equilibration of the pH to the environment can occur.

Most of the experimental support for the application of this theory to algal-bearing symbiotic systems (e.g. hermatypic corals and foraminifera) comes from observations that light stimulates calcification in these systems (e.g. Goreau, 1963; Lee and Zucker, 1969; Röttger, 1974; Erez, 1978). Erez's (1978) stable isotope data seemed to contradict the idea that the bicarbonate splitting theory was applicable to larger foraminifera. Later studies of corals in which low concentrations of the inhibitor of photosystem II (DCMU) were used, suggested that the theory was not applicable to this symbiotic system either (Chalker and Taylor, 1975). Additional experiments on planktonic foraminifera (Bé et al., 1982; Erez, 1983) and larger foraminifera (ter Kuile et al., 1989a) led to the same conclusions. *Amphistegina lessonii*, a perforate larger foraminifer, showed slightly increased calcification rates when photosynthetic CO_2 fixation was inhibited by DCMU. This observation indicated the existence of competition for inorganic carbon between the processes of photosynthesis and calcification. This was also confirmed in *Amphistegina lobifera*, but not in *Amphisorus hemprichii*, an imperforate larger foraminiferan species (ter Kuile et al., 1989b). Hence, it can be concluded that in contrast to calcareous algae (Borowitzka, 1977, 1982a), calcification in algal symbiont-bearing foraminifera is not stimulated by the photosynthetic fixation of CO_2.

2.2 The organic matrix theory

Another theory on calcification that is not necessarily incompatible with the first, is the organic matrix theory. This theory emphasizes the importance of an organic matrix – also called *anlagen* (Hemleben et al., 1977) or "primary organic lining" (Angell, 1979, 1980) – in the initiation and direction of calcification (for a review, see Weiner, 1986). The tertiary structure of this matrix is proposed to be such that it can initiate or inhibit calcification by spatially arranging the calcium and carbonate ions. In planktonic and perforate larger foraminifera, the importance of such an organic matrix has been amply documented (Hemleben et al., 1977, 1986; Anderson and Bé, 1978; Röttger et al., 1984; Weiner and Erez, 1984). The occurrence of calcification on this matrix, and not elsewhere in the organism, may indicate an elevated concentration of the reactants near the matrix (see below). In imperforate species, this mechanism may play a minor role. Calcification in this group occurs in the vesicles, where needles are formed that are later transported to the site of calcification (Angell, 1979, 1980; Hemleben et al., 1986). It is possible that calcification is initiated by organic molecules but, at present, there is little evidence for this.

2.3 Energy-dependent concentration of the reactants

Even though there is considerable evidence that active concentration of calcium and carbonate occur at least in some calcifying systems, no theory including this feature has been proposed. Inhibitor studies have demonstrated the existence of a Ca-ATP-ase in hermatypic (Chalker and Taylor, 1975; Chalker, 1976) and gorgonian corals (Kingsley, 1984). The coccolithosomes of coccolithophorids were shown to contain calcium at a concentration of 6 M, 600 times the concentration of the medium (van der Wal et al., 1982). Concentration to such a degree is energy- (ATP)-dependent. The rapid calcification rates of algal symbiont-bearing foraminifera in the light in the presence of DCMU, when compared with rates in the dark, suggest that ATP synthesized through cyclic phosphorylation using photosystem I can be transferred to the host and used for calcification.

In these foraminifera, the most obvious energy-dependent system in the calcification process is the concentration of carbonate into the inorganic carbon pool of perforate species (ter Kuile and Erez, 1988; ter Kuile et al., 1989a). The inorganic carbon concentration in this pool was estimated to be up to 1000 times that in the medium (ter Kuile and Erez, 1988). The pH of the pool was estimated to be at least one unit above the pH of the cytoplasm (ter Kuile et al., 1989a), suggesting that a pH gradient must also be maintained. The existence of calcium pools in planktonic foraminifera was suggested by the time-course experiments of Anderson and Faber (1984). Calculations of the discrepancy between growth rates of algal symbiont-bearing larger foraminifera estimated using ^{45}Ca tracer methodology and increases in dry weight (J. Erez, unpublished), suggested the probability of an internal calcium pool, that may contain up to 3 M calcium. In addition, the inhibition of calcification in *A. lobifera* by vanadate, an inhibitor of the Ca-ATP-ase, suggests that perforate foraminifera actively concentrate calcium. The active concentration of the reactants can be combined with an organic matrix-based mechanism for calcification, explaining the rapid calcification shown to occur on the matrix or primary organic lining (Röttger, 1974; Hemleben et al., 1977; Angell, 1979; Röttger et al., 1984).

2.4 The poison removal theory

Sea water is supersaturated with respect to calcite and aragonite by a factor of 5–6. Spontaneous precipitation of calcium carbonate is thought to be prevented by the presence of other ions in the solution. The poison removal theory (Borowitzka, 1977; Chalker, 1983; Swart, 1983) proposes that the organism removes ions that inhibit calcification (e.g. ammonium, phosphate and magnesium) from the fluid at the site of calcification. Calcification can then occur spontaneously. This theory is compatible with the organic matrix theory and to some extent with the active concentration theory. In the latter case, inhibiting ions might be removed, while calcium and carbonate are transported in the other direction. In a wider sense, the word "removal" could be used as well to describe the dilution of inhibiting ions due to the concentration of calcium and carbonate.

The poison removal theory is particularly applicable to the imperforate foraminifera, where calcification occurs in the vesicles (Hemleben *et al.*, 1986). The removal of inhibiting ions from these vesicles would be sufficient to initiate calcification. It is possible that some organic macromolecules or matrices serve as crystallization points (Angell, 1979; Hemleben *et al.*, 1986). Experimental support for this mechanism in imperforate species comes from the observation that calcification in *A. hemprichii* depends less on ATP than in *A. lobifera* and is rate-limited by diffusion, rather than by an enzymatic (transport?) step (ter Kuile *et al.*, 1989a).

3.0 SCHEMES FOR CALCIFICATION IN FORAMINIFERA

As discussed previously, the calcification mechanisms in perforate and imperforate larger foraminifera differ to a great extent. Figures 1 and 2 represent schemes for the calcification process in each group. These schemes are based on information derived from studies using sea water with variable inorganic carbon levels and pH combined with

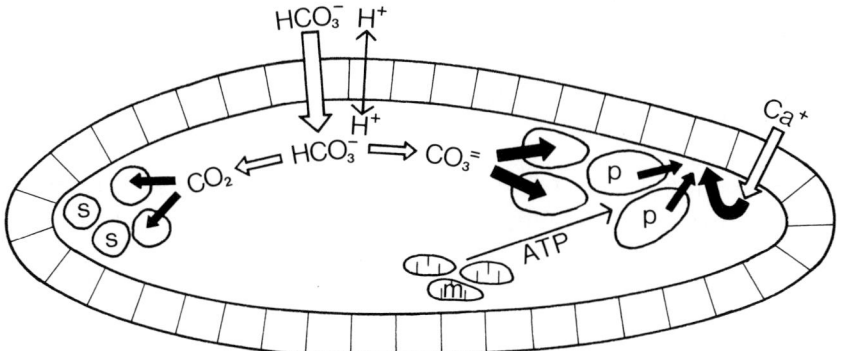

Fig. 1. Scheme for inorganic carbon and calcium uptake in *Amphistegina lobifera*. S, symbionts; P, pool; M, mitochondria; filled arrows, enzyme-mediated uptake; open arrows, uptake by diffusion; thin arrows, transfer of ATP or H^+. For an explanation, see text.

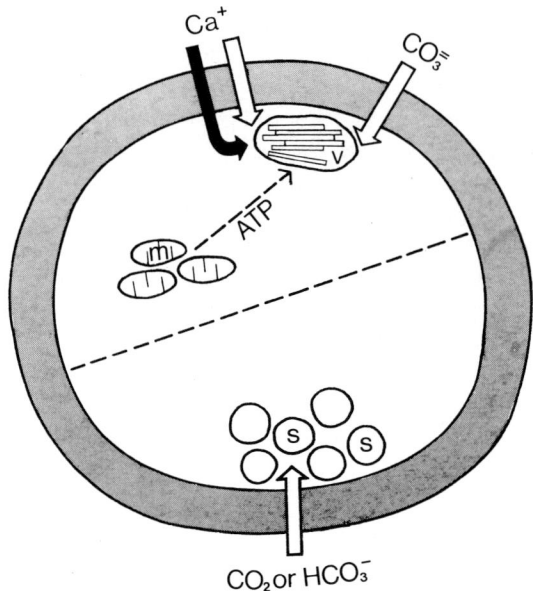

Fig. 2. Scheme similar to that presented in Fig. 1 for *Amphisorus hemprichii*. V, vesicles; other symbols as in Fig. 1.

experiments using specific inhibitors (ter Kuile *et al.*, 1989a,b). *Amphistegina lobifera* was used as a representative of the perforate species and *Amphisorus hemprichii* as a representative of the imperforate group. Studies of the uptake kinetics of inorganic carbon in these and other species suggest that the results can be generalized (ter Kuile and Erez, 1987).

3.1 Amphistegina lobifera

The total uptake of inorganic carbon by *A. lobifera* as a function of external concentrations fits the Hill–Whittingham equation for enzymatic reactions that are provided with their substrate by a diffusion step. The optimum pH was between 8.0 and 8.5. These observations indicate that inorganic carbon is initially taken up in the form of bicarbonate in a diffusion-limited step. Experiments on *Fragilaria shiloi*, one of the common algal symbionts of *A. lobifera*, showed that the symbionts take up inorganic carbon for photosynthesis in the form of CO_2. The rate-limiting step is probably fixation by the

Rubisco enzyme (Michaelis-Menten Kinetics, $K_m = 400\,\mu\text{M}$ for total inorganic carbon) (ter Kuile et al., 1989a). The inorganic carbon for calcification is not taken up directly from the cytoplasm, but first concentrated in an inorganic carbon pool. Carbon from this pool is used primarily for calcification and not for photosynthesis. The carbon in this pool is derived from (1) the cytoplasm, (2) sea water and (3) metabolic carbon that was initially photoassimilated by the symbionts and afterwards respired ($\sim 10\%$ C in skeleton). Experiments using the external pH as a variable showed that the uptake rates into the pool for calcification were a function of the carbonate concentration and fit Michaelis-Menten kinetics ($K_m = 500\,\mu\text{M}$ for total inorganic carbon). This suggests that inorganic carbon is concentrated in the pool in the form of carbonate by means of a transport enzyme.

The nature of this transport enzyme is open to speculation, because to our knowledge no transport enzyme for carbonate has been described. A variety of mechanisms has been proposed for the transport of bicarbonate in aquatic plants, such as electrogenic transport (uniport), HCO_3^-/H^+ co-transport and HCO_3^-/OH^- anti-port (for a review, see Lucas, 1983). It has been speculated (ter Kuile and Erez, 1988) that the pool is located in the electron light vesicles observed with the electron microscope in perforate, but not in imperforate, foraminifera (Leutenegger, 1977; Koestler et al., 1985). If this is indeed the case, it is possible that there is a simple mechanism for concentrating carbonate in the pool. A vesicle could be formed initially containing sea water, while the transport mechanism could consist of a Cl^-/CO_3^{2-} anti-porter that gradually changes the sodium chloride into sodium carbonate. In order to maintain the charge balance and the ionic strength of the solution, Na^+ would have to be transported inwards as well. The resulting concentration of inorganic carbon in the pool would be approximately 500 mM, comfortably in the estimated range. The collapse of the pool as a result of treatment with weak acids (ter Kuile et al., 1989a), suggests that the pH of the pool is at least one unit more basic than the cytoplasm. Therefore, it is reasonable to assume that the membrane of the pool contains a pH-regulating mechanism as well.

The pH curve for photosynthesis of the isolated symbionts and the symbionts in situ was very similar (ter Kuile et al., 1989a), suggesting that the pH of the cytoplasm can exchange freely with the external pH. As discussed above, this explains our earlier conclusions that the bicarbonate splitting theory is not applicable to foraminifera, because the absence of pH exchange with the environment is essential for this mechanism to work. In fact, when pH exchange is more rapid than the consumption of CO_2, at a pH of sea water five times less than CO_3^{2-}, competition for inorganic carbon may occur between the processes of photosynthesis and calcification. It is particularly likely to occur when the external inorganic carbon concentration is lowered artificially. Such competition was, in fact, demonstrated using kinetic experiments in the presence and absence of low concentrations of DCMU (ter Kuile et al., 1989b). The enzyme carbonic anhydrase that catalyses the reversible conversion of dissolved CO_2 into bicarbonate is relatively inactive in corals (Graham and Smillie, 1976; Isa and Yamazato, 1984). However, it was thought to have carbonate concentrating function and hence to stimulate calcification. The activity of carbonic anhydrase (CA) could not be demonstrated in foraminifera (ter Kuile et al., 1989b). The addition of CA to the medium stimulated photosynthesis in *A. lobifera*, primarily at inorganic carbon levels lower than those of sea water. Surprisingly, CA had a strong inhibiting effect on calcification. This inhibition can be explained by assuming leakage of the uncharged CO_2 molecule, present in traces even at high pH, out of the pool. Usually, this leakage of CO_2 would hardly affect the total inorganic carbon in the pool, but in the presence of CA it is replaced rapidly, causing the collapse of the pool. The absence of natural CA in foraminifera can readily be understood, because it would change the balance between the consumption of inorganic carbon for photosynthesis and calcification and because it inhibits the concentration of inorganic carbon into the pool. The CA antagonite ethoxyzolamide stimulated photosynthesis and inhibited calcification in the calcareous alga *Halimeda tuna* (Borowitzka and Larkum, 1976). This suggests that in calcareous algae, CA may have a role in balancing the inorganic carbon flows for photosynthesis and calcification.

From a mechanistic point of view, the main function of the pool may be to provide large quantities of carbonate when rapid calcification occurs during chamber formation. Chamber formation, and hence calcification, is a discontinuous process in perforate foraminifera. Storage of inorganic carbon in the pool destined for calcification allows the organism to maintain a steady inflow of inorganic carbon while calcifying. On this view, the pool functions as a capacity buffer. Experimental support for this comes from kinetic experiments measuring ^{14}C uptake and incorporation into the skeleton in specimens whose pools had been emptied by preincubation in inorganic carbon-free sea water (ter Kuile et al., 1989b). Under these conditions, no time-lag for calcification occurred and the pool started filling up only after calcification reached its maximum. In specimens that have been preincubated in sea water with normal inorganic carbon levels, incorporation of ^{14}C into the skeleton was measured only after the pool was filled up (ter Kuile and Erez, 1987, 1988). This suggests that when organisms are first starved of inorganic carbon, it is used immediately for calcification instead of being stored. However, an alternative explanation is also possible: During preincubation in inorganic carbon-free sea water, the pool is completely emptied. If the pool is indeed located in the vesicles as suggested, these vesicles may collapse as a result of the absence of inorganic carbon. If this is the case, time will be needed to renovate the vesicles, which is reflected in the delay in filling the pool.

3.2 Amphisorus hemprichii

Carbon cycling in imperforate foraminifera is much simpler than in the perforate species. Kinetic experiments indicate that imperforate species obtain inorganic carbon for calcification directly from sea water and not through a pool (ter Kuile and Erez, 1987). This is similar to Angell's (1979) conclusion that neither calcium nor carbonate are stored in the cytoplasm (prior to calcification) in *Rosalina floridana*, a non-symbiont-bearing small perforate species. The absence of a pool was later confirmed for *A. hemprichii* (ter Kuile and Erez, 1987, 1988). Experiments with variable external inorganic carbon concentrations suggested that *A. hemprichii* takes up inorganic carbon for photosynthesis and calcification in two separate flows. The photosynthetic uptake fits the Hill–Whittingham equation, indicating that, as in the case of the total uptake of *A. lobifera*, an enzymatic step is preceded by a rate-limiting diffusion step (ter Kuile et al., 1989a). Unfortunately, we did not have the isolated symbionts of *A. hemprichii* at our disposal, and hence it could not be determined which inorganic carbon species is taken up. Diffusion of inorganic carbon was the rate-limiting step for calcification in the range of external concentrations examined. Experiments with variable pH at constant inorganic carbon levels demonstrated that CO_3^{2-} is the species taken up. Calcification of *A. hemprichii* was less affected by uncouplers and inhibitors than *A. lobifera*, indicating that it depends less on energy. This was expected, because diffusion is rate-limiting in *A. hemprichii*, whereas an enzymatic step is limiting in *A. lobifera* (ter Kuile et al., 1989a). Experiments using DCMU and carbonic anhydrase (CA) suggested that competition for inorganic carbon between photosynthesis and calcification, as observed in *A. lobifera*, does not occur in *A. hemprichii*, nor does mutual enhancement. Photosynthesis by the symbionts of *A. hemprichii* was not affected by CA, but calcification was slightly inhibited when CA was added to the medium (ter Kuile et al., 1989b).

These observations suggest that *A. hemprichii* takes up inorganic carbon for photosynthesis and calcification in two separate flows that do not interfere with each other. As mentioned above, calcification in imperforate species occurs in the vesicles (Hemleben et al., 1986). Considering all of the experimental evidence available, calcification in the imperforate foraminifera can best be explained by the poison removal theory.

4.0 CALCIFICATION AS PART OF THE CARBON BUDGET

The importance of calcification as a carbon sink in foraminifera varies as a function of size (age). Young specimens, especially in the perforate group, use a greater part of the incoming inorganic carbon for calcification than older specimens. Figures 3 and 4 present the carbon budgets of two larger foraminifera. These budgets reflect the mechanistic differences discussed above, and incorporate feeding on algae and respiration as additional parameters. It is possible that the relations between the flows of the schemes for *Amphistegina lobifera* and *Amphisorus hemprichii* can be generalized to the perforate and imperforate larger foraminifera respectively.

4.1 Carbon budget for *Amphistegina lobifera*

One of the main points of dispute with respect to carbon cycling in foraminifera and other symbiotic systems is the contribution of feeding as a carbon source. Lee and Bock (1976) estimated that in *Archaias angulatus* and *Sorites marginalis*, two imperforate species, feeding exceeds photosynthesis by more than 10 times. Their experiments were short-term experiments, designed to enhance feeding rates. Later, time-course and pulse-chase experiments suggested that most of the carbon taken up in food by *Amphistegina lobifera* is rapidly egested and not incorporated or respired (ter Kuile et al., 1987). Feeding mainly served as a source of nutrients, rather than reduced carbon. Little or none of the carbon taken up through feeding ended up in the skeleton. Carbon initially photoassimilated by the symbionts and respired by the host was incorporated into the skeleton, accounting for $\sim 10\%$ of the skeletal carbon in total (ter Kuile and Erez, 1987). Foraminifera are extremely conservative with respect to carbon (Hallock, 1978; ter

Kuile and Erez, 1987): only 20% of the carbon fixed during a 2-h labelled incubation was released to the environment during a cold chase of 1 week. Corals, if we can generalize from one experiment, lose more than half of their carbon fixed during 2-h incubations in 40 h on the reef (Crossland et al., 1980). This difference may be caused by the excretion of mucus by corals (Crossland, 1987).

The numbers given in Fig. 3 are valid for relatively young specimens of *A. lobifera*, weighing on average between 66 and 72 µg. For this size class, the incorporation of inorganic carbon into the skeleton is almost double the photoassimilation by the symbionts. Very light (young) specimens had either smaller or greater ratios under comparable experimental conditions (ter Kuile and Erez, in press). Heavier specimens showed a general trend towards more photosynthesis than calcification, but a declining total uptake.

4.2 Carbon budget for *Amphisorus hemprichii*

The budget for *A. hemprichii* (Fig. 4) and, in principle, other imperforate larger foraminifera, is much simpler than that of the perforate species. Uptake for photosynthesis and calcification occurs in two separate flows that appear not to affect each other (ter Kuile et al., 1989a,b). The organisms used for compiling the budget presented here had average weights of between 242 and 523 µg. The uptake of inorganic carbon for photosynthesis was approximately equal to the incorporation into the skeleton. Younger specimens had slightly higher calcification: photosynthesis ratios, whereas older specimens had slightly lower ratios (ter Kuile and Erez, in press). The differences, however, were much smaller than in the case of *A. lobifera*. The contribution of feeding to the carbon budget is problematic. Carbon/phosphorous double-labelling pulse-chase experiments suggested that *A. hemprichii* may use feeding as a source for reduced carbon (ter Kuile et al., 1987). More research is needed to determine the relative importance of feeding and photosynthesis. The estimated respiration rates were strongly dependent on feeding rates.

4.3 Carbon budgets for other species

Complete carbon budgets as drawn up for *A. lobifera* and *A. hemprichii* (Figs 3 and 4) have not been presented for other species. Observations on the perforate species *Heterostegina depressa* (Erez, 1978; Röttger et

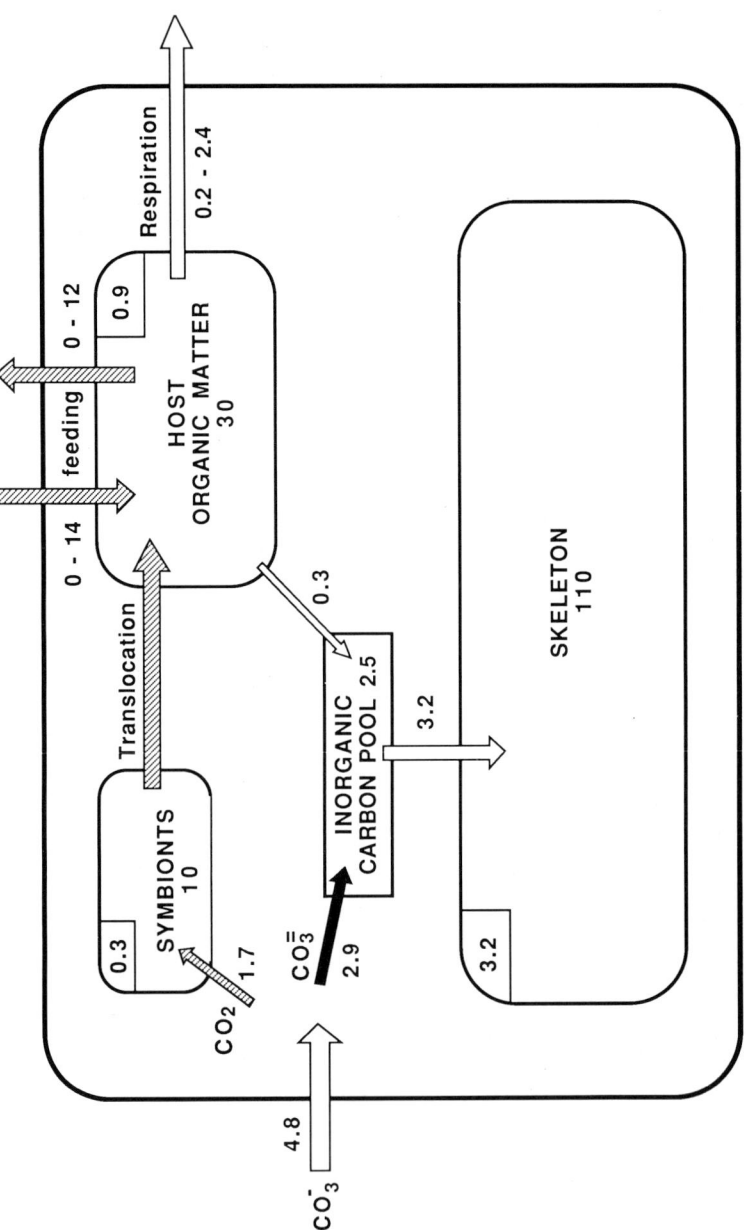

Fig. 3. Carbon budget for *Amphistegina lobifera*. The names and the sizes of the different compartments are given in capital letters and numbers; the names of the processes and the amounts of carbon transferred are given in lower-case letters. The open arrows indicate the transfer of inorganic carbon, the hatched arrows indicate the transfer of organic carbon and the solid arrow indicates the active transport of carbonate. The numbers framed in the corners of the compartments indicate the daily increase in size of that compartment. *Units*: sizes of the compartments in µg C/mg foram, rates of fixation and transfer and daily increase in µg C/mg foram/24 h.

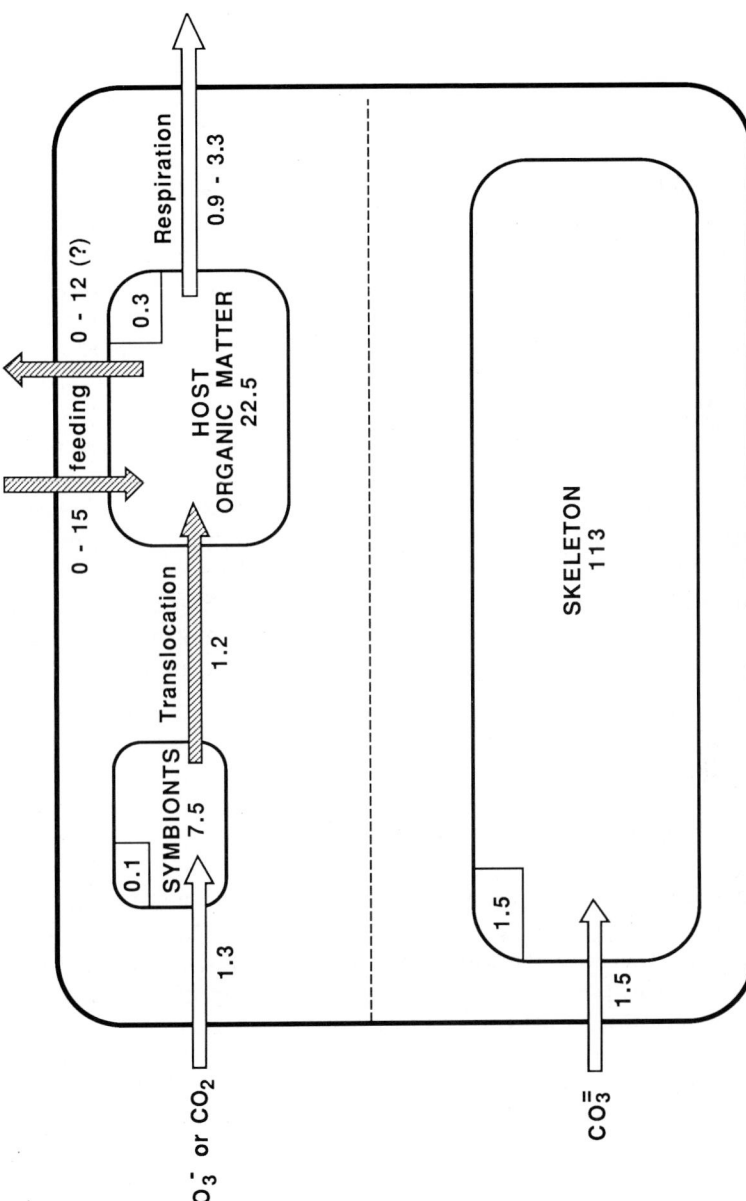

Fig. 4. Carbon budget for *Amphisorus hemprichii*. See Fig. 3 for details.

al., 1980; ter Kuile and Erez, 1987) suggest that inorganic carbon uptake rates per weight are slightly lower than those of *A. lobifera*. Lee and co-workers (1980) found that *H. depressa* did not feed on algae, but did ingest bacteria. Feeding did not enhance the growth rate of *H. depressa*, whereas the addition of nutrients to the medium did (Röttger *et al.*, 1980). This suggests that feeding in *H. depressa* may provide nutrients rather than reduced carbon, just as in *A. lobifera*. Another perforate species *Operculina ammonoides*, showed much higher inorganic carbon uptake rates into the skeleton (ter Kuile and Erez, 1987; ter Kuile *et al.*, 1987). Measurements of the photosynthetic and calcification rates of the imperforate species *Archaias angulatus* (Duguay and Taylor, 1978; Duguay, 1983), *Sorites marginalis, Cyclorbiculina compressa* (Duguay, 1983) and *Borelis schlumbergeri* (ter Kuile and Erez, 1987) indicate that the ratio of inorganic carbon uptake by these processes is constant for each species, though not necessarily 1:1, as in *A. hemprichii*. In general, feeding rates were higher in the imperforate species studied – *A. angulatus, S. marginalis* (Lee and Bock, 1976), *A. hemprichii* (Lee *et al.*, 1980; ter Kuile *et al.*, 1987) and *B. schlumbergeri* (ter Kuile *et al.*, 1987) – than the perforate species – *A. lobifera, H. depressa* (Lee *et al.*, 1980; ter Kuile *et al.*, 1987) and *O. ammonoides* (ter Kuile *et al.*, 1987). This suggests that the earlier conclusion that feeding in *A. lobifera* is used primarily as a source of nutrients and in *A. hemprichii* to provide reduced carbon as well, may be a general character of their respective groups.

4.4 Carbon cycling of planktonic species

The carbon cycling of planktonic foraminifera resembles that of the perforate symbiont-bearing species. For instance, they possess carbon and calcium pools as well (Anderson and Faber, 1984). In other respects, there are vast differences. Planktonic species deposit their skeletons in isotopic equilibrium (Erez and Luz, 1983), indicating that no carbon of metabolic origin is incorporated into the skeleton as it is in larger symbiont-bearing perforate species. Due to the much shorter life-cycle (2–4 weeks) of planktonic foraminifera compared to benthonic species (6–12 months), the photosynthetic and calcification rates of the former exceed those of the latter by a factor of 10 (Erez, 1983; Anderson and Faber, 1984; Jorgensen *et al.*, 1985). Feeding is essential in order to maintain growth, and the growth rate may even be determined by the feeding rate (Caron *et al.*, 1981).

REFERENCES

Anderson, O.R. and Bé, A.W.H. (1978). Recent advances in foraminifera fine structural research. In *Foraminifera* (eds R.H. Hedley and C.G. Adams), Vol. 3, pp. 122–202. Academic Press, London.

Anderson, O.R. and Faber, W.W., Jr (1984). An estimation of calcium carbonate deposition rate in the planktonic foraminifer *Globigerinoides sacculifer* using ^{45}Ca as a tracer: A recommended procedure for improved accuracy. *J. Foram. Res.* **14**:303–308.

Angell, W. (1979). Calcification during chamber development in *Rosalina floridana*. *J. Foram. Res.* **9**:341–353.

Angell, W. (1980). Test morphogenesis (chamber formation) in the foraminifer *Spiroloculina hyalina* Schulze. *J. Foram. Res.* **10**:89–101.

Bé, A.W.H., Spero, H.J. and Anderson, O.R. (1982). Effects of symbiont elimination and reinfection on the life processes of the planktonic foraminifer *Globigerinoides sacculifer*. *Mar. Biol.* **70**:73–86.

Borowitzka, M.A. (1977). Algal calcification. *Oceanogr. Mar. Biol. Ann. Rev.* **15**:189–223.

Borowitzka, M.A. (1982a). Mechanisms in algal calcification. *Prog. Phycol. Res.* **1**:137–177.

Borowitzka, M.A. (1982b). Morphological and cytological aspects of algal calcification. *Int. Rev. Cyt.* **74**:127–162.

Borowitzka, M.A. and Larkum, A.W.D. (1976). Calcification in the green alga *Halimeda*. III: The sources of inorganic carbon for photosynthesis and calcification and a model of the mechanism of calcification. *J. Exp. Bot.* **279**:879–893.

Caron, D.A., Bé, A.W.H. and Anderson, O.R. (1981). Effects of variations in light intensity on life processes of the planktonic foraminifer *Globigerinoides sacculifer* in laboratory culture. *J. Mar. Biol. Ass. U.K.* **62**:435–451.

Chalker, B.E. (1976). Calcium transport during skeletogenesis in hermatypic corals. *Comp. Biochem. Physiol.* **54A**:455–459.

Chalker, B.E. (1983). Calcification by corals and other animals on the reef. In *Perspectives on Coral Reefs* (ed. D.J. Barnes), pp. 29–45.

Chalker, B.E. and Taylor, D.L. (1975). Light enhanced calcification, and the role of oxidative phosphorylation in calcification of the coral *Acropora cervicornis*. *Proc. R. Soc. Lond.* **B190**:323–331.

Cohn, F. (1863). *Jahresber. Schles. Ges. Vaterländ. Kultur* **40**:65–67.

Crossland, C.J. (1987). *In situ* release of mucus and DOC-lipid from the corals *Acropora variabilis* and *Stylophora pistillata* in different light regimes. *Coral Reefs* **6**:35–42.

Crossland, C.J., Barnes, D.J., Cox, T. and Devereux, M. (1980). Compartmentation and turnover of organic carbon in the staghorn coral *Acropora formosa*. *Mar. Biol.* **59**:181–187.

Duguay, L.E. (1983). Comparative laboratory and field studies on calcification and carbon fixation in foraminiferal–algal associations. *J. Foram. Res.* **13**:252–261.

Duguay, L.E. and taylor, D.L. (1978). Primary production and calcification by the soritid foraminifer *Archaias angulatus*. *J. Protozool.* **25**:356–361.

Erez, J. (1978). Vital effect on stable-isotope composition seen in foraminifera and coral skeletons. *Nature, Lond.* **273**:199–202.

Erez, J. (1983). Calcification rates, photosynthesis and light in planktonic foraminifera. In *Biomineralization and Biological Metal Accumulation* (eds P. Westbroek and E.J. de Jong), pp. 307–313. Reidel, Dordrecht.

Erez, J. and Luz, B. (1983). Experimental paleotemperature equation for planktonic foraminifera. *Geochim. Cosmochim. Acta* **47**:1025–1031.

Goreau, T.F. (1963). Calcium carbonate deposition by coralline algae and corals in relation to their role as reef builders. *Ann. N.Y. Acad. Sci.* **109**:127–167.

Graham, D. and Smillie, R.M. (1976). Carbonate dehydratase in marine organisms of the Great Barrier Reef. *Aust. J. Plant Physiol.* **3**:113–119.

Hallock, P. (1978). Carbon fixation and loss in a foraminiferal–algal symbiont system. *J. Foram. Res.* **8**:35–41.

Hemleben, Ch., Bé, A.W.H., Anderson, O.R. and Tuntivate, S. (1977). Test morphology, organic layers and chamber formation of the planktonic foraminifer *Globorotalia menardii* (d'Orbigny) *J. Foram. Res.* **7**:1–25.

Hemleben, Ch., Anderson, O.R., Berthold, W. and Spindler, M. (1986). Calcification and chamber formation in Foraminifera: A brief overview. In *Biomineralization in Lower Plants and Animals* (eds B.S.C. Leadbeater and R. Riding), pp. 237–249. The Systematics Association, Spec. Vol. 30. London.

Isa, Y. and Yamazato, K. (1984). The distribution of carbonic anhydrase in a staghorn coral, *Acropora hebes. Galaxea* **3**:25–36.

Jorgensen, B.B., Erez, J., Revsbech, N.P. and Cohen, Y. (1985). Symbiotic photosynthesis in a planktonic foraminiferan, *Globigerinoids sacculifer* (Brady), studied with microelectrodes. *Limnol. Oceanogr.* **30**:1253–1267.

Kingsley, R.J. (1984). Spicule formation in the invertebrates with special reference to the Gorgonian *Leptogorgia virgulata. Amer. Zool.* **24**:883–891.

Koestler, R.J., Lee, J.J., Reidy, J., Sheryll, R.P. and Xenophontos, X. (1985). Cytological investigation of digestion and reestablishment of symbiosis in the larger benthic foraminifer *Amphistegina Lessonii. Endocyt. C. Res.* **2**:21–54.

ter Kuile, B. and Erez, J. (1987). Uptake of inorganic carbon and internal carbon cycling in benthonic symbiont-bearing foraminifera. *Mar. Biol.* **94**:499–510.

ter Kuile, B. and Erez, J. (1988). The size and function of the internal inorganic carbon pool of the foraminifer *Amphistegina lobifera. Mar. Biol.* **99**:481–487.

ter Kuile, B. and Erez, J. (1991). Carbon budgets for two species of benthonic symbiont-bearing foraminifera. *The Biological Bulletin*, in press.

ter Kuile, B., Erez, J. and Lee, J.J. (1987). The role of feeding in the metabolism of larger symbiont-bearing foraminifera. *Symbiosis* **4**:335–350.

ter Kuile, B., Erez, J. and Padan, E. (1989a). Mechanisms for the uptake of inorganic carbon by two species of symbiont-bearing foraminifera. *Mar. Biol.* **103**:241–251.

ter Kuile, B., Erez, J. and Padan, E. (1989b). Competition for inorganic carbon between the processes of photosynthesis and calcification in the symbiont-bearing foraminifer *Amphistegina lobifera. Mar. Biol.* **103**:253–259.

Lee, J.J. and Bock, W.D. (1976). The importance of feeding in two species of soritid foraminifera with algal symbionts. *Bull. Mar. Sci.* **26**:530–537.

Lee, J.J. and Zucker, W. (1969). Algal flagellate symbiosis in the Foraminifer *Archaias. J. Protozool.* **16**:71–81.

Lee, J.J., McEnery, M.E. and Garrison, J.R. (1980). Experimental studies of larger foraminifera and their symbionts from the Gulf of Elat on the Red Sea. *J. Foram. Res.* **10**:31–47.

Leutenegger, S. (1977). Ultrastructure de Foraminiféres perforés et imperforés ainsi que leurs symbiotes. *Cahiers de Micropaléontologie* **3**:1–52.

Lucas, W.J. (1983). Photosynthetic assimilation of exogenous HCO_3^- by aquatic plants. *Ann. Rev. Plant Physiol.* **34**:71–104.

Pentecost, A. (1985). Photosynthetic plants as intermediary agents between environmental HCO_3^- and carbon deposition. In *Inorganic Carbon Uptake by Aquatic Photosynthetic Organisms* (eds W.J. Lucas and J.A. Berry), pp. 459–480. American Society of Plant Physiologists.

Pringsheim, N. (1888). Uber die entstehung von Kalkinkrustationen von süswasserpflanzen. *Jb. Wiss. Bot.* **19**:138–154.

Röttger, R. (1974). Larger foraminifera: Reproduction and early stage of development in *Heterostegina depressa*. *Mar. Biol.* **26**:5–12.

Röttger, R., Irwan, A., Schmaljohann, R. and Franzisket, L. (1980). Growth of the symbiont-bearing foraminifera *Amphistegina lessonii* D'Orbigny and *Heterostegina depressa* D'Orbigny (Protozoa). In *Endosymbiosis and Cell Biology* (eds W. Schwemmler and H.E.A. Schenk), Vol. 1, pp. 125–132. Walter de Gruyter, Berlin.

Röttger, R., Spindler, M., Schmaljohann, R., Richwien, M. and Fladung, M. (1984). Functions of the canal system in the rotaliid foraminifer, *Heterostegina depressa*. *Nature, Lond.* **309**:789–791.

Swart, P.K. (1983). Carbon and oxygen isotope fractionation in scleractinian corals: A review. *Earth Sci. Rev.* **19**:51–80.

Wal, P. van der (1984). Calcification in two species of coccolithophorid algae. *Gua Papers of Geology* **20**:113 pp.

Wal, P. van der, de Jong, L., Westbroek, P. and de Bruijn, W.C. (1982). Calcification in the Cocclithophorid Alga *Hymenomonas carterae*. *Ecological Bulletin*, **35**:251–258.

Weiner, S. (1986). Organization of extracellularly mineralized tissues: A comparative study of biological crystal growth. *CRC Crit. Rev. Biochem.* **220**:365–408.

Weiner, S. and Erez, J. (1984). Organic matrix of the shell of the foraminifer, *Heterostegina depressa*. *J. Foram. Res.* **14**:206–212.

5
The motility of foraminifera

JEFFREY L. TRAVIS and SAMUEL S. BOWSER

1.0	Introduction	91
2.0	Pseudopodial networks	92
	2.1 General morphology	92
	2.2 Foraminiferan locomotion	95
	2.3 Motility	99
3.0	Intrathalamous movements	112
4.0	Reticulopodial components	113
	4.1 Ultrastructure	113
	4.2 Molecular composition of the reticulopodial cytoskeleton	127
5.0	Mechanisms of foraminiferan motility	130
	5.1 Mechanism of reticulopod movements	130
	5.2 The streaming of reticulopodial contents: Membrane domain transport	139
	5.3 Network withdrawal	141
	5.4 Mechanism of pseudopodial tension and foraminiferan locomotion	143
6.0	Future prospects	144
	Acknowledgements	145
	References	146
	Glossary of terms	153

1.0 INTRODUCTION

The motile activities of granuloreticulose pseudopods (i.e. pseudopodial networks or reticulopods) are intimately involved in many foraminiferan life processes. In benthic foraminiferans, for example, locomotion to or within a suitable habitat depends on the co-ordinated contractile activities of their reticulopods. Reticulopods also participate in all known foraminiferan trophic mechanisms (reviewed in Lipps, 1983), while their associated cytoplasmic transport undoubtedly contributes to metabolic exchange processes. In addition, test construction in agglutinated and calcareous species is largely mediated by reticulopod deployment and transport (reviewed in Lipps, 1973). From these considerations, it seems clear that, to understand foraminiferan biology, we need detailed knowledge of reticulopod structure and motility.

This chapter reviews the current literature on foraminiferan motility, beginning with a description of the major features of reticulopods, their movements, and associated transport processes. Information regarding reticulopod structural and molecular composition is then presented, followed by a discussion of models that help explain some of their motile activities. Finally, some major questions for future investigations are considered.

It is not our intent to provide comprehensive information regarding the basics of cell motility or cytoplasmic structure as it relates to the foraminifera. Instead, the non-specialist reader is directed to pertinent topical reviews for background or supplemental information. Furthermore, we acknowledge that this chapter is biased toward the motility of small, easily cultured benthic species (especially *Allogromia*) – a necessary limitation as these are the most accessible subjects for laboratory studies. Although the dangers associated with making generalizations from such a small sample are clear, our experience with larger agglutinated and calcareous foraminifers, and related protists (e.g. the freshwater athalamiid *Reticulomyxa*), leads us to assert that the fundamental aspects of reticulopodial motility and underlying mechanisms are applicable to granuloreticuloseans in general. Future studies will undoubtedly reveal interesting adaptations in these basic motile processes that will enrich our understanding of this important protistan group.

2.0 PSEUDOPODIAL NETWORKS

2.1 General morphology

Most light microscopic studies of foraminiferan pseudopods have been made using specimens plated on glass slides or coverslips. Under these conditions, the pseudopods of an adult benthic foraminifer, particularly one that is actively feeding or in motion, appear as an intricate net of branched and anastomosed filopods (thin, cylindrical pseudopods) containing numerous refractile granules in rapid, bidirectional motion. Although these major pseudopodial characteristics are the basis for assigning foraminifera to the class Granuloreticulosea (Lee *et al.*, 1985), numerous authors lament that the pseudopods of only a few dozen of the c. 4000 extant species have ever been observed. Even more disappointing is the fact that pseudopod structure and motility has been intensively studied in only a few species.

All foraminiferans examined to date appear capable of extending a pseudopodial network (Figs 1, 2). Even pelagic species extend typical networks when in contact with a planar substrate (e.g. *Globigerina bulloides*: Grell, 1973; Adshead, 1980) and scanning electron microscopy clearly shows reticulopods vesting their test surfaces (reviewed in Bé et al., 1977). Certain pelagic forms also have specialized pseudopods, such as those forming the "bubble chambers" that are thought to serve as a flotation device (reviewed in Hemleben et al., 1989). Adshead (1980) noted the development of long, thin "axopods" that preceded the formation of spines in juvenile pelagic globigeriniids. It is not yet

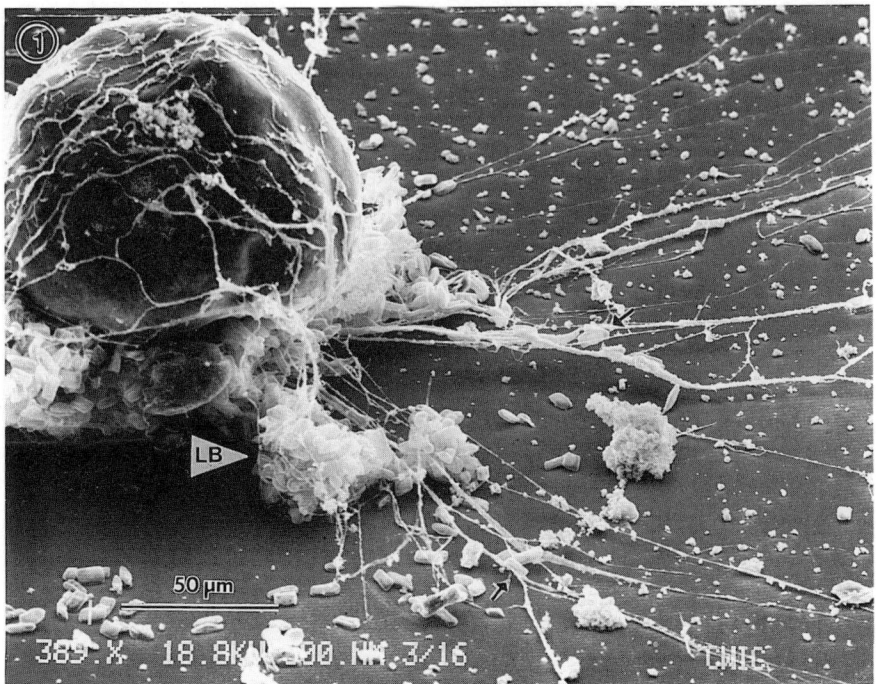

Fig. 1. Survey scanning electron micrograph of a foraging *Allogromia laticollaris*, illustrating the main morphological features of a typical pseudopodial network. The reticulopods (i.e. highly branched and anastomosed filopods) have accumulated diatoms (arrows) in net-like enclosures (LB, "lunch boxes"). Note that reticulopods extend over the test surface. Reproduced with permission from Travis and Allen (1981).

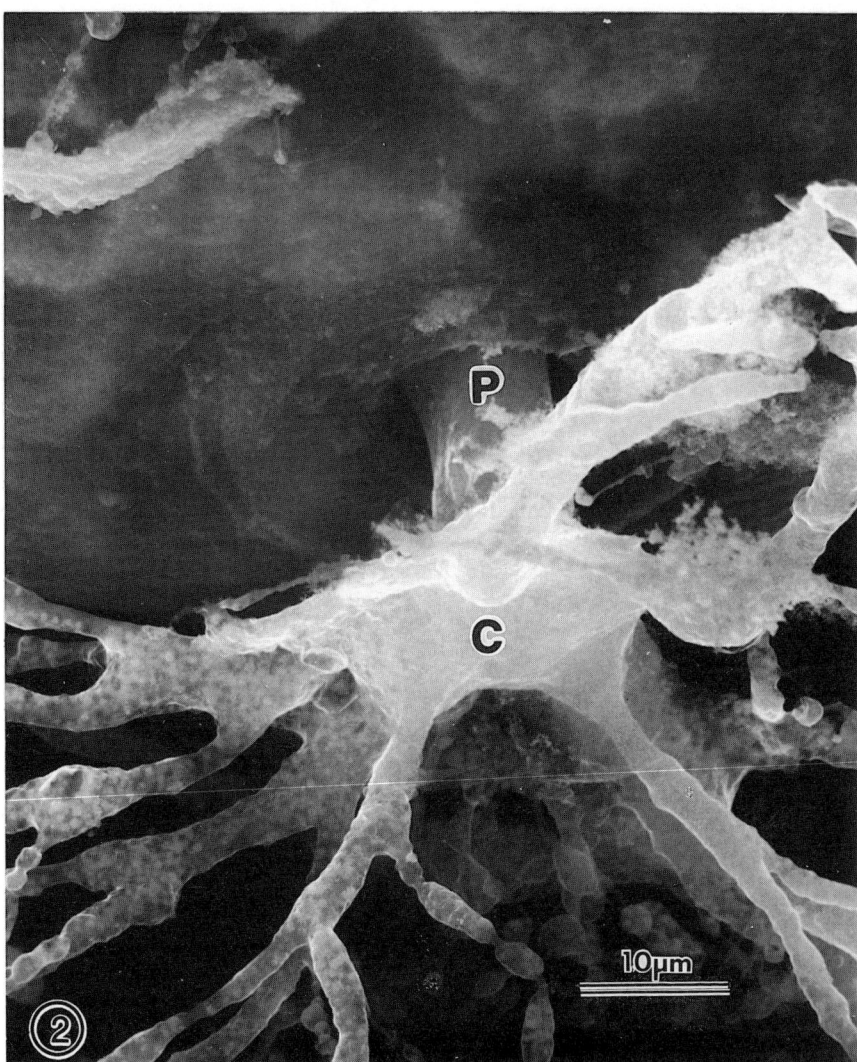

Fig. 2. Scanning electron micrograph of the apertural area of *Allogromia laticollaris*, prepared using the osmium-thiocarbohydrazide-osmium technique, which renders cytoplasmic granules visible through the plasma membrane. The peduncle (P) extends through the test aperture where it fans out to form the circumoral cytoplasm (C), from which numerous trunk pseudopods arise. Ordinarily, this region is obscured by the cell body, but in this preparation part of the reticulopodium is lifted off the substrate to bring the area into view.

known whether these axopods are structurally similar to the relatively stiff and highly retractable axopods of heliozoans.

The exact reticulopodial pattern is complex and constantly changes as new filopods extend, branch and anastomose with others, and old ones retract. On a broader scale, the characteristic deployment of reticulopods varies between foraminiferan species and between individuals in different physiological states (Sheehan and Banner, 1972). Broad, flattened cytoplasmic sheets (sometimes termed lamellipods*) are frequently observed in networks of foraminiferans attached to planar surfaces. In *Allogromia*, flattened pseudopodial sheets are found instead of reticulopods when specimens are plated on positively charged substrates (Travis and Allen, 1981; Bowser and Rieder, 1985). This finding suggests that environmental factors (as well as intrinsic structural factors) may have a profound influence on pseudopodial morphology.

Superficially, the pseudopods of some large, discoidal calcareous species appear distinct. Instead of a radial array of fine reticulopods, these forms possess massive, cone-shaped pseudopods that extend from the test margin and converge on the substrate. However, when examined under light optics at a higher resolution, a given cone-shaped *Elphidium* pseudopod was reported to be composed of a massive collection of unbranched filopods (Sheehan and Banner, 1972). Electron microscopy of similar pseudopods in *Amphisorus* revealed a closely woven array of anastomosed filopods, i.e. reticulopods (Travis *et al.*, 1988). Such examples show that low-resolution light microscopic views of seemingly novel foraminiferan pseudopods may prove deceiving, and deserve careful study by correlative light and electron optics.

To add a final level of complexity, other types of motile appendages occur in foraminifers. Their flagellated or amoeboid gametes or zygotes, which lack reticulopods, are obvious examples. Unfortunately, the information concerning these forms is largely anecdotal; they will not be considered further.

2.2 Foraminiferan locomotion

Benthic foraminifera migrate on two general types of substrate: (1) planar surfaces, such as an algal frond or mollusc shell, and (2)

---

*These pseudopodial sheets are not to be confused with the structurally and functionally distinct lamellipods (leading lamellae) of metazoan tissue cells (see Abercrombie, 1980, for a description of the latter).

sediments, ranging from fine silt to coarse sand. Detailed comparative studies of foraminiferan locomotion on these two substrates were made by Kitazato (1988) and Wetmore (1988). Consistent with the earlier literature (reviewed in Boltovskoy and Wright, 1976), these authors found that the speeds attained by related or diverse species were quite variable and did not correlate well with test size. They also found that the same species moved faster on glass than through sediment. Both authors suggested that the glass and surrounding water offered minimal resistance to movement; Kitazato (1988) added that the foraminifers were perhaps displaying an avoidance reaction to the strange environment, an impression Wetmore (1988) attributed to the absence of nutrients on the glass surface. An explanation not considered by these authors is that, when compared to loose sediment, solid planar surfaces facilitate the application of pseudopodial traction forces.

It is widely accepted that foraminiferans crawling across planar surfaces use their pseudopods to attach to and exert a rearward traction force along a substrate (e.g. Arnold, 1953; Röttger, 1973; Boltovskoy and Wright, 1976). This traction force is thought to be mediated by pseudopodial contraction; it serves to drag the test forwards, in a manner similar to the locomotion of testate amoebae (e.g. Wohlman and Allen, 1968). Unfortunately, the evidence for this mode of locomotion is based largely on casual observations. Indeed, Kitazato (1988) noted that a foraminifer's movement was quite smooth, indicating that any attachment and pulling processes occur continuously, instead of in discrete phases as displayed by testaceans.

In an effort to provide more detailed information on the movement of larger foraminifers along planar substrates, Travis et al. (1988) used a deformable rubber substrate assay (Harris et al., 1980) to visualize rearward traction forces in *Amphisorus* pseudopodial networks. In this study, the specimens were plated on a highly flexible rubber sheet. Wrinkles in the sheets (i.e. traction forces) were frequently observed at points where pseudopodial trunks initially contacted the substrate (Fig. 3). However, the correlation between pseudopodial traction and locomotion was not made. It remains unclear whether these foraminifers used the observed pseudopodial traction to move or to adhere firmly to the substrate.

Similar observations were made with *Amphistegina*, *Astrammina* and several larger Pacific calcariniids (Travis and Bowser, unpublished observations). However, smaller species that were examined (*Allogromia laticollaris*, *Allogromia* sp.) only rarely deformed the rubber sheets, even while moving rapidly. Either *Allogromia* generates insufficient

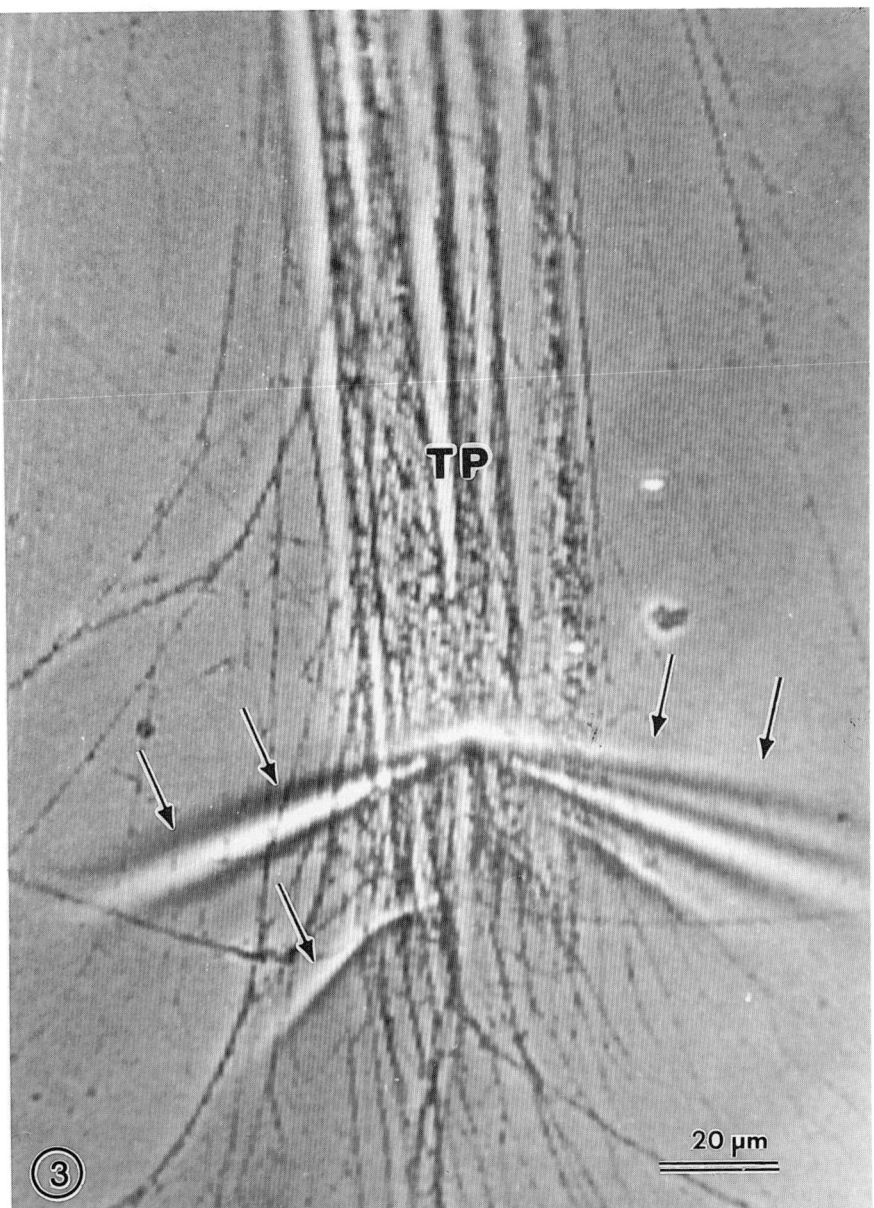

Fig. 3. Tension exerted along an *Amphisorus* trunk pseudopod (TP) deforms a flexible silicone rubber substrate. Arrows point to compression wrinkles in the rubber sheet. The pseudopods angle down from the test margin to contact the rubber substrate, causing the more proximal portions of the pseudopod (top) to appear out of focus. Reproduced with permission from Travis et al. (1988).

pseudopodial tension to deform the rubber sheet, or these smaller foraminifers migrate in a different way. It is noteworthy that the juveniles of *Allogromia* sp. initially disperse by riding along the surface of the parental pseudopodial network, rather than by moving autonomously (Bowser *et al.*, 1984b). Perhaps reticulopodial surface transport, a process considered in greater detail in Section 2.3.2, plays a more general role in foraminiferan locomotion than currently appreciated.

Studies of infaunal locomotion are compounded by the difficulty of observing foraminifers buried in opaque sediments. Two general approaches have been used to address this problem: mathematical model-building and direct observation. Severin and co-workers (Severin and Erskian, 1981; Severin, 1987) have championed the first approach, using *Quinqueloculina impressa* as a model infaunal foraminifer. Their studies helped define environmental factors, such as sediment overburden pressure, that affect foraminiferal migration through sediments. In contrast, direct observations on foraminiferal burrowing were made by Kitazato (1988), who used a side-mounted binocular microscope to view specimens in a sediment-filled beaker, and Wetmore (1988), who made clever use of a parallel-plate glass viewing chamber (the foraminiferal equivalent of an "ant farm"). Their studies provided much-needed information regarding the behaviour of individual foraminiferans moving through sediment.

Foraminiferans that move on, or burrow through, sediments face two major problems not encountered on planar substrates: (1) clearing away sediment as the organism moves forward and (2) achieving firm attachment to the irregular, loose surfaces. Pseudopodial traction forces probably serve to dislodge sediment particles and, perhaps in conjunction with pseudopodial surface transport (discussed below), temporarily displace these particles as they burrow. Such a tunnelling activity was reported by Kitazato (1988), who observed that *Pseudorotalia gaimardii* used its pseudopods to first clear a leading cavity, and then drag its test forward. Kitazato (1988) stated that this species "anchored their pseudopodia to the wall of the resulting cavity with viscous granules". It is not clear from this description whether the foraminifer secreted a cement material to bind sediment and thus form a surface firm enough to pull against. Such secretions would help explain the persistence of foraminiferal burrows, which may even form trace fossils (Severin *et al.*, 1982). Much more work must be done, at both the comparative and mechanistic levels of analysis, on the locomotion of infaunal foraminifera.

It seems appropriate here to point out that traction forces generated by reticulopods may serve other functions besides locomotion. For example, distal pseudopods in *Allogromia* are able to rend soft food materials, an activity (termed skyllocytosis) probably mediated by locally generated pseudopodial tension (Bowser, 1985). Related activities may enable herbivorous foraminiferans to disperse clumps of algal prey, or carnivorous species to tear apart metazoan prey (e.g. Anderson and Bé, 1976; Spindler *et al.*, 1984). In addition, pseudopodial tension probably mediates simple forms of reproduction in foraminiferans, e.g. during cytotomy in *Allogromia* sp. (Lee and Pierce, 1963). Later, we discuss how pseudopodial tension is also involved in reticulopodial morphogenesis and shape maintenance (see Sections 2.3.1.1 and 2.3.4).

2.3 Motility

2.3.1 *Reticulopod movements*

The three major types of reticulopodial movements – extension, bending to form branches and anastomoses, and withdrawal – can occur simultaneously at different points within a fully developed network. Predomination of extension or withdrawal results in the expansion or retraction of the network respectively, whereas the co-ordination of the different reticulopodial movements results in the expression of more complex activities, such as locomotion or skyllocytosis. Reticulopodial movements occur at rates considerably faster than pseudopodial motions in vertebrate tissue cells. For example, foraminiferans extend pseudopods at speeds in excess of $1\,\mu m/s$, even in Antarctic species living at sub-zero temperatures (Bowser and Delaca, 1985). In contrast, neurite outgrowth from neurons cultured at 37°C (albeit only a superficially similar process) occurs at approximately $10\,\mu m/h$ (Cohan and Kater, 1986).

The incessant pseudopodial movements are crucial to network function, for it is through these activities that the network effectively scours a surface of particulates (e.g. bacterial or algal prey). For instance, anastomosed branches frequently move bidirectionally along their major trunks (see Koonce *et al.*, 1986, Fig. 2a–d, for such a sequence in *Reticulomyxa*). In *Allogromia*, these movements sweep particulates towards the veil of hyaline cytoplasm located at the juncture of these mobile branches – a principal site of phagocytosis in pseudopodial networks (McGee-Russell, 1974; Bowser *et al.*, 1985). This food-collecting "sweeping behaviour" seems inefficient when witnessed by light microscopy, due to the network's two-dimensional appearance. However, McGee-Russell (1974) rated its true efficiency higher after examining, from the perspective of a prey item, the network's three-dimensional detail by scanning electron microscopy.

2.3.1.1 *Network morphogenesis.* In benthic species, the formation of the pseudopodial network begins with the extension of several long, thick filopods. These initial trunk filopods originate from the apertural cytoplasm (Fig. 2), or from an "extramural" layer of cytoplasm enveloping the test (Jepps, 1942; McGee-Russell and Allen, 1971; Sheehan and Banner, 1972). From here the filopodia may extend straight into the surrounding sea water for several millimetres before contacting a solid surface – an observation interpreted by some to indicate that these pseudopods are highly rigid structures (e.g. Jepps, 1942). However, this notion is inconsistent with the observation that pseudopodial trunks bend freely in response to the slightest applied force (e.g. fluid movement or manipulation with extremely fine glass needles: Bowser, unpublished observations, on *Astrammina rara* and *Allogromia*). A more likely (but untested) hypothesis is that the neutral buoyancy of pseudopodial trunks accounts for the extreme lengths they sometimes attain, and their apparent rigidity.

Allen (1964) used *Allogromia* sp. to provide the most detailed account of the next stages of network formation: pseudopod branching and fusion. His frame-by-frame cinemicrographic analysis revealed that a given trunk filopod eventually contacted and apparently adhered to a planar substrate (i.e. glass microscope slide). The filopod then extended upwards, followed by a pivoting motion at the point of substrate contact. This bent filopod typically formed an acute angle relative to the long axis of its proximal portion. After extending several micrometres, the filopod tip again contacted the substrate and the adhesion/pivot/extension sequence was repeated several more times.*
The pseudopod, now bent into small segments, next developed tension and rapidly straightened between its most proximal and distal substrate contacts. As a result, cytoplasmic threads trailed from the interposed substrate contacts (thus confirming them as sites of pseudopod/substrate adhesion) as the bent pseudopod straightened and pulled away. These threads either persisted or lost adhesion and immediately "snapped back" to become reincorporated into the trunk filopod. A persisting (i.e. firmly attached) thread often recruited cytoplasm to form a branch

---

*Curiously, in *Allogromia laticollaris* and *Allogromia* sp., most reticulopods bend clockwise (Travis, unpublished observations).

pseudopod, which then extended a new filopod from its adhering tip and began the same adhesion/bend/extension behaviour.

In unpublished studies, we confirmed this general sequence of events in *Allogromia* using interference reflection optics, which allowed us to examine the sites of pseudopod/substrate contact (for more information regarding this imaging mode, Verschueren, 1985). However, other routes for the production of pseudopodial branching were also noted. For example, trunk pseudopods suspended free in the sea water frequently extended small branch filopods. In contrast to branches formed as described above, these branches appeared unstable and were quickly withdrawn unless they contacted the substrate.

Irrespective of their mode of formation, most pseudopodial branches inevitably encountered neighbouring pseudopods. Immediately upon contact, the apposing pseudopodial membranes fused to establish cytoplasmic continuity (i.e. anastomose), as evidenced by the free exchange of intracellular organelles between the two pseudopods. Provided divalent cations (Ca^{2+}, Mg^{2+}) are present in the sea water, plasma membrane fusion readily occurs between any pseudopods that touch each other, irrespective of their diameters or contact angles (Bowser and Travis, unpublished observations). Such rapid fusion implies that the reticulopodial membrane is highly fluid (Poste and Allison, 1974). Considering the biotechnological significance of membrane fusogens, this unusual feature of reticulopods deserves much more attention.

2.3.1.2 *Network withdrawal.* Pseudopodial nets are vital to foraminiferan survival and represent a considerable investment in energy (in terms of both their synthesis and metabolic activity). It comes as little surprise, therefore, that foraminifers possess a mechanism to quickly withdraw their networks within the protective confines of the test. The withdrawal response in *Allogromia* sp. (Strain NF) can be induced by a number of factors. For experimental studies, the simplest method is to treat the organisms with hypertonic sea water (e.g. with osmolarities expected for hypersaline tide pools). Extreme environmental factors, such as temperature (<6 or $>45°C$), pH (<3 or >9) or mechanical trauma, can also induce network withdrawal (Koury, 1982). Using microperfusion methods, McGee-Russell and Allen (1971) observed several distinct phases of the withdrawal response:

1. Initially, most cytoplasmic particles switched from their normal bidirectional saltatory transport to a more uniform inward motion.
2. As the inward transport progressed, the reticulopods thinned distally, lost substrate attachment, and appeared flaccid; concurrently,

refractile patches of cytoplasm formed varicosities that also moved inwards.
3. The inward flow of cytoplasm and organelles gradually slowed.
4. After approximately 3 min, all cytoplasmic transport stopped, with only a few thick, highly refractile pseudopods remaining outside the test. This "stabilized" state was fully reversible: subsequent treatment with plain sea water quickly resulted in pseudopod re-extension and normal motility.

The molecular mechanism for triggering the withdrawal response is poorly understood. In *Allogromia* treated with an isotonic sea water substitute with a reversed Na:K ratio, networks retracted for approximately 2 min, after which time they adapted to the new conditions and re-extended (Koury et al., 1985). This patterned response suggests that membrane depolarization induces withdrawal, but much more detailed study is required to test this hypothesis.

2.3.2 Intracellular organelle streaming and extracellular surface transport

Several quantitative studies of transport in *Allogromia* (Allen, 1964; Rinaldi and Jahn, 1964; Bowser et al., 1984a; Travis and Bowser, 1988), *Astrammina rara* (Bowser and DeLaca, 1985) and *Reticulomyxa* (Koonce et al., 1986) provide the basis for the following account of reticulopodial organelle and membrane surface movements. All but the thinnest branches within a network display rapid (up to 25 μm/s), bidirectional transport of intracellular organelles (seen by light microscopy as various "granules"). Plasma-membrane domains, visualized by the motions of membrane surface markers,* are similarly transported at all points in the network. The trajectories of both types of movement are typically linear and directed parallel to the pseudopodial long axis.

In general, the movements of intracellular organelles and membrane surface markers share common properties, including the following, which partly fulfill Rebhun's (1972) definition of saltatory transport:

1. Particles undergo instantaneous acceleration, and generally move at nearly constant velocities during any one excursion.
2. They may start, stop or reverse direction at any time.

---

*A microsphere will bind to and locally immobilize a patch of membrane surface components. It is thought that microsphere movement reveals the transport of this immobilized membrane domain within the two-dimensional plane of the fluid lipid bilayer (reviewed in Bloodgood, 1988).

3. They may travel considerable distances (over 50 μm) during a given excursion – well beyond that expected of random (Brownian) motion.

Furthermore, the displacements occur over a wide range of velocities within a given pseudopod (Figs 4, 5). Allen (1964, 1968) noted that in thicker reticulopods there was a tendency for particle velocities to occur in multiples of the lowest velocity.

At least two characteristics distinguish reticulopodial transport from Rebhun's classic definition of saltation. First, collisions between transported particles are common, and these are frequently accompanied by changes in particle speed or direction. Secondly, the transported elements can form clusters or tandem arrays that display common motion (McGee-Russell, 1974; Bowser, 1983; Bowser et al., 1984a;

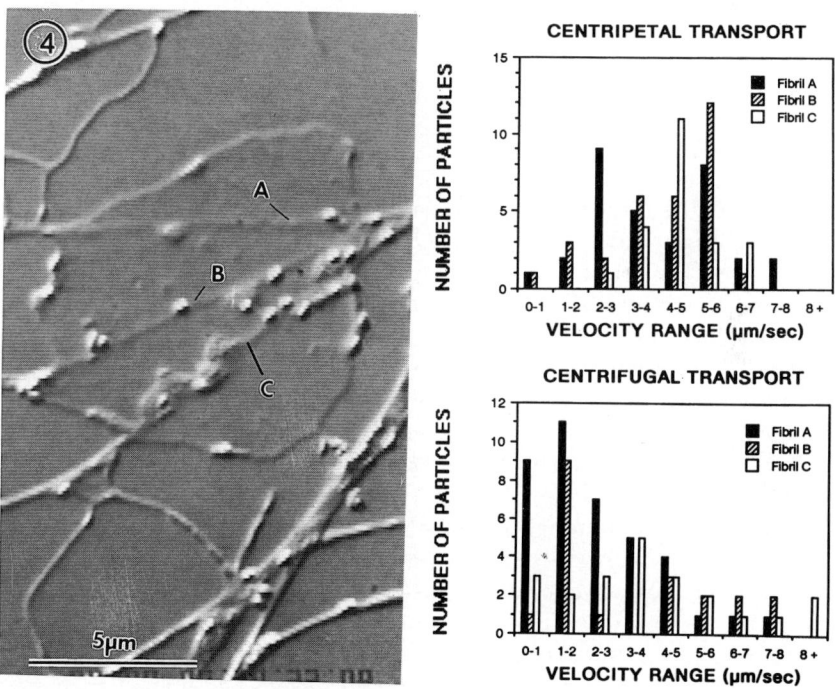

Fig. 4. Organelle transport along pseudopodial fibrils in an *Allogromia* lamellipod. (Left) A video light micrograph of a lamellipod containing three prominent fibrils, labelled, A, B and C. Organelle transport along each fibril was recorded on video-tape and analysed by photokymography as described by Travis and Bowser (1988); these data are presented as histograms of organelle velocity profiles for centripetal (top right) and centrifugal (bottom right) transport. The micrograph is reproduced with permission from Travis et al. (1983).

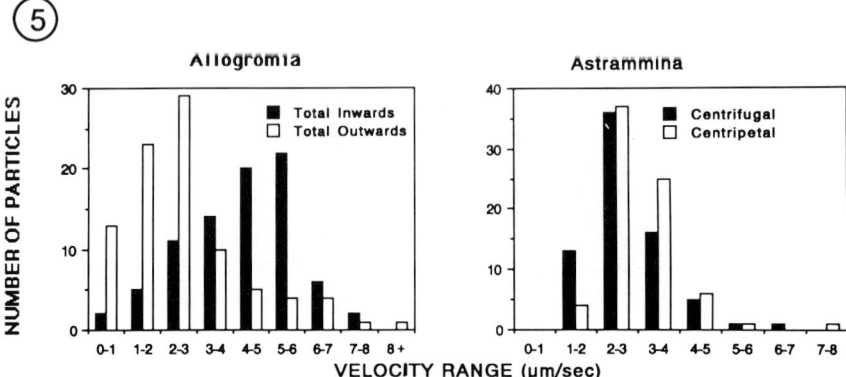

Fig. 5. Histograms comparing organelle transport in *Allogromia laticollaris*, studied at room temperature, and *Astrammina rara*, observed at 0°C. In *Allogromia* (left), 85% of the particles analysed ($n = 164$) moved between 1 and 6 µm/s, with centripetal particles moving more rapidly than centrifugal particles. In contrast, no velocity bias is seen in *Astrammina* (right). The bar chart for *Allogromia* represents pooled organelle transport velocities from the three fibrils pictured in Fig. 4. The bar chart for *Astrammina* was replotted using data presented in Bowser and DeLaca (1985).

Bowser and Rieder, 1985). In addition, videomicroscopy revealed the extremely rapid and uninterrupted movement of barely detectable granules within reticulopods, a phenomenon termed "continuous transport" to distinguish it from saltation (Allen *et al.*, 1982). Continuous transport has not been observed with membrane surface markers (Bowser, unpublished). Finally, Travis *et al.* (1983) illustrated the bulk flow of hyaline cytoplasm along microtubule fibrils in optically favourable regions of *Allogromia* reticulopods, a process not unlike the bulk streaming seen in the intrathalamous cytoplasm (discussed in Section 3).

Foraminifers transport extracellularly attached materials both centripetally and centrifugally along reticulopodial surfaces (Jahn and Rinaldi, 1959; Bowser *et al.*, 1984a; Bowser and DeLaca, 1985). Certain foraminifers employ surface transport to aid in the dispersal of young (e.g. Bowser *et al.*, 1984b) or to collect sediment for building agglutinated tests (Sandon, 1957). However, probably all species use surface transport as a principal means of collecting food (bacteria, algae and other nutritive particulates). Banner and Culver (1978) infer that surface transport in *Haynesina* forces aggregates of detritus past the sharp tubercles lining specific regions of the test (the sutural lacunae), where they become fragmented and sieved in preparation for the subsequent ingestion of the digestible fraction. Alexander and Banner (1984) noted that *Haynesina* could use the forces generated by surface transport to break open diatom frustules against the tubercules and liberate chloroplasts for incorporation as endosymbionts (e.g. Lopez, 1979). Arnold (1964) hinted that a similar rasping activity might occur at the apertural tooth of *Spiroloculina hyalina*. As illustrated by these examples, the use of surface transport in feeding may have provided selective pressure for the evolution of certain test ornamentations.

The properties and mechanism of surface transport in *Allogromia* sp. were examined in a series of recent studies in which reticulopods were labelled with polystyrene microspheres, which served as membrane surface markers (Bowser and Bloodgood, 1984; Bowser *et al.*, 1984a; Bowser and Rieder, 1985, 1986). It was found that microspheres of various sizes, or with net positive, negative or neutral surface charges, or with varying degrees of hydrophobicity, reversibly bound to and moved along the reticulopodial surface. In contrast to surface-bound bacteria, which were quickly enveloped to form a blister-like phagosome, none of the membrane markers tested were phagocytosed in the reticulopodium (suggesting that phagocytosis is triggered by specific molecules, e.g. those associated with the outer envelopes of bacterial prey).

Membrane surface markers were found to move bidirectionally along even the thinnest branch pseudopods, and microspheres frequently passed close by other microspheres moving in the opposite direction (Fig. 6a–d). Groups of particles occasionally moved in tandem, as though linked to a common motile element; that the particles were not merely attached to one another was evidenced by the fact that any member of the group could spontaneously change speed or direction (Bowser and Rieder, 1985). Such changes occurred most frequently at branch points in the network. Microspheres were also seen transiently linked in common motion to underlying cytoplasmic vesicles (Bowser *et al.*, 1984a).

Although membrane surface markers moved either centripetally or centrifugally at nearly identical velocities, approximately 70% of the excursions were directed centripetally. This inward directional bias resulted in the collection of surface-bound microspheres at the circumoral cytoplasm – mirroring the frequent observation that certain foraminifers become embedded within a bolus of algae and detritus (Fig. 6e). As mentioned previously, *Allogromia's* reticulopodial surface transport serves to collect and deliver larger food items to the circumoral cytoplasm, where some selection apparently takes place and digestible material is phagocytosed (Bowser *et al.*, 1985).

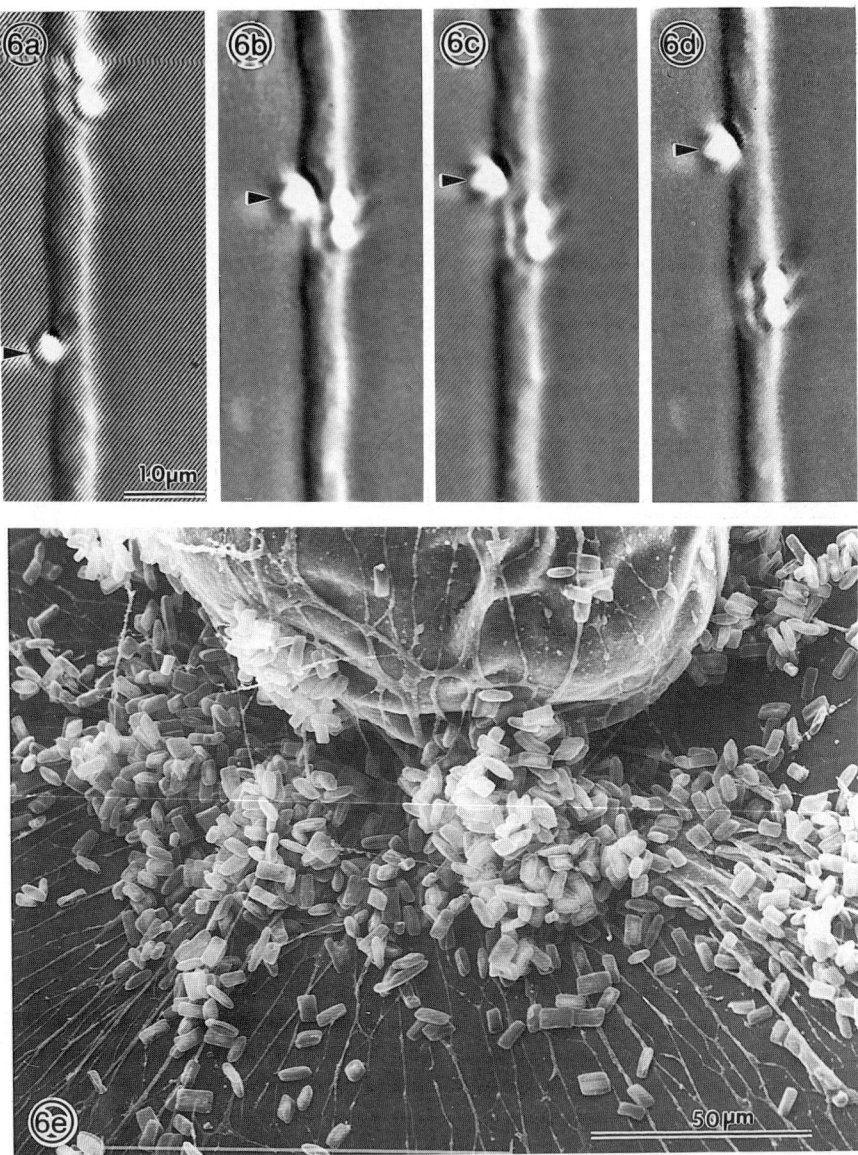

Fig. 6. Membrane surface transport associated with *Allogromia* pseudopods. A sequence of Hoffman modulation-contrast videomicrographs (a–d) shows a single polystyrene microsphere of 0.45 μm diameter (arrowhead) moving towards the top of the page, passing a pair of beads moving in the opposite direction. Elapsed time between frames is approximately 0.25 s. Scanning electron microscopy (e) shows an accumulation of algal prey near the oral region, which resulted from the inward bias associated with surface transport. Figure 6a–d reproduced with permission from Bowser *et al.* (1984).

2.3.3 Cytoplasmic fibrils

Historically, the presence of fibrillar supportive structures in foraminiferan pseudopods was the subject of considerable controversy. The pseudopods of other rhizopods (e.g. heliozoans, radiolarians) were supported by cores of readily demonstrable, optically refractile material (stereoplasm). Earlier workers described such stereoplasmic cores in the pseudopods of foraminifera (reviewed in Sandon, 1963), and Schmidt (1937) clearly demonstrated the positive axial birefringence of *Miliola* (= *Iridia*) pseudopods. However, Jahn and Rinaldi (1959) were unable to detect stereoplasmic elements in *Allogromia*, and Le Calvez (1953) and Sandon (1963) noted them in other species only under certain physiological stresses, such as starvation or hyperosmotic shock. Allen (1964) reported that cytoplasmic fibrils were visible by phase-contrast microscopy only in the thin veils located at pseudopod bifurcations in *Allogromia*.

The unambiguous demonstration of intracellular fibrillar elements in pseudopods of living foraminifers (Fig. 7) was made possible by two simultaneous advances. The first was the development of video-enhanced contrast light microscopy (videomicroscopy), which generated sufficient contrast to image very weak phase objects (Allen *et al.*, 1981a,b; Inoué, 1981). The second was the use of positively charged glass to flatten the normally cylindrical pseudopods into very thin (c. 0.5 μm) sheets (Fig. 7; see also Fig. 10). Under these conditions, the elusive reticulopodial stereoplasm was readily seen as an elaborate intracellular network of birefringent cytoplasmic fibrils. This intracellular fibrillar system, originally described in allogromiids (Travis *et al.*, 1983; Bowser and Rieder, 1985), has subsequently been identified in every other species we have examined (e.g. *Astrammina rara*: Bowser and DeLaca, 1985). Flattening the pseudopods proved to be the key technical improvement because it spatially separated the fibrils from extreme phase gradients situated at the pseudopod margins; these gradients obscured the fibrils in cylindrical reticulopods.

2.3.3.1 *Association between fibrils and cytoplasmic transport.* One of the most significant aspects of reticulopodial fibrils is their association with cytoplasmic transport. Travis *et al.* (1983) noted that cytoplasmic organelles moved only when in contact with the fibrils; elsewhere, they were essentially motionless. Membrane surface transport was similarly found to occur only in association with underlying cytoplasmic fibrils (Bowser and Rieder, 1985). In both cases, movement followed precisely the paths defined by the fibrils. Organelles typically align

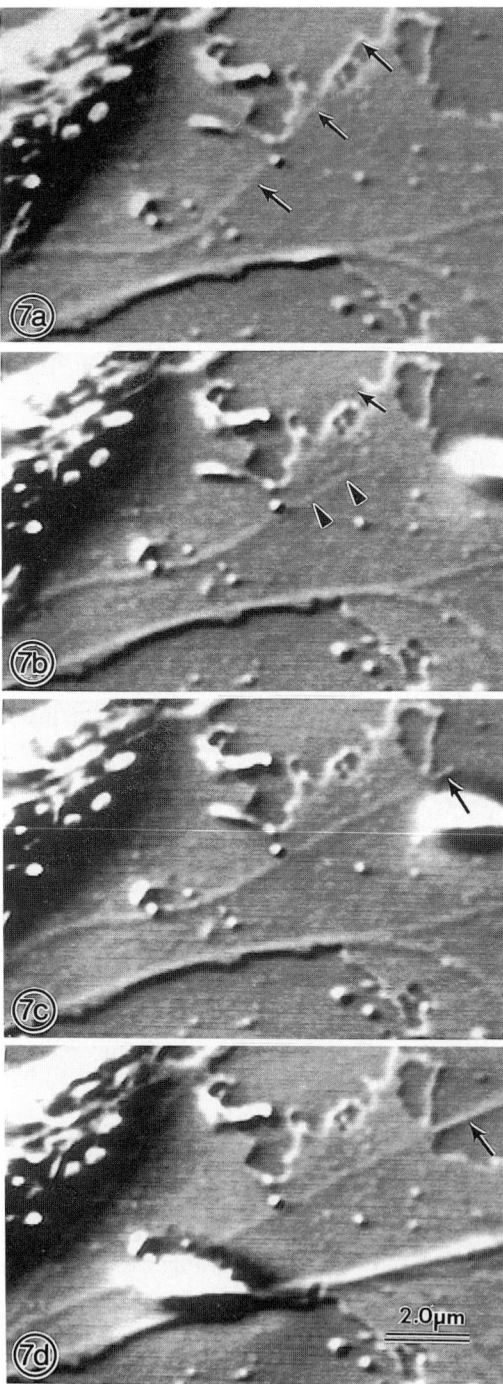

their long axes parallel to the fibrils during transport, especially large vacuoles that become flattened where they contact the fibrils, and often assume a tear shape as they translocate through the cytoplasm (Kachar et al., 1987). Such observations provide evidence that organelles interact with the fibrils at multiple sites. Similar shape changes have been described for transported organelles in animal cell types (Hard and Cloney, 1971; Martz et al., 1984). As detailed in Section 4.1.3.1, the intracellular fibrils correspond to bundles of microtubules. The clear demonstration of microtubule-associated transport in *Allogromia* prompted the work that led to similar discoveries in metazoan cells (Allen, 1987).

2.3.3.2 *Dynamic properties of cytoplasmic fibrils.* The distribution of the cytoplasmic fibrils seen within extensive flattened pseudopodial sheets was very similar to the branching and anastomosing pattern of normal cylindrical reticulopods, suggesting that the fibrils played a role in determining network morphology. Video-recordings of such preparations further revealed that the fibres were highly dynamic, displaying rapid axial (lengthening/shortening) and lateral (bending) movements (Travis et al., 1983). A given fibril was also seen to split to form two thinner filaments (terminology of Allen, 1964), each of which could split to form even thinner filaments. Filaments were seen to loop out from the sides of fibrils. When an extending fibril contacted a neighbouring fibril, or one fibril bent or a filament looped sufficiently to contact an adjacent fibre, the two invariably "zipped" together to form a single, thicker fibre. Thus, the axial and lateral movements of the fibrils also mirrored the extension/retraction and bending movements of reticulopods.

Interactions between moving cytoplasmic fibrils and the plasma membrane are readily observed at the lateral margins of lamellipods (reviewed in Travis and Bowser, 1988). It was found that, when an extending fibril collided with the plasma membrane, a new filopodium emerged from the site of contact (Fig. 7a–e). Filopods were never seen extending from the lamellipod margin in the absence of such

Fig. 7. Filopod formation as a result of microtubule poking. In (a), a microtubule fibril (arrows) courses through the lamellipodial cytoplasm and terminates in a short filopod. In (b), the fibril withdrew from the original filopod (arrow), straightened, and began to extend in a new direction. The newly extended portion of the fibril (arrowheads in b) appears thinner than the original fibril. In (c), the tip of the extending fibril has reached the lateral margin of the lamellipod, where it poked out a nascent filopod (arrow). At this stage, the newly extended portion of the fibril has thickened, so that the diameter of the entire fibril appears uniform. The new filopod (arrow d) continued to extend and thicken. Elapsed time: (a) 0 s; (b) 5 s; (c) 8 s; (d) 27 s. Reproduced with permission from Travis and Bowser (1988).

interactions. Furthermore, when such a filopod subsequently withdrew from the margin, the fibril was seen concomitantly to shorten. These observations provide convincing evidence that fibril lengthening and shortening drives filopod extension and withdrawal.

Lateral interactions between bending or looping fibrils and the pseudopodial membrane resulted in complementary pseudopodial shape changes (Travis and Bowser, 1990). When bending or looping fibrils collided against the membrane, a protuberance emerged from the site of contact. The shape of the bent fibril mirrored precisely the shape of the protuberance, and if the bent fibril subsequently straightened, the protuberance concomitantly receded. In such cases, only the region of membrane in contact with the fibril was deformed; the remainder of the lamellipod remained firmly anchored to the charged glass substrate. It seems very likely that in the absence of such a highly adhesive substrate, i.e. in a normal cylindrical pseudopod, such interactions between bending fibrils and the plasma membrane would result in pseudopod bending.

2.3.4 *Autonomous motility of isolated networks*

There is considerable local control over pseudopodial motility and organization in reticulopods, as evidenced by the formation of "secondary centres of organization" within the network. For example, Liedy (1879) noted that "spindle like accumulations of protoplasm occur in the course of the pseudopodial threads. Sometimes, through the conjunction and spreading of several of the latter together, islet-like expansions occur, and become the centres of secondary nets". These secondary centres have been reported in the networks of diverse foraminifers (Jepps, 1942; Allen, 1964; Sheehan and Banner, 1972; Knight, 1986).

The potential for local control of pseudopodial activity is probably best demonstrated by severing networks with a microneedle (Fig. 8). Motility continues unabated in an isolated network, which slowly reorganizes to form a cytoplasmic mass (i.e. satellite) centralized within a radial array of reticulopods (for concise accounts of this process, see Cushman, 1920; Sandon, 1944; Jahn and Rinaldi, 1959). In *Allogromia*, several satellites may initially arise when a large portion of the network is severed from the cell body, but these eventually fuse to form a single larger satellite. Similarly, when two separate satellites are brought into contact, they combine to form one larger satellite. The cytoplasmic reorganization associated with satellite formation is an active process, being reversibly inhibited by uncouplers of oxidative phosphorylation (Travis and Rosenbaum, unpublished observation).

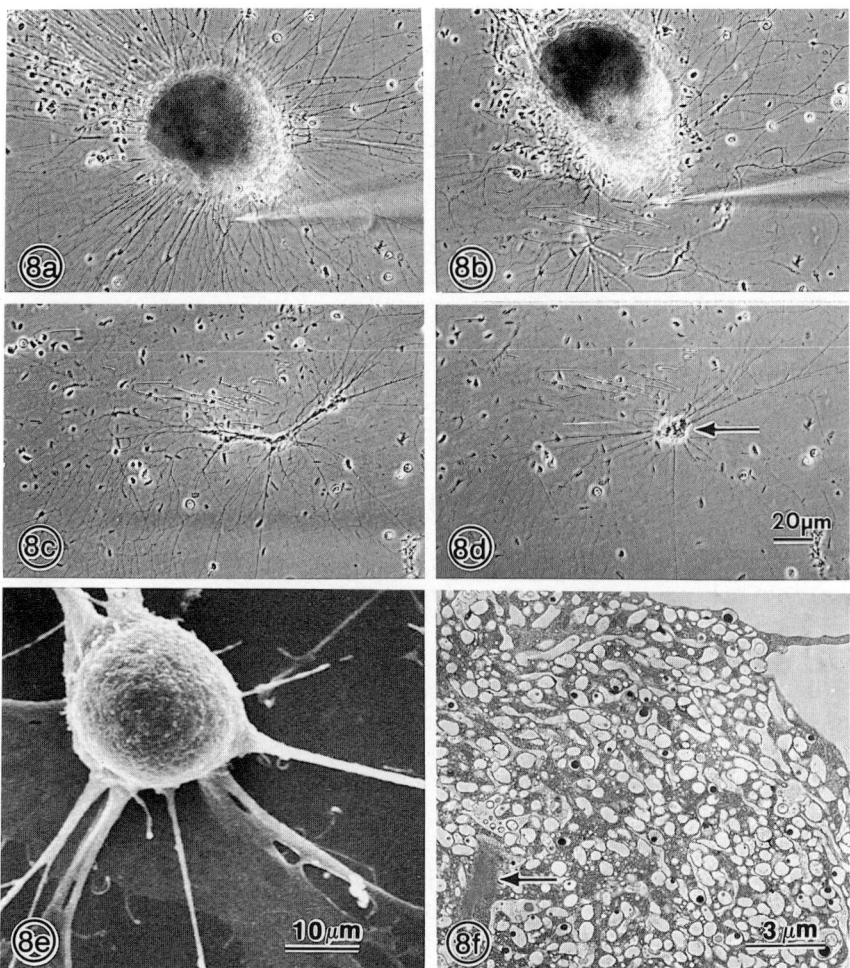

Fig. 8. Pseudopodial satellites (autonomous pseudopodial fragments) in *Allogromia laticollaris*. In this series of phase-contrast light micrographs (a–d), pseudopods were severed from the cell body using a glass micropipette (a, b), and the cell body was pushed aside. Scratches on the polystyrene substrate made by the microneedle serve as reference marks. Pseudopodial cytoplasm streams centripetally over the next few minutes to form a central mass (arrow, d). (e) Scanning electron micrograph and (f) thin-section transmission electron micrograph showing, respectively, the rough surface relief and the highly vacuolated cytoplasm of a satellite. The arrow in (f) points to a paracrystalline array of helical filaments common to satellite formation. Figure 8a–d reproduced with permission from Travis and Bowser (1986).

It has been suggested that the balance of tensile forces within the intact network is disrupted when it is severed, and that satellite formation reflects the re-equilibration of these forces (Travis and Bowser, 1986b).

At least one subtle change distinguishes satellite motility from that of normal networks: the centripetal (inward) bias in surface transport, characteristic of intact networks, is not seen in satellites (Lawrence and Travis, unpublished observations). This finding suggests that the cell body plays an undefined role in establishing pseudopodial polarity.

Pseudopodial satellites remain active for several hours, but eventually motility slows and finally ceases. If contact with the "donor" specimen is re-established, the satellite cytoplasm becomes re-incorporated with the donor network. Such pseudopodial fusion typically does not occur between species, or even between two individuals of the same species (e.g. Cushman, 1920; Allen, 1964). However, in laboratory cultures of *Allogromia laticollaris* (Travis and Centonze, 1983), and in specimens of *Myxotheca arenilega* that have been stripped of their shells (Schwab and Schwab-Stey, 1980), satellites from one organism can fuse with the network of a different specimen of the same species to form a chimera.

Satellite formation appears to be a general feature of foraminiferan reticulopods, being observed in all species we have examined (including allogromiids, *Amphistegina, Amphisorus, Textularia,* calcariniids, *Spiroloculina, Astrammina, Pyrgo* and *Elphidium*). In nature, these benthic foraminifers are frequently exposed to forces (e.g. bioturbation or wave/current action) that could easily fragment their networks. In this regard, the ability of severed networks to form satellites and be subsequently rescued may be adaptive, allowing the organism to recoup the energetic investment represented by the reticulopodial cytoplasm.

3.0 INTRATHALAMOUS MOVEMENTS

Several granuloreticuloseans, e.g. *Reticulomyxa* (Nauss, 1949), *Allogromia* (Arnold, 1955) and *Spiroloculina* (Arnold, 1964), have tests that are transparent/translucent enough to permit the direct light microscopic examination of intrathalamous cytoplasm. Whereas reticulopodial organelles typically move independently (or in small groups, each of which displays independent motion), bulk cytoplasmic flow predominates in the intrathalamous cytoplasm of these species. The pattern of intrathalamous movement is exceedingly complex, composed of a three-dimensional maze of numerous "currents" or "streams", each of which defines the linear or curvilinear path taken by groups of particles. The currents move simultaneously in close proximity, often in opposite directions or with significantly different average velocities. Individual streams are not restricted in space but rather "wander with relative freedom throughout the chamber spaces" (Arnold, 1964). McGee-Russell (1974) described such intrathalamous movements of *Allogromia* sp. as "writhing columnar motions – as seen in stirring a viscous sediment in a liquid".

Unfortunately, few studies have focused on the physiological significance or force-generating mechanism of bulk streaming and related intrathalamous processes. It could be argued that giant cells, like the larger foraminifera, cannot depend on the simple diffusion of metabolites, and the dynamic pattern of bulk streaming serves to mix the intrathalamic contents (see Fenchel, 1987). The jumbled array of currents leaves one with the impression that this "cytoplasmic mixing" is a random process. This may be true locally, but the specific transport of materials (e.g. those targeted for secretion or elimination) between chambers or between intrathalamic and extrathalamic compartments undoubtedly requires a more directed process. In this respect, at least one class of non-random intrathalamous movements has been reported: defecation vacuoles or granules containing more concentrated wastes move in a directed fashion from within the cell body, through the apertural region and into the network, where they are subsequently expelled (see Section 4.1.2).

4.0 RETICULOPODIAL COMPONENTS

4.1 Ultrastructure

4.1.1 *Plasma membrane*

Based on his observations of prey capture and surface transport, Sandon (1934) suggested the presence of a resilient surface layer or pellicle on foraminiferan pseudopods. Jahn and Rinaldi (1959), unable to reconcile reticulopodial surface transport with the accepted unit membrane model, suggested that *Allogromia* pseudopods were composed of "naked gel threads" that lacked a plasma membrane. Electron microscopy settled this controversy. The studies of Wohlfarth-Bottermann (1961), Hedley *et al.* (1967) and Marszalek (1969) showed clearly that all reticulopods were enclosed in a typical trilaminar unit membrane (Fig. 9).

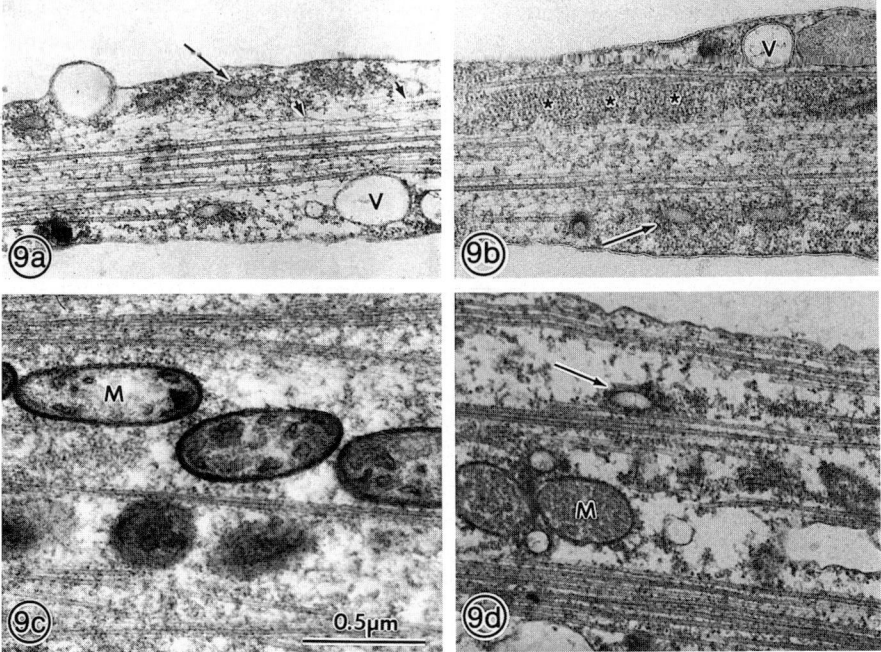

Fig. 9. Transmission electron micrographs of longitudinal sections through pseudopods, demonstrating ultrastructural similarities between four benthic foraminiferans: (a) *Allogromia* sp., strain NF; (b) *Allogromia laticollaris*; (c) *Spiroloculina hyalina*; (d) *Amphisorus hemprichii*. Common features include a prominent microtuble (MT) cytoskeleton, mitochondria (M), electron-lucent vesicles (V) and elliptical, fuzzy-coated vesicles (arrows). Not illustrated are dense bodies and clathrin-coated pits and vesicles. Helical filaments (*), often aggregated in patches of paracrystalline material, and thin filaments (arrowheads), are also commonly seen in the pseudopodial cytoplasm. Figure 9a is reproduced with permission from Koury et al. (1985).

Earlier workers also suggested that reticulopods were vested by an extracellular coat of "mucus, which is left behind as a trail after their withdrawal" (Jepps, 1942; see also Cushman, 1920). More recent ultrastructural studies, using polycationic contrasting agents, failed to demonstrate such a mucus layer in *Allogromia*, revealing instead a conventional glycocalyx (Bowser et al., 1984a). Similar results were obtained with *Astrammina* (Bowser et al., 1986), except that the reticulopods also appeared studded with short, hair-like projections of surface coat material (see also Marszalek's, 1969, description of *Iridia*). Light-microscopic studies using fluorescent lipophilic probes, together with whole-mount transmission electron microscopy, showed that, as reticulopods withdrew from a substrate, they frequently trailed behind plasma-membrane fragments (Bowser and Travis, unpublished; see also Koury et al., 1985, Fig. 1b). It seems likely that these plasma membrane remnants were mistaken for mucus trails in early light microscopic studies.

4.1.2 *Membrane-bound organelles: Identification and distribution*

As mentioned at the onset, a major diagnostic feature of reticulopods is the presence of refractile cytoplasmic "granules", which for many years remained uncharacterized. To this day, these granules remain difficult to identify by their light microscopic appearance alone: correlative light and electron microscopy reveals that a single granule seen with the light microscope may in fact consist of a group of several different membrane-bound organelles (Bowser, 1983; Bowser and Rieder, 1985). Nevertheless, video-enhanced light microscopy, electron microscopy and cytochemical techniques have documented several major types of reticulopodial organelles, the most conspicuous of which include mitochondria, dense bodies, phagosomes, defecation vacuoles, clathrin-coated pits and vesicles, and a unique class of elliptical fuzzy-coated vesicles. Many of these organelles are illustrated in Fig. 9.

Mitochondria are among the most conspicuous reticulopodial organelles. Electron microscopy shows that the mitochondria of various foraminiferan species are oblate spheroids ($c.$ 0.5 μm diameter, up to 5 μm long), possessing the tubular cristae typical of most protists. We have used fluorescent dyes (e.g. Rhodamine 123: Chen et al., 1982) to specifically stain mitochondria in living *Allogromia* and *Reticulomyxa*, permitting their differentiation from other reticulopodial granules (Bowser and Travis, unpublished observations). These studies confirmed that mitochondria are transported bidirectionally through the pseudopodial cytoplasm. Such transport may serve to distribute metabolic energy (ATP) throughout the network – an absolute requirement for all aspects of network motility.

Dense bodies are prominent refractile particles that also contribute significantly to the granular appearance of reticulopods. Like mitochondria, they appear dark with positive phase optics. By electron microscopy, they appear as membrane-bound vesicles possessing an electron-lucent cortex and a homogeneous, electron-opaque core. In some species, e.g. *Allogromia* sp. (strain NF), the dense bodies lack the electron-lucent cortex and the inner membrane leaflet is indistinguishable from the electron-opaque core material (McGee-Russell and Allen, 1971). Different preparative methods might account for this variability. The composition and function of dense bodies are not yet known. Anderson and Bé (1978) proposed that they represent concretions of indigestible material. However, the observation that cold shock causes the osmiophilic core to transform into classic myelin figures (Bowser, unpublished observation) is consistent with the idea that they represent phospholipid stores involved in reticulopodial membrane assembly and turnover. Other highly motile cells, such as *Amoeba proteus* (Szubinska, 1971) and mammalian white blood cells (Caulfield et al., 1987), also have membrane phospholipid storage granules.

In allogromiids, small food items like bacteria may be endocytosed to form blister-like phagosomes either at the hyaline veil that forms at the juncture of bifurcating reticulopods, or at the circumoral cytoplasm. McGee-Russell (1974) reported that bacterial prey were trapped in a loose basket of pseudopodial extensions that gradually closed as the pseudopodia fused around it to form a phagosome. He noted that during engulfment, motile bacteria had about a 50% chance of escaping. In stark contrast to the bidirectional movement of other organelles, phagosomes, often clustered to form large bulges in the reticulopods, typically move unidirectionally towards the cell body.

There seems to be some interspecific variability regarding the presence of lysosomes in reticulopods, as defined by cytochemical methods used to detect acid phosphatase. For example, lysosomes were reported in the reticulopods of the planktonic foraminifer *Hastigerina pelagica* (Anderson and Bé, 1976). In contrast, lysosomes are not observed in *Allogromia* reticulopods. In fact, even those reticulopodial phagosomes containing bacterial prey lack detectable acid phosphatase activity. However, once in the cell body, phagosomes fuse with numerous primary lysosomes and assume a digestive function (Bowser et al., 1985).

Defecation vacuoles, containing the indigestible remains of digestive vacuoles, are transported from the cell body into the reticulopods, whereafter their contents are emptied. In *Allogromia* sp., the specific inward movement of phagosomes and the outward movement of defecation vacuoles are superimposed on the typical bidirectional transport of other organelles (Bowser et al., 1985). This observation indicates that certain organelles can be targeted for specific transport to or away from the cell body. Indeed, Jepps (1942) reported that defecation vacuoles in *Polystomella* (= *Elphidium*) were transported to the distal pseudopodial tips for exocytotic release of their contents, suggesting that foraminifers can target organelles to specific regions within the network.

Coated vesicles c. 0.1 μm in diameter are commonly seen in electron micrographs of reticulopods. In favourable sections, the pentagonal/hexagonal substructure characteristic of clathrin can be seen surrounding vesicles (= coated vesicles), vesting plasma membrane invaginations (= coated pits), or underlying portions of the plasma membrane (Travis and Allen, 1981; Schwab and Schwab-Stey, 1981). In other cell types, these clathrin-coated pits and vesicles mediate the internalization of certain molecules (e.g. polypeptide hormones, low-density lipoproteins, vitellogenins) bound to specific membrane receptors (reviewed in Goldstein et al., 1985). In the foraminifera, the role of coated pits/vesicles is not yet defined.

A final type of membraneous organelle found in reticulopods is a class of small (50–70 nm diameter; up to 300 nm long) elliptical vesicles, which are vested with a dense coat of fibrillar material very distinct from clathrin. These vesicles are barely visible by light microscopy (e.g. see Bowser and Rieder, 1985, Fig. 8). Elliptical vesicles are common to every foraminiferan species we have examined, and Koonce et al. (1986) illustrated them in Reticulomyxa. To our knowledge, these vesicles are unique to the Granuloreticulosea, and may therefore prove to be a valuable taxonomic character. Their ubiquity within the class suggests that they play a fundamental role in the organization or maintenance of reticulopods, but their function remains unknown. In an ultrastructural study, McGee-Russell and co-workers (1982) noted that the fuzzy coat of these vesicles structurally interacted with microtubules, the plasma membrane and other organelles. These authors suggested that the elliptical vesicles served to physically link various organelles to form discrete cytoplasmic domains, and thus coordinate reticulopodial motility. Their observations prompted the term Motility Organizing Vesicles (MOVs) for the elliptical vesicles. Other possible functions of these vesicles include membrane recycling or ion homeostasis.

Foraminifers restrict the passage of certain organelles (e.g. nuclei, Golgi elements, ribosomes and in some cases lysosomes) from the cell body into the pseudopodial cytoplasm (Doyle, 1936; Arnold, 1964; McGee-Russell and Allen, 1971; Koury, 1982; Bowser et al., 1985). The basis of organelle segregation in foraminifera is unknown. Earlier workers suggested that the pores in the tests of calcareous species might serve to restrict the passage of larger organelles, such as nuclei, from the pseudopods (Arnold, 1954; reviewed in Boltovskoy and Wright, 1976). However, such segregation is well documented in non-calcareous species lacking sieve plates, e.g. *Allogromia*. A cytoplasmic structural barrier that might account for organelle segregation, such as a filament network, was specifically sought for but not detected by serial-section electron microscopy of *Allogromia* juveniles – a particularly favourable material for such a study due to their small size (Bowser et al., 1984b). The absence of a structural barrier suggests that steric hindrance does not account for organelle segregation. It seems more likely that the selective transport of only certain membraneous organelles into the reticulopods results in their asymmetric distribution.

4.1.3 *Reticulopodial cytoskeleton*

4.1.3.1 *Microtubules.* Electron microscopy shows that microtubules are the most prominent component of the reticulopodial cytoskeleton (Hedley et al., 1967; Marszalek, 1969; McGee-Russell and Allen, 1971; Schwab and Schwab-Stey, 1973; Hottinger and Dreher, 1974; Nyholm and Nyholm, 1975; Schwab, 1977; reviewed in Anderson and Bé, 1978). This impression is confirmed by light microscopic techniques which provide a better overview of the network's cytoskeleton. For example, the extensive system of birefringent cytoplasmic fibrils seen in flattened pseudopods by light microscopy (Fig. 7) have been identified as microtubule bundles by antitubulin immunofluorescence, and more directly by correlative light and high-voltage electron microscopy (reviewed in Travis and Bowser, 1986b, 1988).

Most reticulopodial microtubules occur in discrete, loosely organized bundles, although single microtubules are occasionally seen (Fig. 10c). The microtubules within a bundle typically assume straight trajectories and are aligned with their long axes parallel to the pseudopodial long axis. However, this basic configuration may be modified by some foraminifers as required for specific life strategies. For example, three-dimensional views of trunk pseudopods showed that microtubules coil tightly around one another in *Astrammina* (Bowser et al., 1986). This coiling may serve to increase the tensile strength of trunk pseudopods in this carnivorous species, which is known to ensnare relatively large metazoan prey.

The average microtubule density (number per unit cross-sectional area) can be quite high, ranging from c. 20 microtubules/μm^2 in *Allogromia* sp. (Koury et al., 1985) to over 650 microtubules/μm^2 in *Amphisorus hemprichii*. In *Allogromia*, microtubules are closely spaced, with a wall-to-wall separation averaging about 15 nm. Microtubule separation is significantly smaller in *Astrammina's* tightly coiled bundles (Jensen and Bowser, unpublished observations). Other densely packed microtubule arrays, e.g. the axonemes of heliozoans or the cytopharyngeal baskets of ciliates, are thought to be held in precise alignment by microtubule–microtubule crossbridging structures (reviewed in

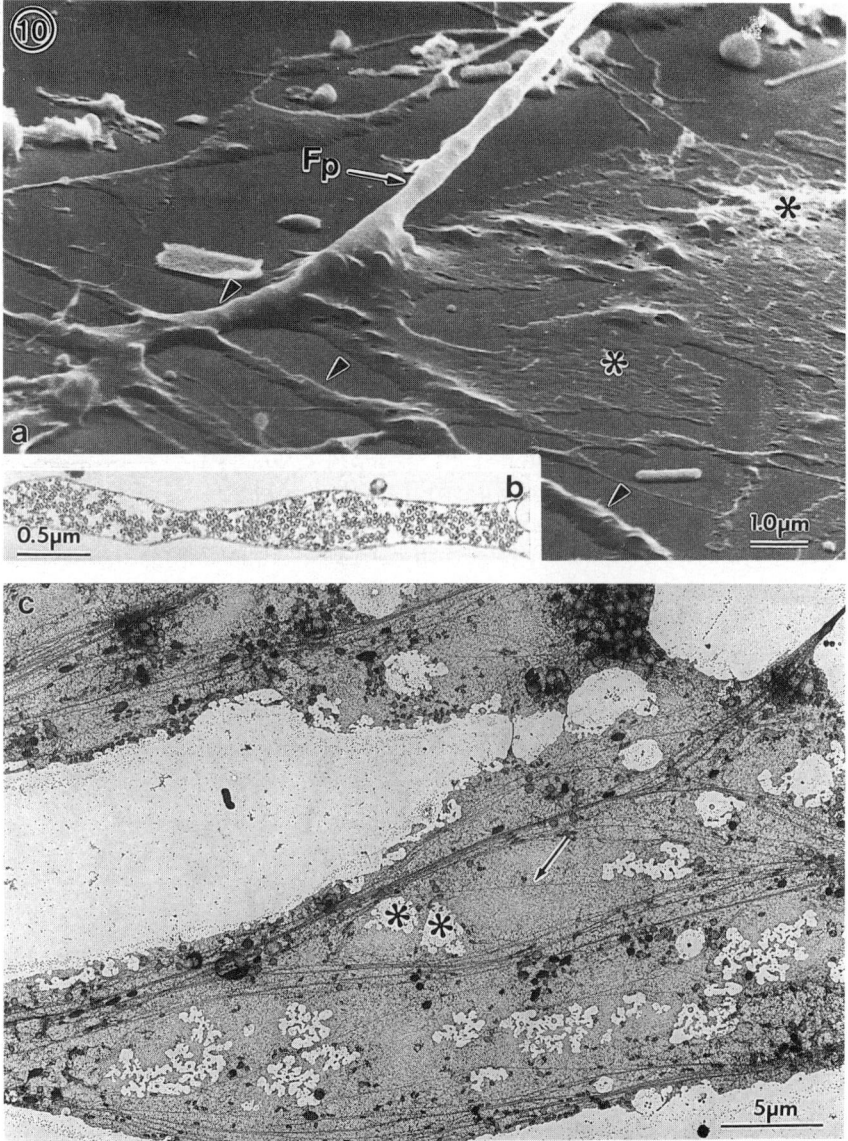

Fig. 10. Two-dimensional reticulopods on cationized surfaces. (a) Scanning electron microscopy reveals that cylindrical filopods (Fp) flatten considerably after contacting the charged substrate (arrowheads), in some regions forming extensive lamellipods (*). Cross-sections through such areas (b) show that the lamellipods are usually just a few tenths of a micrometre thick. As a consequence, fixed specimens may be critical point-dried and viewed as whole-mount in the transmission electron microscope. A typical *Allogromia* lamellipod prepared in this manner (c) is seen to contain numerous microtubules, occurring singly (arrow) or in bundles that branch and anastomose to form an interconnected network. This micrograph also shows numerous membraneous organelles contacting the microtubules. Note that the margins of the lamellipodia are typically irregular, and that lacunae or openings may form within the boundaries of these sheets. Figure 10b is reproduced with permission from Travis et al. (1983).

Tucker, 1982). Similar crossbridges are seen linking adjacent microtubules in foraminiferan pseudopods (Fig. 11; McGee Russell and Trautwein, 1977; Travis and Allen, 1981; Golz and Hauser, 1986; Bowser *et al.*, 1986). Two types of microtubule crossbridge are seen in detergent-extracted (i.e. demembranated) reticulopods: one with the periodicity and helical pitch resembling that of vertebrate MAP_2 (Fig. 11b,c), and the other displaying a linear periodicity resembling that of ciliary dynein (Fig. 11d,e). A quantitative structural analysis of untreated *Allogromia* reticulopods confirmed these findings, and showed that the two classes of bridging structures occur in mutually exclusive domains (Jensen *et al.*, 1990). Both types of crossbridge can also be seen linking microtubules to membrane-bound organelles or to the plasma membrane (Travis and Allen, 1981; Kachar *et al.*, 1987). A preliminary comparative study of *Allogromia* and *Astrammina* (Jensen and Bowser, unpublished) found that crossbridge length correlates positively with microtubule wall-to-wall separation distance. This finding strengthens

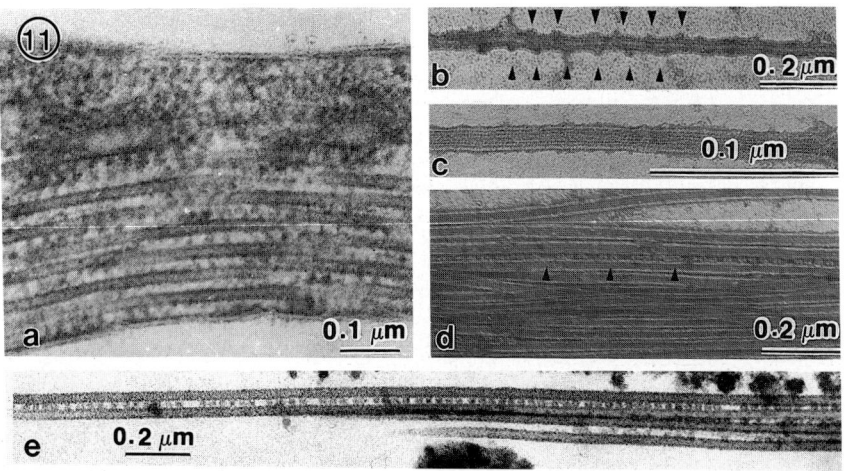

Fig. 11. Reticulopodial microtubules and crossbridging structures. A photographic two-step lateral translation of a transmission electron micrograph (a) enhances the appearance of periodic structures linking adjacent microtubules, and similar structures linking microtubules to the plasma membrane, in this *Allogromia* filopod. Negatively stained preparations of lysed *Allogromia* filopods reveal helically arranged projections decorating some microtubules (arrowheads, b), as well as a microtubule protofilament substructure (c) and occasional globular crossbridges between bundled microtubules (arrowheads, d). (e) Microtubules from a detergent-lysed and stabilized *Allogromia* filopod, demonstrating crossbridges displaying a 24-nm dynein-like periodicity. All parts are reproduced with permission: (a) from Travis and Bowser (1988); (b–d) from Golz and Hauser (1986); (e) from Travis and Allen (1981).

the argument that inter-microtubule crossbridges define the spatial organization of microtubules in reticulopods.

4.1.3.2 *Helical filaments and paracrystals.* Schwab and Schwab-Stey (1972) first described the presence of helical filaments 5 nm in diameter in *Allogromia* cytoplasm. These authors noted that, following some harsh fixation regimens, microtubules were not preserved; instead, the cytoplasm contained numerous helical filaments. Through studies with various pharmacological agents, the notion that foraminiferan microtubules might transform into helical filaments was subsequently advanced by several groups (Hauser and Schwab, 1974; McGee-Russell, 1974; Schwab-Stey and Schwab, 1979). In all of these studies, it was found that helical filaments could align laterally to form paracrystalline aggregates (Fig. 12a). In addition, Koury et al. (1985) demonstrated that, during reticulopod withdrawal, microtubules decreased in number and the frequency of helical filaments and paracrystals concomitantly increased. Conversely, Bowser et al. (1984b) reported that paracrystals

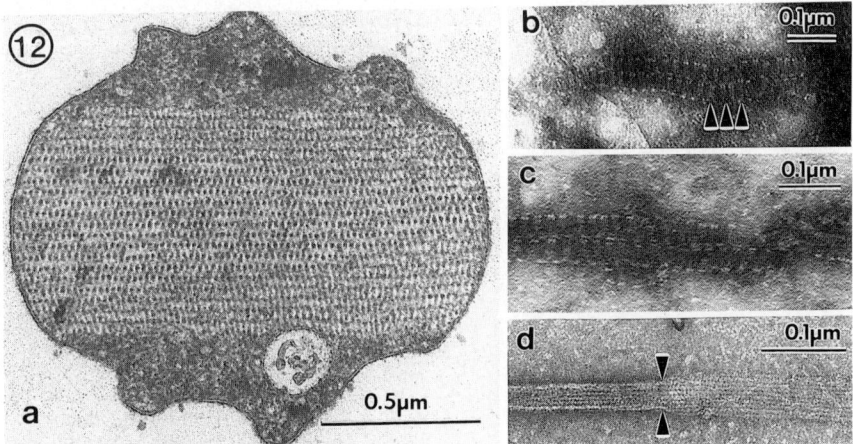

Fig. 12. Transmission electron micrographs of tubulin-containing paracrystals and helical filaments from *Allogromia laticollaris* reticulopods. In (a), a paracrystal is seen in a pseudopod treated with ruthenium red prior to fixation. When networks are lysed in a microtubule-stabilizing buffer and then treated with ruthenium red, the microtubules transform into helical filaments, as seen in negative stain (arrowheads, b). Each helical filament appears to be composed of a pair of adjacent subfilaments, in contrast to intact microtubules, which are composed of 13 longitudinally oriented protofilaments (cf. c, d). These protofilaments are particularly evident in (d), where the microtubule has split open and flattened to the right of the arrowheads. Reproduced with permission from Golz and Hauser (1986).

in the cell bodies of *Allogromia* juveniles disappeared as they assembled reticulopodial microtubules.

More recently, the close relationship between microtubules, helical filaments and paracrystals was confirmed using immunocytochemical techniques. Rupp *et al.* (1986) used correlative fluorescence immunocytochemistry and electron microscopy to show that paracrystals in retracted networks contained tubulin (Fig. 13). Golz and Hauser (1986) showed that microtubules obtained from detergent-extracted *Allogromia* pseudopods transformed into helical filaments upon the addition of the polycationic dye ruthenium red (cf. Figs 12b–d). However, the molecular details regarding the transition between microtubules, helical filaments and paracrystals are obscure, and the mechanisms regulating them are completely unknown. It is unclear, for example, whether microtubules disassemble into dimers or ring-like oligomers, which then assemble to form helical filaments, or whether the microtubules directly "uncoil" to form helical filaments. The latter possibility was presented as a

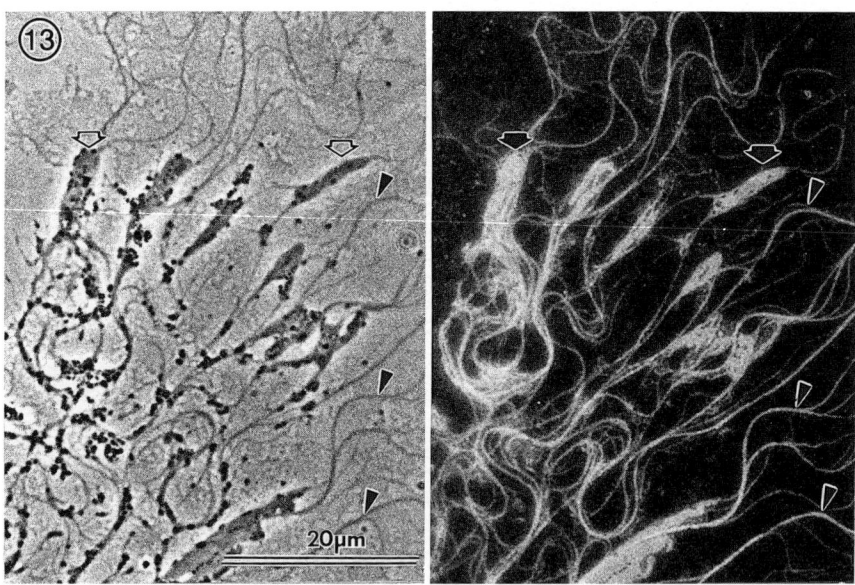

Fig. 13. Phase-contrast (a) and antitubulin immunofluorescence (b) micrographs of an *Allogromia* sp. (strain NF) pseudopod, flattened on a highly charged glass coverslip, following withdrawal induced by high Mg^{2+}-seawater treatment. The thin, wavy fibrils (solid arrowheads) and pools of phase-dense material (squat, open arrowheads) stain intensely with tubulin antibodies (cf. a, b). Note the accumulation of organelles within the pools. Reproduced with permission from Rupp *et al.* (1986).

formal model by Hauser and Schwab (1974), and is supported by the *in vitro* work of Golz and Hauser (1986). Such a transition is difficult to explain in the light of current concepts of microtubule structure (Mandelkow *et al.*, 1986; Mandelkow and Mandelkow, 1989). The structure of foraminiferan microtubules, helical filaments and paracrystals requires detailed study by molecular resolution electron microscopy before any model can be accepted. Helical filaments and paracrystals were found in all other foraminifers examined to date, and similar paracrystals were reported in other microtubule-rich structures, most notably in *Sticholonche* axopods following various treatments (Cachon and Cachon, 1981a, and references therein).

4.1.3.3 *Actin filaments and their association with microtubules.* The explosive growth in the field of cell motility was fuelled over the past two decades by technical developments such as refinements in immunofluorescence of cytoskeletal proteins and fine structural preservation, and the application of contractile protein biochemistry to non-muscle cells. During this time, familiar sarcodines such as *Amoeba proteus* (Pollard and Korn, 1971), *Chaos carolinensis* (Comly, 1973) and *Difflugia* (Eckert and McGee-Russell, 1973) were shown to possess an actomyosin-based cytoskeleton. Actin was even found in the microtubule-rich axopods of heliozoans (Edds, 1975). Were the foraminifera an exception?

Despite early reports of microfilaments in foraminiferan pseudopods (Schwab and Schwab-Stey, 1972, 1973), initial attempts to identify actin as a component of *Allogromia* pseudopods using heavy meromyosin or myosin subfragment 1 decoration techniques were unsuccessful (Travis and Allen, 1981). Later, actin was identified by immunostaining protein blots of cell body and pseudopodial extracts from *Astrammina rara* (Travis *et al.*, 1985). In *Allogromia*, actin-containing filaments were localized, using the fluorescent actin-specific probe rhodamine-phalloidin, to discrete tear-shaped plaques (Fig. 14; Travis and Bowser, 1986a,b; Bowser *et al.*, 1988). These actin filament plaques were restricted to the ventral cell surface, and were especially prominent in regions where trunk pseudopods initially contacted the substrate (Bowser *et al.*, 1988). Electron microscopy demonstrated that these plaques contained a feltwork of roughly parallel microfilaments. In contrast to *Allogromia*'s highly restricted distribution, actin microfilaments were found throughout the pseudopodial network in *Reticulomyxa* (Koonce *et al.*, 1986; Hauser *et al.*, 1989). Because microfilaments play a role in contractile vacuole activity in diverse freshwater protists (reviewed in Patterson, 1980), it is possible that *Reticulomyxa*'s pervasive actin cytoskeleton is a modification necessary for the functioning of the numerous contractile

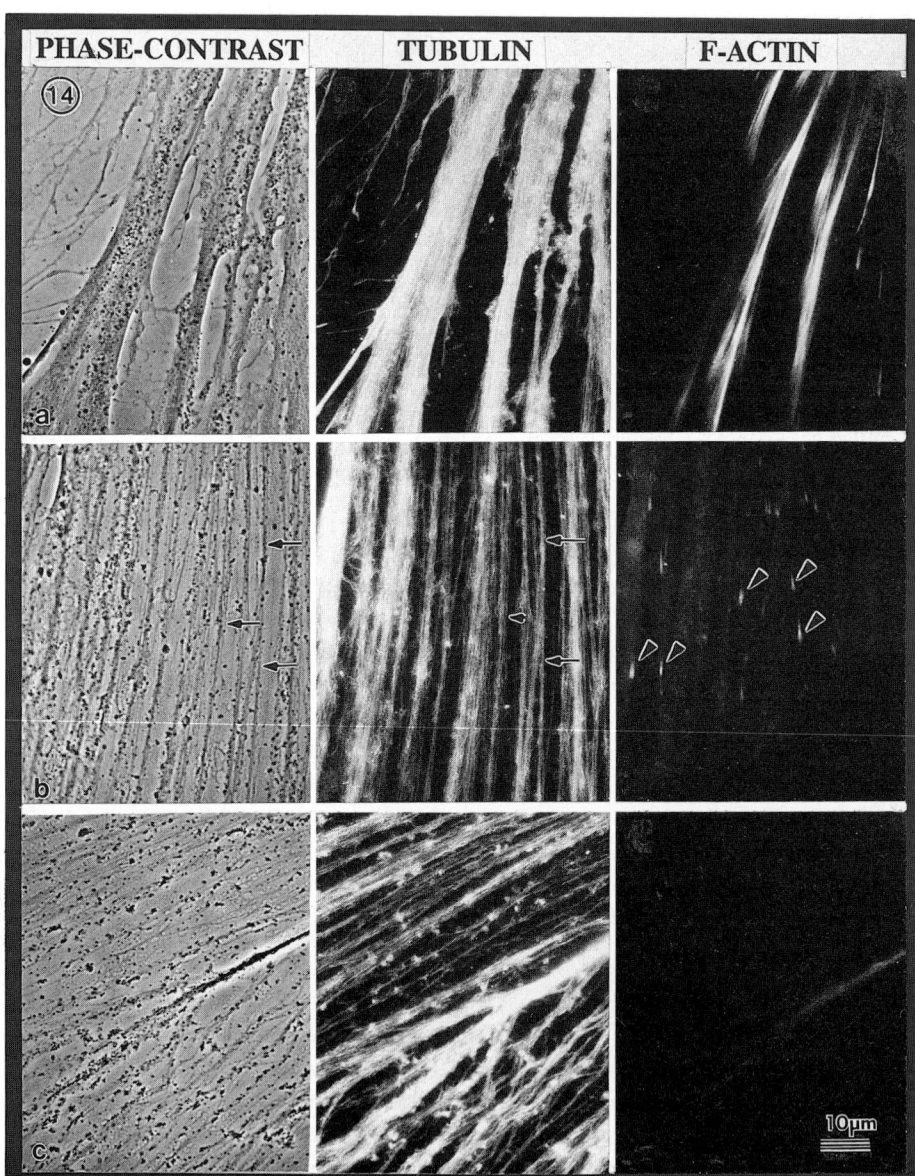

vacuole complexes seen streaming through its pseudopodial network (Bowser, unpublished).

There is considerable evidence that structural interactions occur between microtubules and actin microfilaments in reticulopods. Correlative immunofluorescence and high-voltage electron microscopy of serial thick sections through *Allogromia* pseudopodial trunks revealed that a subset of microtubules closely associates with actin plaques at cell/substrate contact sites (Fig. 15; Bowser *et al.*, 1988). Similar results were also obtained with *Amphisorus* (Travis *et al.*, 1988). In *Reticulomyxa*, Koonce *et al.* (1986) showed the close, co-axial alignment of actin microfilaments and microtubules, and Hauser *et al.* (1989) reported 25-nm-long crossbridging structures between the two filament systems.

4.1.3.4 *Other filament systems.* It remains an open question whether other cytoskeletal elements are present in reticulopods. Electron microscopy and immunolabelling studies of *Allogromia* have failed to reveal intermediate-sized filaments, the third major fibrous protein system of eukaryotic cells. Travis and Allen (1981) found that filaments 5 nm in diameter, which appeared distinct from actin by several criteria, were co-linear with reticulopodial microtubules in *Allogromia laticollaris*. These authors suggested that the 5-nm filaments belonged to a newly recognized class of contractile filaments that occur widely in protistan taxa (reviewed in Cachon and Cachon, 1981b; Roberts, 1987). An alternative notion is that the 5-nm filaments represent tubulin protofilaments fixed at some intermediate stage of microtubule assembly or disassembly (Golz and Hauser, 1986).

Fig. 14. Different areas of the same reticulopodium, imaged by phase-contrast (left column) and fluorescence optics (centre, right columns). Double-label fluorescence microscopy using tubulin antibodies and tetramethylrhodamine (TMR)-labelled phalloidin allows the simultaneous localization of microtubules (TUBULIN, centre column) and actin-containing microfilaments (F-ACTIN, right column) in the same specimen. The spatial distribution of microtubules and microfilaments differs: microtubules extend throughout reticulopodial trunks (a), whereas F-actin is restricted to a few linear fibres that taper proximally. (b) Approximately 30 μm distal to the above region, the trunks flatten to form an extensive lamellipod on the cationized glass coverslip. The fibrils visible with phase optics stain intensely for tubulin (small arrows), whereas F-actin is restricted to only a few foci (arrowheads). At the periphery of the lamellipod (c), the fibrils continue to stain intensely with tubulin antibodies, but no F-actin staining is seen. Reproduced with permission from Bowser *et al.* (1988).

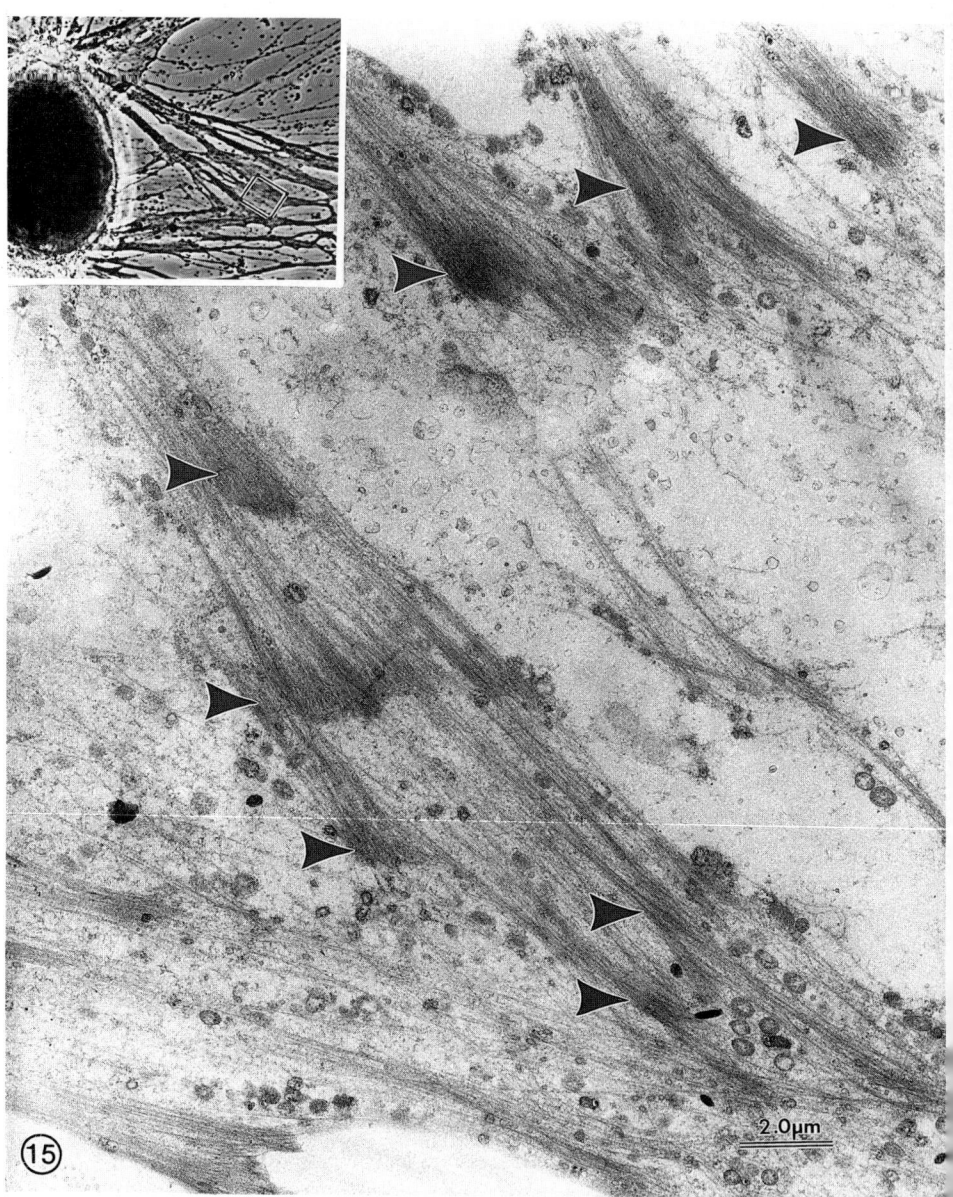

Fig. 15. This high-voltage electron micrograph of a thick (0.25 μm) section through the ventral surface of an *Allogromia* sp. pseudopodial trunk shows dense filament plaques (large arrowheads) interspersed among microtubule bundles. The section was the first cut from the boxed region of the organism pictured in the inset. Reproduced with permission from Bowser et al. (1988).

4.2 Molecular composition of the reticulopodial cytoskeleton

Although foraminifera and allied protists are excellent systems for structural or biophysical studies, they remain difficult subjects for biochemical analysis. This is largely due to contamination of typical laboratory cultures of granuloreticuloseans with other eukaryotic organisms (e.g. algal endosymbionts, ectocommensals, or prey). Therefore, the characterization of any protein isolated from such crude cultures must be fully substantiated by other methods, such as the exclusive immunocytochemical localization of the polypeptide in question to reticulopods. As a result of this constraint, very little reliable information is available regarding the molecular composition of the reticulopodial cytoskeleton. Nevertheless, some biochemical data are available for reticulopodial microtubules, primarily those of *Allogromia* sp. (strain NF) and *Astrammina rara* – species which are grown or collected, respectively, under conditions that exclude all other eukaryotic organisms.

Microtubules are composed of two distinct polypeptides, alpha and beta tubulin (Luduena *et al.*, 1971). In mammalian tissue cells, tubulins are subject to a number of post-transitional modifications, including phosphorylation, tyrosylation and acetylation (reviewed in Greer and Rosenbaum, 1989). Preliminary electrophoretic and immunochemical studies suggest that foraminiferan tubulins differ substantially from those of vertebrate cells. For example, Rupp *et al.* (1986) found that *Allogromia* tubulins display considerably slower electrophoretic mobilities than vertebrate brain tubulins, and Koonce *et al.* (1986) documented a similar electrophoretic behaviour for *Reticulomyxa* tubulins. It also remains questionable whether post-transitional modification of tubulin occurs in *Allogromia* or *Reticulomyxa*. For example, Centonze and Travis (1983), Koonce *et al.* (1986) and Lindenblatt *et al.* (1988) were unable to detect tyrosylated alpha tubulin in *Allogromia* or *Reticulomyxa* using highly sensitive immunological methods. Hauser *et al.* (1989) did, however, report weak staining of reticulopods in juvenile *Allogromia* gamonts using an antibody against tyrosinylated tubulin.

In other cell types, three general types of microtubule-associated proteins (MAPs) interact with the microtubule lattice:

1. *Mechanochemical transducers (mechanoenzymes)*, which convert the energy liberated by the hydrolysis of adenosine triphosphate (ATP) into mechanical work.

2. *Crosslinkers*, which connect adjacent microtubules or link microtubules to membranes or other cytoskeletal filaments.
3. *Assembly/disassembly regulators*, proteins that might facilitate or inhibit the nucleation or elongation phases of microtubule assembly (reviewed in Warner and McIntosh, 1989; Warner *et al.*, 1989).

The specific interactions between microtubules, MAPs and other cellular structures define the functional role of a given set of microtubules.

Ultrastructural data on inter-microtubule crosslinks seen in *Allogromia* reticulopods indicate that at least two classes of MAPs are present: those related to structural MAPs (i.e. presumably microtubule crosslinkers) and those related to dynein mechanoenzymes (see Section 4.1.3.1). Microtubule–membrane crosslinks (Fig. 11a) also have a dynein-like linear repeat. However, the biochemical characterization of these reticulopodial MAPs is only in its infancy. Preliminary immunocytochemical studies using an antibody directed against vertebrate brain MAP_2 suggested the presence of a related polypeptide in *Allogromia* reticulopods. When challenged with this antiserum, protein blots of *Allogromia* whole-cell lysates stained a doublet of approximately 70 kD (Rupp, Travis and Bowser, unpublished observations). Euteneuer *et al.* (1986) reported the isolation of a MAP-like protein from *Reticulomyxa* that promoted the formation of needle-shaped microtubule bundles *in vitro*.

The sliding movement (Figs 7, 16a,b) and packing order of foraminiferan microtubules suggests that a dynein-like molecule is present in reticulopods. Hauser (personal communication) recently detected a reticulopodial component that is immunologically related to axonemal 14S dynein. The dynein-like component stains in a punctate pattern (Fig. 16d,e), consistent with the discontinuous distribution of dynein-like crossbridges seen by electron microscopy. We used the techniques

---

Fig. 16. Differential interference-contrast videomicrographs of microtubule sliding in a cytoskeletal model of the athalamiid *Reticulomyxa* (a–c), and the distribution of dynein-like molecules in *Allogromia* revealed by indirect immunofluorescence (d, e). *Reticulomyxa* pseudopods were lysed with non-ionic detergent and the residual cytoskeleton was fragmented with a microneedle, resulting in short microtubule bundles (a). In (b), ATP was added to the preparation, resulting in the extrusion of microtubules from both ends of the bundle (cf. b, c). Note that these microtubules also displayed dramatic shape changes including looping and bending. Arrows point to microtubule ends. (d, e) The same *Allogromia* pseudopod doubly stained with antibodies against tubulin (d) and ciliary 14S dynein (e). The dynein staining appears stippled, consistent with the discontinuous distribution of dynein-like crossbridges seen in electron micrographs. All parts are reproduced with permission: (a–c) from Koonce *et al.* (1987); (d, e) unpublished micrographs courtesy of Dr M. Hauser, Ruhr-Universitat, Bochum.

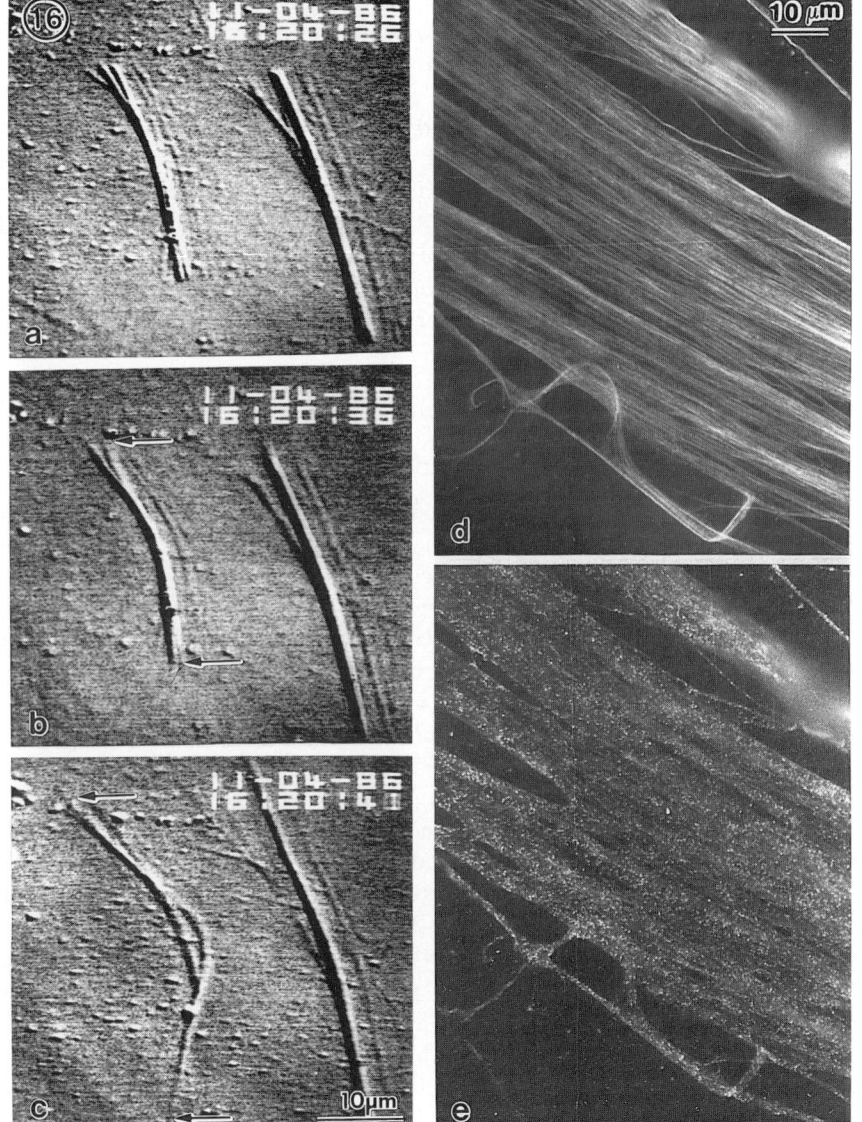

of Telzer and Haimo (1981) to label *Allogromia* microtubules with *Tetrahymena* 21S ciliary dynein and observed that they became decorated with up to six dynein arms. This finding suggested that an analogous molecule in *Allogromia* could account, at least in part, for the highly labile and roughly hexagonal bundling of microtubules (as would be expected if six bifunctional dynein-like molecules bound each microtubule). It is important to note that Euteneuer *et al.* (1988) recently described the presence of dynein-like ATPase in *Reticulomyxa* extracts. This enzyme had latent dynein-like ATPase activity, and exhibited the characteristic vanadate-sensitive UV cleavage pattern of its high-molecular-weight (>300 kD) polypeptide. Hopefully, with the production of specific antibodies to this microtubule-associated ATPase, we will be able to learn its precise intracellular location within reticulopods, and whether it cross-reacts with any native foraminiferan crossbridges.

5.0 MECHANISMS OF FORAMINIFERAN MOTILITY

We recently presented a brief historical account of progress made towards understanding the mechanism of reticulopodial motility (Travis and Bowser, 1988). In short, earlier investigators attempted to explain various facets of reticulopodial motility within the framework of a single mechanism (e.g. Jahn and Rinaldi, 1959; Kavanau, 1963; Allen, 1964). However, studies based on modern ultrastructural and video-microscopic methods, together with various experimental manipulations, support McGee-Russell's (1974) contention that reticulopodial motility is too complex to be adequately explained by one mechanism. As discussed in the following section, it seems likely that the following four major facets of foraminiferan motility are mechanistically distinct: (1) filopod extension/retraction and bending movements; (2) network withdrawal; (3) intracellular and surface transport; and (4) pseudopodial tension and whole-cell locomotion. Although the available information is inadequate to formulate detailed models, in the following sections we argue that microtubules play an integral role in each of these processes.

5.1 Mechanism of reticulopod movements

The morphological integrity and motility of reticulopods depend on an intact microtubule system. The principal evidence for this assertion is the loss of pseudopodial shape and inhibition of all movement when reticulopodial microtubules are disassembled by treatment with antimitotics or cold (Fig. 17; Travis and Bowser, 1986a). Furthermore, Travis et al. (1983) observed that the movements of cytoplasmic fibrils, corresponding to microtubule bundles, are essentially identical to reticulopod movements (i.e. both fibrils and filopods lengthen and shorten, bend, split to form branches, and anastomose to form intricate networks). The inescapable conclusion is that microtubule movements drive the pseudopodial movements. The question therefore distills to "what is the mechanism of the microtubule movements?"

5.1.1 Axial microtubule movements

Three major hypotheses help explain the observed axial movements of reticulopodial microtubule fibrils. In the first, microtubules are thought to translocate by a sliding mechanism dependent on a specific mechano-enzyme (i.e. motor molecule). Secondly, it is possible that microtubule-dependent motors are associated with the plasma membrane. Provided that these membrane-associated motors are immobilized, e.g. by transmembrane anchorage to the substrate at the ventral pseudopodial surface, the microtubules would be propelled through the cytoplasm. The conceptual framework for such "gliding" microtubule motility has been established from work on squid axoplasm (reviewed in Allen, 1987). A third possibility is that these axial movements actually correspond to the assembly/disassembly of their constituent microtubules. Evidence that argues against a fourth possibility, namely that some other contractile system (e.g. actomyosin) powers these movements, will also be considered.

The most likely hypothesis is that fibril length changes are driven by the sliding of adjacent reticulopodial microtubules. Microtubule sliding is best understood in the axonemes of cilia and flagella (Summers and Gibbons, 1971) and the axostyles of termite hind-gut flagellates (Woodrum and Linck, 1980). In these highly ordered microtubule systems, dynein transduces the free energy of ATP hydrolysis into mechanical work to effect the ratchet-like movement of adjacent microtubules past one another (Chilcote and Johnson, 1989). The evidence that reticulopodial fibril movements are caused by microtubule sliding is strong, and much of it is discussed in the earlier sections. Briefly, compelling evidence was obtained by Koonce et al. (1987), who demonstrated the rapid (up to 10 μm/s), ATP-dependent ejection of individual microtubules from severed pieces of demembranated *Reticulomyxa* reticulopods (Fig. 16a,b). Their observations are reminiscent of the classic ATP-induced "telescoping" of microtubules from trypsin-treated flagellar axonemes (Summer and Gibbons, 1971). The

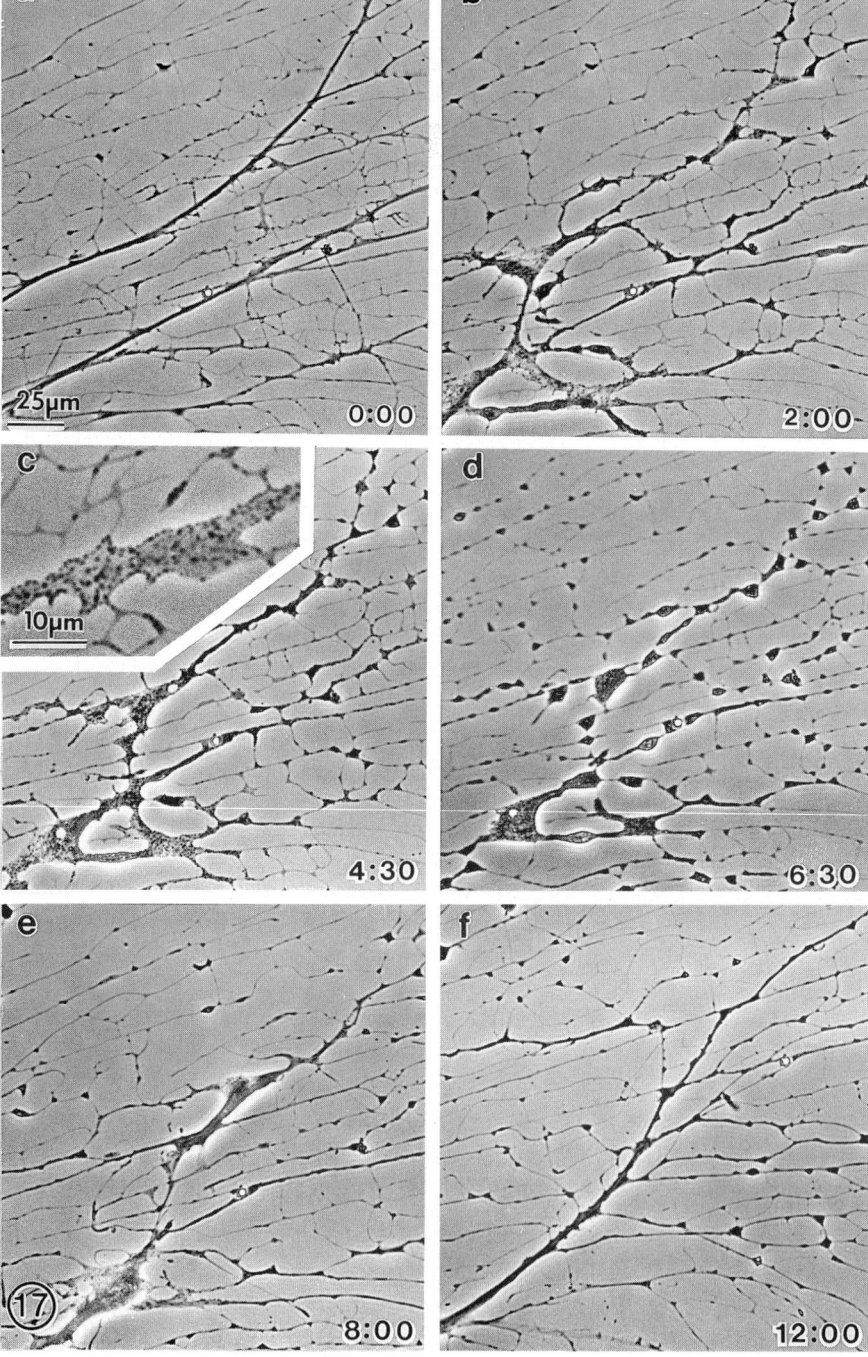

sliding of reticulopodial microtubules *in vivo* remains to be demonstrated unequivocally. Using videomicroscopy, we recently reported the axoneme-like bending of fibrils in flattened *Allogromia* reticulopods (Travis and Bowser, 1990). We also observed short fibrils (i.e. those with both ends in view) translocating along long stationary ones, which were interpreted as short microtubule fragments sliding along longer microtubules (see Travis and Bowser, 1988). Unfortunately, the latter dynamic events proved difficult to document convincingly in still micrographs.

As discussed above, microtubule–microtubule crosslinks displaying a dynein-like periodicity have been demonstrated in both detergent-extracted (Fig. 11d, e; Travis and Allen, 1981; Golz and Hauser, 1986) and native (Jensen *et al.*, 1990) *Allogromia* reticulopods. Furthermore, reticulopods stain with antidynein antibodies (Fig. 16d,e; Hauser, personal communication). It will be interesting to learn whether the dynein-like molecule isolated by Euteneuer *et al.* (1988) from *Reticulomyxa* mediates microtubule sliding *in vivo*.

It seems relevant to ask whether different motor molecules are required to move reticulopodial microtubules (and hence the fibrils) in opposite directions. Although the answer to this question is unknown, information on ciliary beating may provide some clues. During normal ciliary beating, the power (effective) stroke in the mechanochemical cycle of dynein is limited to a single direction (presumably a consequence of its molecular structure). Consequently, an A-subfibre moves forwards (relative to the basal body) along the B-subfibre of the adjacent doublet. However, the direction of the microtubule translocation depends on which microtubule is free to move (reviewed in Satir, 1989). For example, if the A-subfibre is held fast, the B-subfibre will now move backwards (towards the basal body), rather than the A-subfibre moving

---

Fig. 17. Reticulopodial motility is reversibly inhibited by colchicine. These sequential phase-contrast light micrographs show *Allogromia* reticulopods (a) immediately before application of 1 mM colchicine, (b, c) during treatment and (d–f) after removal of the drug. The colchicine response is initially marked by the accumulation of cytoplasm at branch points in the network (b). Pseudopodial movements, intracellular organelle transport and cell surface motility cease within 3–5 min. This inhibition is evidenced by a 2-s timed exposure (c, inset), which shows clear images of accumulated granules in a cytoplasmic varicosity. (In untreated pseudopodia, particles appear blurred due to their motion during such exposures). After the drug is removed (at elapsed time = 5 min), movement resumes and cytoplasm streams between the varicosities (d, e) to re-establish normal network shape (f). Elapsed time (min:sec) is indicated by the numbers in the lower right corner of each panel. Reproduced with permission from Travis and Bowser (1986).

forwards. Therefore, it is possible to explain the apparent bidirectional sliding of reticulopodial microtubules by a single mechanoenzyme, provided that some mechanism exists for differentially tethering the sliding microtubule pair. Evidence that microtubule tethers exist in reticulopods, and additional consequences of such tethering, are discussed below (see Section 5.1.3).

Microtubule motors (i.e. dynein, kinesin), adsorbed to a glass substrate, promote to ATP-dependent translocation, or "gliding", of purified microtubules along the glass (reviewed in Allen, 1987). It is therefore possible that dynein-like or kinesin-like molecules bind to the cytoplasmic face of the reticulopodial membrane and effect microtubule motion. Provided that the membrane to which the motors are attached is immobile, one would expect the microtubules to translocate through the cytoplasm (as described above for glass). Conversely, if the microtubules are anchored, one would expect the motors to generate a shear force within the membrane (as is thought to occur during surface transport; see Section 5.2). Electron microscopy, shows microtubules in close juxtaposition to the plasma membrane, indicating that, structurally, such interactions are possible. As mentioned with respect to network morphogenesis (Section 2.3.1.1), portions of the ventral reticulopodial membrane are firmly attached to the substrate; these same areas might immobilize microtubule motors to effect gliding. One way to document microtubule gliding *in vivo* is to image, by videomicroscopy, both ends of a short microtubule fragment translocating within a flattened network, far removed from other cytoskeletal structures. Such observations are difficult due to the great lengths attained by reticulopodial microtubules.

One cannot rule out that fibril axial movements represent microtubule length changes rather than microtubule movement. It is known that microtubules purified from vertebrate brain display simultaneous phases of assembly and disassembly – a process termed dynamic instability (Mitchison and Kirschner, 1984; reviewed in Cassimeris *et al.*, 1987). Videomicroscopic studies of living vertebrate cells confirm that microtubules display distinct phases of elongation and shortening *in vivo* (Cassimeris *et al.*, 1988). However, the possibility that the axial movement of reticulopodial fibrils actually represents microtubule assembly/disassembly seems unlikely in the light of available data regarding the kinetics of this process. Direct observation of microtubule dynamic instability, both *in vitro* and *in vivo*, indicates that microtubule elongation and shortening rates are very different, showing very slow growth (approximately 4 μm/min) while shortening about 10 times faster

(Cassimeris *et al.*, 1988). In contrast, the elongation and shortening of microtubule fibrils in reticulopods occur at equivalent rates which, more importantly, far exceed (by up to two orders of magnitude) the length changes reported for microtubule dynamic instability in vertebrate cells (reviewed in Travis and Bowser, 1988). Although these dramatic rate differences raise serious doubts that microtubule assembly/disassembly accounts for axial fibril displacements, the evidence is far from conclusive. In particular, the reversible transition between microtubules and helical filaments (Section 4.1.3.2) suggests that microtubule dynamics in reticulopods are operationally different from those of vertebrate cells. Indeed, Rupp *et al.* (1986) observed the rapid (micrometres/second) extension of microtubules from tubulin paracrystals in flattened *Allogromia* networks following recovery from induced withdrawal.

Finally, the possibility exists that the axial movements of fibrils are effected by linkage of their microtubules to some other contractile system, such as actomyosin. However, several observations make actomyosin an unlikely candidate. As discussed above (Section 4.1.3.3), actin microfilaments are restricted to dense plaques decorating the ventral cytoplasmic face of trunk pseudopods; they are not detected in more distal regions of the network, where fibril movements are prominent (Bowser *et al.*, 1988). Furthermore, fibril movements continue unabated when these distal reticulopods are microsurgically isolated from the actin-rich regions (Bowser and Travis, unpublished observations). Fibril movements also continue in *Allogromia* treated with concentrations of cytochalasin D that eliminate the actin filament plaques and cause the loss of pseudopod adhesion (Travis and Bowser, 1986a). It is noteworthy that Koonce and Schliwa (1986a) used the actin filament-severing protein gelsolin to remove experimentally actin microfilaments from their demembranated models of *Reticulomyxa* networks, which nevertheless continued to display ATP-dependent microtubule sliding movements that were identical to control preparations. The potential role of actomyosin in other aspects of reticulopodial motility are discussed below.

5.1.2 *Lateral microtubule movements: Zipping/unzipping*

Intersecting fibrils are frequently observed to "zip" rapidly together. Alternatively, a branched fibril may quickly "unzip". These motions were originally attributed to the making or breaking of lateral associations between microtubules (perhaps mediated by intermicrotubule crossbridges). However, on the basis of unpublished video-recordings, we now recognize that vesicles may also mediate these movements. This behaviour is best illustrated when a vesicle travels down a fibril to encounter a Y-shaped branch. Continued motion of the vesicle serves to draw the divergent filaments together, which subsequently fuse and behave as a single fibril. Conversely, when a vesicle encounters intersecting fibrils (i.e. forming an X-shape), it sometimes jumps tracks, in so doing pulling out a filament like a bowstring. The salient point to be made from these observations is that microtubule distribution within the network may largely be determined by cytoplasmic transport. This is in stark contrast to the situation in metazoan tissue cells, where microtubule deployment (organized by the centrosome) mandates vesicle traffic (reviewed in Cassimeris *et al.*, 1987).

5.1.3 Additional notes on network morphogenesis

5.1.3.1 *Extension and withdrawal of filopods.* Videomicroscopy of flattened reticulopods revealed that filopods emerged where axially travelling fibrils (i.e. microtubules) poked against the plasma membrane (Travis *et al.*, 1983; Fig. 7a–d). This mechanism probably explains how the first filopods are extended through the test aperture, and how some branch filopods later arise. As discussed in the preceding section, axial fibril movements are bidirectional, and the microtubules may reverse and retreat from the filopods. Direct videomicroscopic observations showed that such microtubule withdrawal caused the filopod to retract or vesiculate concomitantly, attesting to the fact that an unsupported fluid cylinder (i.e. a naked plasma membrane strand) longer than π times its diameter is highly unstable (Boys, 1959).

5.1.3.2 *Filopod bending.* The sliding mechanism proposed for axial fibril movements (Section 5.1.1) provides a convenient mechanism for the bending movements of the fibrils and, consequently, the reticulopods. Microtubules bend when their axial displacement encounters some mechanical resistance, as evidenced by wave propagation in beating cilia or flagella. Here, the dynein-powered axial displacement of the outer doublet microtubules (Summers and Gibbons, 1971) is resisted by mechanical linkages (i.e. nexin links or radial spokes) between the doublets. In reticulopods, this role might be played by the observed microtubule crossbridges or immobilized (i.e. substrate-attached, see below) plasma membrane domains. Kamimura and

Takahashi (1981) observed that, when trypsin-treated flagellar axonemes were experimentally tethered at both ends, individual doublet microtubules looped out from the axonemes. Koonce et al. (1987) described similar sliding-related microtubule looping in reactivated fragments of *Reticulomyxa* pseudopods, and we reported looping of *Allogromia* microtubules *in vivo* (Travis and Bowser, 1990). Tethers impeding the axial displacement of sliding microtubules would be expected to cause bending of the microtubule bundle.

5.1.3.3 *Role of reticulopod/substrate adhesion.* As discussed in Section 2.3.1.1, cell-substrate adhesion plays a key role in network morphogenesis. Recall that a nascent pseudopodial trunk typically remains unbranched until it contacts the substratum, whereafter the characteristic adhesion, bending and extension behaviour ensues. Branch pseudopods arise from this bent trunk as it straightens and pulls away from the adhesion sites, trailing behind thin cytoplasmic threads that subsequently recruit cytoplasm. The attachment of these nascent trunks to the substrate apparently depends on the formation of actin filament-containing plaques at the adhesion site. When *Allogromia* are plated in the presence of 10 μM cytochalasin D, which inhibits plaque formation, pseudopodial trunks extend normally but do not adhere to the substrate; consequently, pseudopodial networks do not develop (Travis and Bowser, 1986a). Treating fully developed trunks with cytochalasin D apparently results in the loss of actin-containing plaques and the diminution of reticulopod/substrate adhesion. In cytochalasin-treated networks, microtubule-based movements continue but are now unrestrained by substrate adhesion sites. As a result, cytoplasm coalesces to form irregularly shaped bodies that display chaotic motility (Fig. 18). Travis and Bowser (1986a,b) suggested that actin filaments serve to periodically "spot weld" the network to the substrate, and thus transiently stabilize its morphology. Consistent with this hypothesis is the observation that highly charged substrates, which augment pseudopod/substrate attachment, largely abrogate this cytochalasin response (Bowser and Travis, unpublished observation).

The tethering of reticulopodial microtubules at cell-substrate adhesion sites suggests a mechanism for the formation of branch pseudopods. In it, a subset of reticulopodial microtubules is held in place at the substrate adhesion site (e.g. Bowser et al., 1988). As the trunk pulls away, the attached microtubules remain in place to serve as the cytoskeleton of the new branch and to promote the recruitment of cytoplasm by microtubule-dependent transport.

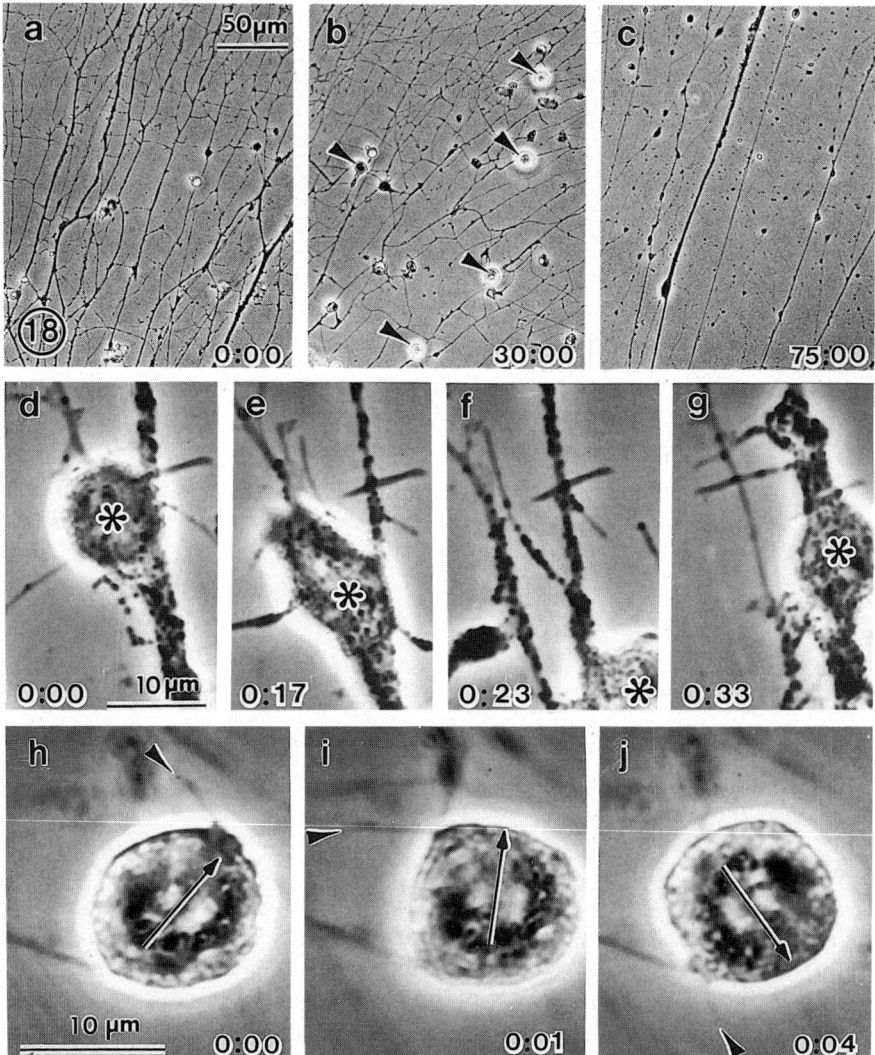

Fig. 18. Effect of cytochalasin D (5×10^{-5} M) on reticulopodial morphology and motility. Sequential phase-contrast light micrographs (a–j) show an *Allogromia laticollaris* network (a) immediately before treatment, (b, d–j) after 30 min of treatment and (c) 45 min after washing out the drug. Within 10–20 min after the drug is applied, the reticulopodial cytoplasm coalesces to form irregularly shaped bodies (b, arrowheads). Higher magnification (d–g) reveals that these bodies (*) translocate through the thinned network and continuously change shape. Often they rotate (up to 12 revolutions/min; h–j, arrow) and may extend and retract minute filopodia (h–j, arrowhead). Upon removal of the cytochalasin, cytoplasm within the bodies gradually streams into the remaining network (c), and normal motility is restored. Elapsed time (min:sec) is indicated by the numbers to the lower right of each panel. Reproduced with permission from Travis and Bowser (1986).

5.2 The streaming of reticulopodial contents: Membrane domain transport

The sum of available data indicates that membrane surface and cytoplasmic transport share a common, cytoskeleton-dependent mechanism. In addition, compelling evidence implicates microtubules as the major cytoskeletal element involved. As discussed previously (Section 2.3.2), the basic features of these two transport phenomena are the same: both movements are bidirectional, display certain saltatory characteristics, and have essentially indistinguishable velocity profiles. Disassembly of microtubules with antimitotic drugs or cold shock simultaneously inhibits membrane surface transport and organelle streaming (Fig. 17). The movements of cytoplasmic particles (Travis et al., 1983) and surface-attached microspheres (Bowser and Rieder, 1985) occur only in close association with microtubules and precisely follow the paths defined by these microtubules. Furthermore, as microtubules move laterally through the cytoplasm (i.e. zipping movements), the associated organelles move laterally through the cytoplasm; similarly, membrane-associated microspheres are laterally dragged across the reticulopodial surface. These observations indicate that cytoplasmic organelles and plasma membrane domains are somehow tethered to the microtubules. This notion is further supported by the observation that surface-attached microspheres and cytoplasmic granules remain bound to reticulopodial microtubules following extraction of the plasma membrane and soluble cytoplasmic components using non-ionic detergents (Bowser and Rieder, 1985; Koonce and Schliwa, 1986a; Bowser et al., 1987). Finally, structural crossbridges, which might serve as physical tethers, are readily seen linking microtubules to organelles or the plasma membrane in electron micrographs of both intact (Travis and Allen, 1981; Bowser and Rieder, 1985; Koonce et al., 1986; Kachar et al., 1987) and detergent-treated reticulopods (Bowser and Rieder, 1985; Bowser and Travis, unpublished observations).

Philosophically, these common characteristics should not be surprising, because in both cases the cytoplasmic face of a membrane interface (a closed vesicle or a plasma membrane patch) is transported along the length of a microtubule; the difference being that organelle transport occurs within a cytoplasmic volume, whereas surface transport is restricted to the two-dimensional plane of the highly fluid plasma membrane. Therefore, for the remainder of this discussion, organelle and cell surface transport are considered manifestations of the same basic process, here termed "membrane domain transport".

There are a number of ways in which a membrane domain might be moved by or along microtubules (see Allen *et al.*, 1982). For reticulopods, the two most likely mechanisms will be considered: (1) specific mechanoenzymes move membrane domains along the lengths of static microtubules, and (2) membrane domains are statically linked to motile microtubules and are thus translocated by microtubule movements. The hypothesis that membrane domains are driven along microtubules by actomyosin is not supported by the available experimental data, as already discussed with respect to the axial translocation of microtubule fibrils (Section 5.1.1; see also Travis and Bowser, 1986a, 1988).

Because the microtubule surface lattice is structurally asymmetrical, movement along a microtubule in one direction is a different molecular process from movement in the opposite direction (reviewed in Kirschner and Mitchison, 1986; Cassimeris *et al.*, 1987). In accord with this concept, work on axoplasmic transport has identified two distinct classes of motor molecules – kinesins and dyneins – each of which is specific for one direction of transport along the microtubule surface (reviewed in Warner and McIntosh, 1989; Warner *et al.*, 1989). Because reticulopodial membrane domain transport is bidirectional along a given microtubule bundle, which in *Reticulomyxa* is of essentially uniform polarity (Euteneuer *et al.*, 1989), one would expect reticulopods to have both kinesin-like and dynein-like mechanoenzymes. Therefore, by this model, a given reticulopodial membrane domain would require both mechanoenzymes in order to move bidirectionally. As a corollary to this concept, unidirectionally transported components (e.g. centripetally transported phagosomes or centrifugally transported defecation vacuoles) would be so targeted by possessing only one type of motor molecule. To date, only a dynein-like mechanoenzyme has been detected by conventional biochemical approaches in extracts of *Reticulomyxa*. Although the picture is far from complete, Euteneuer and co-workers (1989) raise the possibility that only a dynein-like mechanoenzyme, which can be uniquely regulated to operate in either direction along a microtubule, is present in this system.

Videomicroscopy of flattened reticulopods showed that organelles could translocate through the cytoplasm by being associated with axially or laterally moving microtubule fibrils (reviewed in Travis and Bowser, 1988). These observations are consistent with a mechanism whereby organelles or plasma membrane domains are statically linked to sliding microtubules. As already discussed in terms of bidirectional axial microtubule movement (Section 5.1.1), a sliding microtubule mechanism could operate with only a single dynein-like motor. Such a mechanism would help explain the tandem motion of numerous organelles or membrane surface markers. It also helps explain the wide range of velocities displayed by membrane domain transport. Consider a membrane domain attached to one microtubule, which in turn was sliding along another. If the latter microtubule was moving along still another, the particle would have a final velocity equal to the sum of the two component velocities. Iterative addition within a bundle – essentially Allen's (1964) "piggy-back" concept – would give rise to a wide velocity spread, and considerably faster upper speeds.

Allen (1964) also noted that particles often change velocity or direction at branch points and sites of substrate attachment – the same sites where microtubule bundles branch, fuse or intersect. It is unlikely that the particle will be interacting with the same number of sliding microtubules at either side of these sites; in turn, this might explain the velocity changes frequently seen here. A final requisite for such a model is that the linkages between membrane domains and microtubules are labile, enabling the domains to "switch tracks" readily. Evidence for such lability comes from the direct observation of particles readily associating/dissociating from microtubule fibrils *in vivo* (Travis *et al.*, 1983), a process which might be regulated by protein phosphorylation (Koonce and Schliwa, 1986b).

Of course, these two models are not mutually exclusive: membrane domains could translocate autonomously along microtubules, and the microtubules in turn could slide relative to each other. At present, this hybrid model appears to be the best to explain the highly complex patterns of reticulopodial membrane domain transport. Sorting out these possibilities remains a major challenge.

5.3 Network withdrawal

The mechanism responsible for the slow inward flow of cytoplasm during reticulopod withdrawal is unknown, but the available data indicate that it is distinct from that responsible for normal bidirectional organelle transport. Rupp *et al.* (1986) took advantage of the improved resolution afforded by videomicroscopy, and the flattening of pseudopods by charged glass coverslips, to re-examine this phenomenon at the resolution of just a few MTs. It was found that, 10–15 s after adding a hypertonic sea water stimulus, the inward flow occurred only in association with cytoplasmic fibrils. This inward movement was remarkably uniform ($0.6 \pm 0.02\ \mu m/s$), while the distances between organelles did not change, as though the organelles were riding along a conveyor belt. Later during the withdrawal, the fibrils become wavy and coalesced to form phase-dense pools, which slowly moved inwards; these phase-dense pools correspond to the refractile droplets of cytoplasm described by McGee-Russell and Allen (1971). When the retraction solution was replaced by normal sea water, Rupp et al. (1986) noted the rapid growth of fibrils from the phase-dense pools and, subsequently, the resumption of normal, bidirectional organelle transport along the new fibrils.

The observation that only organelles in contact with the wavy, paracrystalline-rich fibrils (Fig. 13) move inwards – and those not in contact are immobile – rules out mechanisms invoking global, isometric contraction of the cytoplasmic matrix (e.g. Byers and Porter, 1977). The active contraction of helical filaments also seems unlikely to cause withdrawal because neighbouring organelles maintain fixed distances along a given fibril. The most straightforward interpretation of the available data is that foraminifera are able to quickly transform their microtubules into helical filaments (Fig. 12). With the loss of normal microtubule structure, organelles no longer can translocate but remain locked in position along the helical filaments. The helical filaments are reeled in, perhaps by an actomyosin contractile system or by their entropy-driven aggregation into paracrystals, carrying with them bound organelles and cytoplasm.

In a preliminary report, Koonce and Schliwa (1986b) noted that cyclic AMP-dependent protein kinase activity induced the unidirectional centripetal transport of organelles in the *Reticulomyxa* lysed reticulopodial model. This finding suggests that protein phosphorylation plays a role in determining the directionality of organelle movement. Given the striking similarity between this *in vitro* result and the withdrawal response in the intact organism (Bowser, unpublished observation), it will be interesting to learn whether protein phosphorylation induces the microtubule-helical filament transition in *Reticulomyxa*.

During recovery from the withdrawal response, microtubules quickly grow from the paracrystals and organelles subsequently resume normal bidirectional transport. In this view, the paracrystals represent a temporary reservoir of microtubule precursors for their local assembly. As mentioned earlier, fully extended networks are very dynamic structures, continually changing size and shape, primarily through the extension of new pseudopods in some regions and the retraction of old ones elsewhere. These continual changes require local microtubule assembly in the extending regions, and disassembly in the retracting ones. Simple diffusion of microtubule subunits is clearly inadequate to supply local demands for pseudopod growth, particularly when it occurs many millimetres from the cell body (where the synthesis of subunits occurs). By first packaging microtubules into a paracrystalline intermediate, and then rapidly transporting this intermediate to the sites where microtubule growth is required, the foraminifera have evolved an energetically efficient mechanism to supply local subunit demands.

5.4 Mechanism of pseudopodial tension and foraminiferan locomotion

When crawling across a hard planar surface, benthic foraminifera, like many other cell types (reviewed in Abercrombie, 1980), appear to employ a "grip and tug mechanism" in which they firmly attach portions of their pseudopodia to the substratum and subsequently apply a rearward traction force (see Section 2.2). The mechanism generating this force, manifested as pseudopodial tension, remains in doubt. Pharmacological studies strongly suggest that pseudopodial tension depends on intact microtubules and actin-containing filament plaques, while electron microscopy provides evidence for a structural interaction between these two filament systems at the sites where traction force is applied (Bowser et al., 1988; Travis et al., 1988). Given these observations, at least two hypotheses for the generation of pseudopodial tension can be proposed.

In the first, tension is generated by microtubule sliding, presumably through the agency of their lateral crossbridges (Section 5.1.1). This force would then be transmitted to the substrate by the subset of microtubules structurally interacting with (i.e. tethered to) the actin-containing adhesion plaques. In this regard, the plaques would be functionally analogous to the z-lines of striated muscle or the dense bodies of smooth muscle.

Alternatively, contraction of the actin-containing plaques, mediated by an as yet unidentified myosin-like mechanoenzyme, could generate tension locally which would then be transmitted along the attached microtubules to pull the cell body. In this context, one could compare the actin-containing filament plaques of the ventral reticulopodial surface to the contractile stress fibres of vertebrate tissue cells. However, if the plaques did contract in a manner analogous to stress fibres, one would expect to observe in the rubber sheet assay the development of compression wrinkles immediately beneath the adhesion plaques. However, micromanipulation studies indicate that the plaques alone do not deform the rubber sheet, but do allow the transmission of an externally applied force against the substratum (Travis et al., 1988). A more definitive answer to the question of how pseudopodial tension is generated and applied to effect locomotion requires further study (e.g. the application of methods such as the microinjection of specific antibodies to the various mechanoenzymes, to rule out one mechanochemical system or the other).

The "grip and tug" model described above provides a simple mechanism for locomotion along a cohesive surface, but it is not known if it applies to infaunal locomotion as well. Logically, one would expect that when pseudopods attach to fine, loose sediment and exert a rearward traction force, the particles would dislodge and move rearwards, rather than the bulky shell move forwards. Whether this actually occurs may be a matter of degree: perhaps the distal portions of the network pervade the sediment sufficiently to provide enough anchorage for the forward locomotion of the body. In this respect, it is noteworthy that *Astrammina rara*, an infaunal species, coats its pseudopods with sediment – a behaviour expected to increase the frictional resistance of its pseudopods (Bowser *et al.*, 1986). Similarly, Wetmore (1988) reported that sediments adhere to the pseudopods of other infaunal species.

6.0 FUTURE PROSPECTS

Research during the past century has meticulously catalogued various aspects of foraminiferan motility. Recent observations made with correlative microscopic methods, coupled with studies using specific pharmacological agents, have shown the importance of microtubules in reticulopodial structure and function, and testable models that help explain some facets of foraminiferan motility have been formulated.

Despite these gains, we are still left with serious gaps in our knowledge of foraminiferan motility. For example, one topic not addressed in this chapter is the vertical movement of planktonic foraminifers within the water column. The regulation of buoyancy in planktonic species is clearly among the least understood "motility" processes and, considering its importance to their life habits, deserves intensive study.

With respect to reticulopodial motility, answers to the following partial list of questions are very much needed in order to substantiate the models presented here:

1. Do reticulopodial microtubules slide relative to each other *in vivo*? If so, what role does this process play in cytoplasmic transport or pseudopod motion?

2. What are the identities and properties of the mechanoenzymes associated with reticulopodial microtubules? Are different motors responsible for organelle transport, surface motility, microtubule bending and whole-cell locomotion?
3. Do reticulopodial microtubules assemble spontaneously or is their growth "seeded' from an organizing centre? What factors regulate their assembly and disassembly? What are the molecular details of the microtubule – helical filament transformation?
4. Is myosin present and, if so, what is the function of the reticulopodial actomyosin system?
5. How do foraminifers generate the pseudopodial tension required for locomotion? Does locomotion across a planar substrate differ fundamentally from locomotion through sediment?

The *in vitro* reactivation of reticulopodial transport in *Reticulomyxa* (Koonce and Schliwa, 1986a) is an exciting advance that will undoubtedly lead to further tests of the ideas presented here. On the other hand, allogromiids remain the most accessible system for many *in vivo* studies. For example, the microinjection of fluorescently labelled cytoskeletal proteins will permit studies of *in situ* cytoskeletal dynamics using photobleaching or resonance energy-transfer techniques. Finally, molecular genetic strategies are now available that can be used to amplify, into experimentally manipulable amounts, the picogram quantities of DNA or RNA typically extracted from foraminifera. Using related approaches, the genes encoding their cytoskeletal proteins could be cloned and expressed *en masse*, permitting detailed molecular and biochemical studies.

ACKNOWLEDGEMENTS

This paper is submitted in memory of Dr Harold Sandon, whose pioneering work influenced all students of foraminiferan motility. We thank Drs M. Hauser and M. Schliwa for the use of figures and sharing with us unpublished findings, and Drs S.P. Alexander, J.M. Bernhard and K.P. Severin for enlightening discussions. S.S.B. extends special gratitude to Drs S.M. McGee-Russell, C.L. Rieder and T.E. Delaca for support and encouragement during the course of his studies. Previously unpublished aspects of work reported here were funded by grants from the National Science Foundation (DCB 85–02875 awarded to J.L.T. and DPP 89–17375 awarded to S.S.B.).

REFERENCES

Abercrombie, M. (1980). The crawling movement of metazoan cells. *Proc. R. Soc.* **B207**:129–147.

Adshead, P.C. (1980). Pseudopodial variability and behavior of globigerinids (Foraminiferida) and other planktonic Sarcodina developing in cultures. In *Studies in Marine Micropaleontology and Paleocology* (eds W.V. Slater et al.), pp. 96–126. Special Publication No. 19, Cushman Foundation for Foraminiferal Research, Lawrence, Kansas.

Alexander, S.P. and Banner, F.T. (1984). The functional relationship between skeleton and cytoplasm in *Haynesina germanica* (Ehrenberg). *J. Foram. Res.* **14**:159–170.

Allen, R.D. (1964). Cytoplasmic streaming and locomotion in marine foraminifera. In *Primitive Motile Systems in Cell Biology*, (eds R.D. Allen and N. Kamiya), pp. 407–432. Academic Press, London.

Allen, R.D. (1968). Differences of a fundamental nature among several types of amoeboid movement. *Symp. Soc. Exp. Biol.* **20**:151–168.

Allen, R.D. (1985). New observations on cell architecture and dynamics by video-enhanced contrast optical microscopy. *Ann. Rev. Biophys. Chem.* **14**:265–290.

Allen, R.D. (1987). The microtubule as an intracellular engine. *Sci. Am.* **256**:42–49.

Allen, R.D., Allen, N.S. and Travis, J.L. (1981a). Video-enhanced contrast differential interference contrast (AVEC-DIC) microscopy: A new method capable of analyzing microtubule-related motility in the reticulopodial network of *Allogromia laticollaris*. *Cell Motil.* **1**:291–302.

Allen, R.D., Travis, J.L., Allen, N.S. and Yilmaz, H. (1981b). Video-enhanced contrast polarization (AVEC-POL) microscopy: A new method applied to the detection of birefringence in the reticulopodial network of *Allogromia laticollaris*. *Cell Motil.* **1**:275–289.

Allen, R.D., Travis, J.L., Hayden, J.H., Allen, N.S., Bruer, A.C. and Lewis, L.J. (1982). Cytoplasmic transport: Moving ultrastructural elements common to many cell types revealed by video-enhanced microscopy. *Cold Spring Harb. Symp. Quant. Biol.* **46**:85–87.

Anderson, O.R. and Bé, A.W.H. (1976). A cytochemical fine structure study of phagotrophy in a planktonic foraminifer *Hastigerina pelagica* (d'Orbigny). *Biol. Bull.* **151**:437–449.

Anderson, O.R. and Bé, A.W.H. (1978). Recent advances in foraminiferal fine structure research. In *Foraminifera* (eds R.H. Hedley and C.G. Adams), Vol. 3, pp. 121–202. Academic Press, London.

Arnold, Z.M. (1953). An introduction to the study of movement and dispersal in *Allogromia laticollaris* Arnold. *Contr. Cushman Found. Foram. Res.* **4**:15–21.

Arnold, Z.M. (1954). A note on foraminiferan sieve-plates. *Contr. Cushman Found. Foram. Res.* **5**:77.

Arnold, Z.M. (1955). Life history and cytology of the foraminiferan *Allogromia laticollaris*. *Univ. Calif. Publ. Zool.* **61**:167–252.

Arnold, Z.M. (1964). Biological observations on the foraminifer *Spiroloculina hyalina* Schulze. *Univ. Calif. Publ. Zool.* **72**:1–93.

Banner, F.T. and Culver, S.J. (1978). Quaternary *Haynesina* n. gen. and

Paleogene *Protelphidium* Haynes; their morphology, affinities and distribution. *J. Foram. Res.* **8**:1–10.

Bé, A.W.H., Hemleben, C., Anderson, O.R., Spindler, M., Hacunda, J. and Tuntivate-Choy, S. (1977). Laboratory and field observations of living planktonic foraminifera. *Micropaleontol.* **23**:155–179.

Bloodgood, R.A. (1988). The use of microspheres in the study of cell motility.In *Microspheres: Medical and Biological Applications* (eds A. Rembaum and Z. Tokes), pp. 165–192. CRC Press, Boca Raton, Florida.

Boltovskoy, E. and Wright, R. (1976). *Recent Foraminifera.* W. Junk, The Hague.

Bowser, S.S. (1983). Fully correlative LM and HVEM analysis of motile fibrils and particles in the reticulopods of *Allogromia* sp. (Protozoa, Foraminiferida). In *Proceedings of the 41st Annual Meeting of the Electron Microscopy Society of America* (ed. G.W. Bailey), pp. 538–539. San Francisco Press, San Francisco, Calif.

Bowser, S.S. (1985). Invasive activity of *Allogromia* pseudopodial networks: Skyllocytosis of a gelatin/agar gel. *J. Protozool.* **32**:9–12.

Bowser, S.S. and Bloodgood, R.A. (1984). Evidence against surf-riding as a general mechanism for surface motility. *Cell Motil.* **4**:305–314.

Bowser, S.S. and DeLaca, T.E. (1985). Rapid intracellular motility and dynamic membrane events in an Antarctic foraminifer. *Cell Biol. Int. Rep.* **9**:901–910.

Bowser, S.S. and Rieder, C.L. (1985). Evidence that cell surface motility in *Allogromia* is mediated by cytoplasmic microtubules. *Can. J. Biochem. Cell Biol.* **63**:608–620.

Bowser, S.S. and Rieder, C.L. (1986). Microtubule-dependent reticulopodial surface motility: Reversible inhibition on plasma membrane blebs. *Ann. N.Y. Acad. Sci.* **466**:933–935.

Bowser, S.S., Israel, H.A., McGee-Russell, S.M. and Rieder, C.L. (1984a). Surface transport properties of reticulopodia: Do intracellular and extracellular motility share a common mechanism? *Cell Biol. Int. Rep.* **8**:1051–1063.

Bowser, S.S., McGee-Russell, S.M. and Rieder, C.L. (1984b). Multiple fission in *Allogromia* sp., strain NF (Foraminiferida): Release, dispersal and ultrastructure of offspring. *J. Protozool.* **31**:272–275.

Bowser, S..S., McGee-Russell, S.M. and Rieder, C.L. (1985). Digestion of prey in foraminifera is not anomalous: A correlation of light microscopic, cytochemical and HVEM technics to study phagotrophy in two allogromiids. *Tiss. Cell* **17**:823–839.

Bowser, S.S., DeLaca, T.E. and Rieder, C.L. (1986). Novel extracellular matrix and microtubule cables associated with pseudopodia of *Astrammina rara*, a carnivorous Antarctic foraminifer. *J. Ultrastruct. Mol. Struct. Res.* **94**:149–160.

Bowser, S.S., Koonce, M.P. and Schliwa, M. (1987). Reactivated membrane surface transport in *Reticulomyxa*. *J. Cell Biol.* **105**:129a.

Bowser, S.S., Travis, J.L. and Rieder, C.L. (1988). Microtubules associate with actin-containing filaments at discrete sites along the ventral surface of *Allogromia* reticulopods. *J. Cell Sci.* **89**:297–307.

Boys, C.V. (1959). *Soap Bubbles. Their Colors and Forces Which Mould Them.* Dover, New York.

Byers, H.R. and Porter, K.R. (1977). Transformation in the structure of the cytoplasmic ground substance in erythrophores during pigment aggregation and dispersion. I. A study using whole-cell preparations in stereo high voltage electron microscopy. *J. Cell Biol.* **75**:541–558.

Cachon, J. and Cachon, M. (1981a). Polymorphism of tubulin reassembly in the microtubular system of a heliozoan. II. Cytochemical effects of mechanical shocks and ultrasounds. *Biol. Cell.* **40**:23–33.

Cachon, J. and Cachon, M. (1981b). Movement by non-actin filament mechanisms. *BioSystems* **14**:313–326.

Cassimeris, L.U., Walker, R.A., Pryer, N.K. and Salmon, E.D. (1987). Dynamic instability of microtubules. *BioEssays* **7**:149–154.

Cassimeris, L., Pryer, N.K. and Salmon, E.D. (1988). Real-time observations of microtubule dynamic instability in living cells. *J. Cell Biol.* **107**:2223–2231.

Caulfield, J.P., Hein, A., Schmidt, R.E. and Ritz, J. (1987). Ultrastructural evidence that the granules of human natural killer cell clones store membrane in a nonbilayer phase. *Am. J. Pathol.* **127**:305–316.

Centonze, V.E. and Travis, J.L. (1983). Immunofluorescence of *Allogromia* reticulopodia. *Biol. Bull.* **165**:489.

Chen, L.B., Summerhayes, I.C., Johnson, L.V., Walsh, M.L., Bernal, S.D. and Lampidis, T.J. (1982). Probing mitochondria in living cells with rhodamine 123. *Cold Spring Harb. Symp. Quant. Biol.* **46**:141–155.

Chilcote, T.J. and Johnson, K.A. (1989). Microtubule-dynein cross-bridge cycle and the kinetics of 5'-adenylyl imidodiphosphate (AMP-PNP) binding. In *Cell Movement*, Vol. 1: *The Dynein ATPases* (eds F.D. Warner, P. Satir and I.R. Gibbons), pp. 235–243. Alan R. Liss, New York.

Cohan, C.S. and Kater, S.B. (1986). Suppression of neurite elongation and growth cone motility by electrical activity. *Science* **232**:1638–1640.

Comly, L.T. (1973). Microfilaments in *Chaos carolinensis*: Membrane association, distribution, and heavy meromyosin binding in the glycerinated cell. *J. Cell Biol.* **58**:230–237.

Cushman, J.A. (1920). Observations on living specimens of *Iridia diaphana*, a species of foraminifera. *Proc. U.S. Natl Mus.* **57**:153–158.

Doyle, Wm.L. (1935). Distribution of mitoochondria in the foraminiferan, *Iridia diaphana*. *Science* **81**:387.

Eckert, B.S. and McGee-Russell, S.M. (1973). The patterned organization of thick and thin microfilaments in the contracting pseudopod of *Difflugia*. *J. Cell Sci.* **13**:727–739.

Edds, K.T. (1975). Motility in *Echinosphaerium nucleofilum* II. Cytoplasmic contractility and its molecular basis. *J. Cell Biol.* **66**:156–164.

Euteneuer, U., McDonald, K.L., Koonce, M.P. and Schliwa, M. (1986). Intracellular transport in *Reticulomyxa*. *Ann. N.Y. Acad. Sci.* **466**:936–937.

Euteneuer, U., Koonce, M.P., Pfister, K.K. and Schliwa, M. (1988). An ATPase with properties expected for the organelle motor of the giant amoeba, *Reticulomyxa*. *Nature* **332**:176–178.

Euteneuer, U., Haimo, L.T. and Schliwa, M. (1989). Microtubule bundles of *Reticulomyxa* networks are of uniform polarity. *Eur. J. Cell Biol.* **49**:373–376.

Fenchel, T. (1987). *Ecology of Protozoa*. Science Tech Inc., Madison, Wisconsin.

Goldstein, J.L., Brown, M.S., Anderson, R.G.W., Russell, D.W. and Schneider, W.J. (1985). Receptor mediated endocytosis: Concepts emerging from the LDL receptor system. *Ann. Rev. Cell Biol.* **1**:1–39.

Golz, R. amd Hauser, M. (1986). Polymorphic assembly states of *Allogromia* tubulin under physiologic conditions. *Eur. J. Cell Biol.* **40**:124–129.

Greer, K. and Rosenbaum, J.L. (1989). Post-translational modification of tubulin. In *Cell Movement*, Vol. 2: *Kinesin, Dynein and Microtubule Dynamics* (eds F.D. Warner and J.R. McIntosh), pp. 47–66. Alan R. Liss, New York.

Grell, K.G. (1973). *Protozoology*. Springer-Verlag, New York.

Hard, R. and Cloney, R.A. (1971). Intracellular movements in eosinophilic leucocytes of *Taricha granulosa*. In *Proceedings of the 11th Annual Meeting of the American Society for Cell Biology*.

Harris, A.K., Wild, P. and Stopak, D. (1980). Silicone rubber substrata: A new wrinkle in the study of cell locomotion. *Science* **208**:177–179.

Hauser, M. and Schwab, D. (1974). Mikrotubuli und helikale Mikrofilamente im Cytoplasma der Foraminifere *Allogromia laticollaris* Arnold. Untersuchungen mit Vinblastin und Deuteriumoxid zum Nachweis einer engen Wechselbeziehung. *Cytobiologie* **9**:263–279.

Hauser, M., Lindenblatt, J. and Hulsmann, N. (1989). The cytoskeleton of *Reticulomyxa filosa* reticulopodia contains Glu-tubulin as a main component. *Eur. J. Protistol.* **25**:145–157.

Hedley, R.H., Parry, D.M. and Wakefield, J.St.J. (1967). Fine structure of *Shepherdella taeniformis* (Foraminifera::Protozoa). *J. R. Micr. Soc.* **87**:457–461.

Hemleben, Ch., Spindler, M. and Anderson, O.R. (1989). *Modern Planktonic Foraminifera*. Springer-Verlag, New York.

Hottinger, L. and Dreher, D. (1974). Differentiation of protoplasm in Nummulitidae (Foraminifera) from Elat, Res Sea. *Mar. Biol.* **25**:41–61.

Inoue, S. (1981). Video image processing greatly enhances contrast, quality and speed in polarization-based microscopy. *J. Cell Biol.* **89**:346–356.

Jahn, T.L. and Rinaldi, R.A. (1959). Protoplasmic movement in the foraminiferan, *Allogromia laticollaris*, and a theory of its mechanism. *Biol. Bull.* **117**:100–118.

Jensen, C.G., Bollard, S.M., Jensen, L.C.W., Travis, J.L. and Bowser, S.S. (1990). Microdensitometer-computer correlation analysis of two distinct, spatially-segregated classes of microtubule bridges in *Allogromia* pseudopodia. *J. Struct. Biol.* **105**:1–10.

Jepps, M.W. (1942). Studies on *Polystomella lamarck* (Foraminifera). *J. Mar. Biol. Assoc. U.K.* **25**:607–666.

Kamimura, S. and Takahashi, K. (1981). Direct measurement of the force of microtubule sliding in flagella. *Nature* **293**:566–568.

Katchar, B., Bridgman, P.C. and Reese, T.S. (1987). Dynamic shape changes of cytoplasmic organelles translocating along microtubules. *J. Cell Biol.* **105**:1267–1271.

Kavanau, J.L. (1963). Protoplasmic streaming as a process of jet propulsion. *Dev. Biol.* **7**:22–37.

Kirschner, M. and Mitchison, T. (1986). Beyond self-assembly: From microtubules to morphogenesis. *Cell* **45**:329–342.

Kitazato, H. (1988). Locomotion of some benthic foraminifera in and on sediments. *J. Foram. Res.* **18**:344–349.

Knight, R. (1986). A novel method of dark field illumination for a stereomicroscope and its application to a study of the pseudopodia of *Reophax moniliformis* Siddall (Foraminiferida). *J. Micropalaeontol.* **5**:83–90.

Koonce, M.P. and Schliwa, M. (1986a). Reactivation of organelle movements along the cytoskeletal framework of a giant freshwater amoeba. *J. Cell Biol.* **103**:605–612.

Koonce, M.P. and Schliwa, M. (1986b). Directionality of organelle movement in *Reticulomyxa* may be mediated by phosphorylation. *J. Cell Biol.* **103**:275a.

Koonce, M.P., Euteneuer, U., McDonald, K.L., Menzel, D. and Schliwa, M. (1986). Cytoskeletal architecture and motility in a giant freshwater amoeba, *Reticulomyxa*. *Cell Motil. Cytoskel.* **6**:521–533.

Koonce, M.P., Tong, J., Euteneuer, U. and Schliwa, M. (1987). Active sliding between cytoplasmic microtubules. *Nature* **328**:737–739.

Koury, S.T. (1982). The ultrastructure of *Allogromia* sp. during feeding steady state conditions and during induced retraction of the reticulopodial network. M.Sc. thesis, State University of New York.

Koury, S.T., Bowser, S.S. and McGee-Russell, S.M. (1985). Ultrastructural changes during reticulopod withdrawal in the foraminiferan protozoan, *Allogromia* sp., strain NF. *Protoplasma* **129**:149–156.

Le Calvez, J. (1953). Ordre des foraminifers. In *Traite de Zoologie* (ed. P. Grasse), Vol. 1 (2), pp. 149–265.

Lee, J.J. and Pierce, S. (1963). Growth and physiology of foraminifera in the laboratory; part 4 – monoxenic culture of an allogromiid with notes on its morphology. *J. Protozool.* **10**:404–411.

Lee, J.J., Hutner, S.H. and Bovee, E.C. (1985). *An Illustrated Guide to the Protozoa*. Allen Press, Lawrence, Kansas.

Leidy, J. (1879). Freshwater rhizopods of North America. *U.S. Geol. Surv. Terr. Rep.* **12**:1–324.

Lindenblatt, J., Hulsmann, N. and Hauser, M. (1988). Characterization of the cytoskeletal elements in the reticulopodial network (RPN) of the giant amoeba *Reticulomyxa filamentosa*. *Eur. J. Cell Biol.* **46**:39 (suppl.).

Lipps, J.H. (1973). Test structure in foraminifera. *Ann. Rev. Microbiol.* **27**:471–488.

Lipps, J.H. (1983). Biotic interactions in benthic foraminifera. In *Biotic Interactions in Recent and Fossil Benthic Communities* (eds M.J.S. Tevesz and P.L. McCall), pp. 331–376. Plenum Press, New York.

Lopez, E. (1979). Algal chloroplasts in the protoplasm of three species of benthic foraminifera: Taxonomic affinity, viability, persistence. *Mar. Biol.* **53**:201–211.

Luduena, R.F., Shooter, E.M. and Wilson, L. (1977). Structure of the tubulin dimer. *J. Biol. Chem.* **252**:7006–7014.

Mandelkow, E. and Mandelkow, E.M. (1989). Tubulin, microtubules, and oligomers: Molecular structure and implications for assembly. In *Cell Movement*, Vol. 2: *Kinesin, Dynein and Microtubule Dynamics* (eds F.D. Warner and J.R. McIntosh), pp. 23–45. Alan R. Liss, New York.

Mandelkow, E.-M., Schultheiss, R., Rapp, R., Muller, M. and Mandelkow, E. (1986). On the surface lattice of microtubules: Helix starts, protofilament number, seam, & handedness. *J. Cell Biol.* **102**:1067–1073.

Marszalek, D.S. (1969). Observations on *Iridia diaphana*, a marine foraminifer. *J. Protozool.* **16**:599–611.

Martz, D., Lasek, R.J., Brady, S.T. and Allen, R.D. (1984). Mitochondrial motility in axons: Membraneous organelles may interact with the force generating system through multiple sites. *Cell Motil.* **4**:89–101.

McGee-Russell, S.M. (1974). Dynamic activities and labile microtubules in cytoplasmic transport in the marine foraminiferan *Allogromia*. *Symp. Soc. Exp. Biol.* **28**:157–189.

McGee-Russell, S.M. and Allen, R.D. (1971). Reversible stabilization of labile microtubules in the reticulopodial network of *Allogromia*. *Adv. Cell Mol. Biol.* **1**:153–184.

McGee-Russell, S.M. and Trautwein, R. (1977). Low calcium transmembrane inhibition of the fixation-induced contractility system of *Allogromia* – ultrastructural correlates including cross-bridges. In *Proceedings of the International Congress in Protozoology* (eds S.H. Hutner), abstract 60.

McGee-Russell, S.M., Bowser, S.S. and Koury, S.T. (1982). Motility organizing vesicles (MOVs) in *Allogromia*: A new concept and term in cell motility. *J. Cell Biol.* **95**:329a.

Mitchison, T. and Kirschner, M. (1984). Dynamic instability of microtubule growth. *Nature* **312**:237–242.

Nauss, R.N. (1949). *Reticulomyxa filosa* gen. et sp. nov., a new primitive plasmodium. *J. N.Y. Bot. Garden* **76**:161–173.

Nyholm, K.-G. and Nyholm, P.-G. (1975). Ultrastructure of monothalamous foraminifera. *Zoon* **3**:141–150.

Patterson, D.J. (1980). Contractile vacuoles and associated structures: Their organization and function. *Biol. Rev.* **55**:1–46.

Pollard, T. and Korn, E.D. (1971). Filaments of *Amoeba proteus*. II. Binding of heavy meromyosin by thin filaments in motile cytoplasmic extracts. *J. Cell Biol.* **48**:216–219.

Poste, G. and Allison, A.C. (1973). Membrane fusion. *Biochem. Biophys. Acta* **300**:421–465.

Rebhun, L.I. (1972). Polarized intracellular particle transport: Saltatory movements and cytoplasmic streaming. *Int. Rev. Cytol.* **32**:93–137.

Rinaldi, R.A. and Jahn, T.L. (1964). Shadowgraphs of protoplasmic movement in *Allogromia laticollaris* and a correlation of this movement to striated muscle contraction. *Protoplasm* **58**:369–390.

Roberts, T.M. (1987). Fine (2–5 nm) filaments: New types of cytoskeletal structures. *Cell Motil. Cytoskel.* **8**:130–142.

Röttger, R. (1973). Die Ektoplasmahulle von *Heterostegina depressa* (Foraminifera: Nummulitidae). *Mar. Biol.* **21**:127–138.

Rupp, G., Bowser, S.S., Mannella, C.A. and Rieder, C.L. (1986). Naturally occurring tubulin-containing paracrystals in *Allogromia*: Immunocytochemical identification and functional significance. *Cell Motil. Cytoskel.* **6**:363–375.

Sandon, H. (1934). Pseudopodial movements of foraminifera. *Nature* **133**:761–762.

Sandon, H. (1944). Analogy between pseudopodia and nerve fibres. *Nature* **154**:830–831.

Sandon, H. (1975). Neglected animals – the foraminifera. In *New Biology* (eds M.L. Johnson, M. Abercrombie and G.E. Fogg), Vol. 24, pp. 7–32. Penguin, Harmondsworth.

Sandon, H. (1963). Some observations on reticulose pseudopodia. In *Progress in Protozoology* (eds J. Ludvik, J. Lom and J. Vavra), pp. 166–169. Academic Press, London.

Satir, P. (1989). Structural analysis of the dynein cross-bridge cycle. In *Cell Movement*, Vol. 1: *The Dynein ATPases* (eds F.D. Warner, P. Satir and I.R. Gibbons), pp. 219–234. Alan R. Liss, New York.

Schmidt, W.J. (1937). Ueber den Feinbau der Filopodien insbesondere ihre Doppelbrechung bei *Miliola*. *Protoplasma* **27**:587–598.

Schwab, D. (1977). Light and electron microscopic investigations on the monothalamous foraminifer *Boderia albicollaris* n. sp. *J. Foram. Res.* **7**:189–195.

Schwab, D. and Schwab-Stey, H. (1972). Fibrillare und tubulare Strukturen im Cytoplasma der Foraminifere *Allogromia laticollaris* Arnold. *Cytobiologie* **6**:234–242.

Schwab, D. and Schwab-Stey, H. (1973). Further electronmicroscopic investigations of the foraminifer *Myxotheca arenilega* Schaudinn. Cytoplasmic microtubules and microfilaments. *Cytobiologie* **7**:193–204.

Schwab, D. and Schwab-Stey, H. (1980). Induced cell fusion in Foraminifera. *Protoplasma* **102**:141–146.

Schwab, D. and Schwab-Stey, H. (1981). Hexagonal structures at the plasma membrane of the foraminifer *Allogromia laticollaris* Arnold. *J. Foram. Res.* **11**:212–216.

Schwab-Stey, H. and Schwab, D. (1979). The transformation of cytoplasmic microtubules into helices and paracrystals by Halothane in the foraminifer *Allogromia laticollaris* Arnold. *Z. mikrosk.-anat. Forsch., Leipzig* **93**:751–762.

Severin, K.P. (1987). Laboratory observations on the rate of subsurface movement of a small miliolid foraminifer. *J. Foram. Res.* **17**: 110–116.

Severin, K.P. and Erskian, M.G. (1981). Laboratory experiments on the vertical movement of *Quinqueloculina impressa* Reuss through sand. *J. Foram. Res.* **11**:133–136.

Severin, K.P., Culver, S.J. and Blanpied, C. (1982). Burrows and trails produced by *Quinqueloculina impressa* Reuss, a benthic foraminifer, in fine-grained sediment. *Sedimentology* **29**:897–901.

Sheehan, R. and Banner, F.T. (1972). The pseudopodia of *Elphidium incertum*. *Rev. Espanola Micropaleontol.* **4**:31–63.

Spindler, M., Hemleben, Ch., Salomons, J.B. and Smit, L.P. (1984). Feeding behavior of some planktonic foraminifers in laboratory cultures. *J. Foram. Res.* **14**:237–249.

Summers, K.E. and Gibbons, L.R. (1971). Adenosine triphosphate-induced sliding of tubules in trypsin-treated flagella of sea urchin sperm. *Proc. Natl Acad. Sci. USA* **68**:3092–3096.

Szubinska, B. (1971). "New membrane" formation in *Amoeba proteus* upon injury of individual cells. *J. Cell Biol.* **49**:747–772.

Telzer, B.R. and Haimo, L.T. (1981). Decoration of spindle microtubules with dynein: Evidence for uniform polarity. *J. Cell Biol.* **89**:373–378.

Travis, J.L. and Allen, R.D. (1981). Studies on the motility of the foraminifera. 1. Ultrastructure of the reticulopodial network of *Allogromia laticollaris* (Arnold). *J. Cell Biol.* **90**:211–221.

Travis, J.L. and Bowser, S.S. (1986a). A new model of reticulopodial motility and shape: Evidence for a microtubule-based motor and an actin skeleton. *Cell Motil. Cytoskel.* **6**:2–14.

Travis, J.L. and Bowser, S.S. (1986b). Microtubule-dependent reticulopodial motility: Is there a role for actin? *Cell Motil. Cytoskel.* **6**:146–152.

Travis, J.L. and Bowser, S.S. (1988). Optical approaches to the study of foraminiferan motility. *Cell Motil. Cytoskel.* **10**:126–136.

Travis, J.L. and Bowser, S.S. (1990). Microtubule–membrane interactions *in vivo*: Direct observation of plasma membrane deformation mediated by actively bending microtubules. *Protoplasma* **154**:184–189.

Travis, J.L. and Centonze, V.E. (1983). Intercellular fusion between reticulopodial networks in *Allogromia laticollaris*. *Biol. Bull.* **165**:497.

Travis, J.L., Kenealy, J.F.X. and Allen, R.D. (1983). Studies on the motility of the foraminifera. II. The dynamic microtubular cytoskeleton of the reticulopodial network of *Allogromia laticollaris*. *J. Cell Biol.* **97**:1668–1676.

Travis, J.L., Bowser, S.S., Rupp, G. and Rieder, C.L. (1985). Actin in foraminiferan reticulopodia. *J. Cell Biol.* **101**:398a.

Travis, J.L., Bowser, S.S., Calvin, J.G. and Lee, J.J. (1988). Pseudopodial tension in *Amphisorus hemprichii*, a giant Red Sea foraminifer. *Protoplasma* [suppl. 1]:64–71.

Tucker, J.B. (1982). Microtubule-organizing centres and assembly of intricate microtubule arrays in protozoans. In *Microtubules in Microorganisms* (eds P. Cappuccinelli and N.R. Morris), pp. 15–29. Marcel Dekker, New York.

Verschueren, H. (1985). Interference reflection microscopy in cell biology: Methodology and applications. *J. Cell Sci.* **75**:279–301.

Warner, F.D. and McIntosh, J.R. (eds) (1989). *Cell Movement*, Vol. 2: *Kinesin, Dynein and Microtubule Dynamics*. Alan R. Liss, New York.

Warner, F.D., Satir, P. and McIntosh, J.R. (eds) (1989). *Cell Movement*, Vol. 1: *The Dynein ATPases*. Alan R. Liss, New York.

Wetmore, K.L. (1988). Burrowing and sediment movement by benthic foraminifera, as shown by time-lapse cinematography. *Rev. Paleobiol., Spec. Pub.* **2**:921–927.

Wohlfarth-Bottermann, K.E. (1961). Cytologische Studien. VIII. Zum Mechanismus der Cytoplasmastromung in dunnen Faden. *Protoplasma* **54**:1–26.

Wohlman, A. and Allen, R.D. (1968). Structural organization associated with pseudopod extension and contraction during cell locomotion in *Difflugia*. *J. Cell Sci.* **3**:105–114.

Woodrum, D.T. and Linck, R.W. (1980). Structural basis of motility in the microtubular axostyle: Implication for cytoplasmic microtubule structure and function. *J. Cell Biol.* **87**:404–414.

GLOSSARY OF TERMS

Actin   A protein existing in either monomeric (G-actin) or polymeric (F-actin) form. F-actin comprises the basic structure of microfilaments in eukaryotic cells.

Actomyosin   A complex of actin and myosin forming a cytoplasmic contractile system.

ATP (adenosine 5′-triphosphate)   A triphosphorylated nucleotide possessing relatively high-energy phosphoric anhyride bonds that serves as a chemical energy source to drive cellular metabolism and locomotion.

ATPase   A cellular enzyme that catalyses the hydrolysis of ATP to ADP and inorganic phosphate. The free energy of this hydrolysis reaction is typically used to drive some energetically unfavourable process.

Antimitotics   Compounds that interfere with normal microtubule function (and hence mitosis) in cells. Common examples are colchicine, vinblastine and nocodazole.

Axopod A long, tapering pseudopodium possessing a distinct central core or axial filament (e.g. the radiating pseudopods of the Heliozoa).

Birefringence Optical property of a material displaying different refractive indices for light waves vibrating in different planes; generally indicative of a (non-cubic) crystalline or paracrystalline state.

Cell surface transport Energy-dependent movement of plasma membrane domains or cell surface components within the plane of the plasma membrane. Sometimes referred to as membrane surface transport.

Cilia Short (generally 5–10 μm long) flagella used to propel cells through liquid or move fluids along cell surfaces.

Clathrin Protein complex that constitutes the polyhedral framework of coated pits and vesicles.

Coated pit Bristle-coated invaginations of the plasma membrane that form during the initial stages of endocytosis.

Coated vesicle Small clathrin-vested spherical vesicles formed when a coated pit pinches off completely from a membrane.

Cytochalasin A fungal metabolite that disrupts cellular actin microfilament systems.

Cytoskeleton Intracellular framework consisting chiefly of microtubules, microfilaments and various intermediate and fine-filament systems.

Differential interference contrast microscopy A type of polarizing interference optical system that generates image contrast proportional to the optical path gradients in an object. This results in a shadowcast or pseudo-3D image. Also referred to as Nomarski optics or abbreviated DIC.

Dynein High molecular weight, microtubule-associated mechanoenzyme. In cilia and flagella, the dynein arms mediate the sliding of axonemal microtubules, which is responsible for bend propagation.

Filopod A long, thin cylindrical pseudopodium without anastomoses and generally unbranched.

Flagellum Thin, thread-like organelle protruding from the cell surface, whose undulations serve to propel the cell through the medium. In eukaryotes, it consists of a 9 + 2 array of microtubules and accessory structures (i.e. the axoneme) bounded by a plasma membrane.

Gelsolin One of a family of actin-associated proteins that severs actin microfilaments in the presence of Ca^{2+}.

Glycocalyx Carbohydrate-rich extracellular surface of the plasma membrane in most eukaryotic cells.

Helical filament Alternate assembly polymorph of microtubule proteins seen under certain physiological states in granuloreticulose protists.

Lamellipod A broad and highly flattened sheet-like pseudopod.

Mechanoenzyme A motor molecule; an enzyme that transduces metabolic energy into the force used to perform mechanical work.

Microfilament A thin (6–8 nm diameter) intracellular filament composed of a linear polymer of actin. It may interact with the mechanoenzyme myosin or a variety of "actin binding proteins". Also called an actin filament or F-actin.

Microtubule Hollow, cylindrical cellular filament (outer diameter *c.* 25 nm). Formed by the assembly of tubulin heterodimers.

Microtubule-associated proteins (MAPs) Any of a large family of diverse intracellular proteins that bind to microtubules. Various MAPs perform structural and mechanochemical functions, while others may regulate microtubule assembly.

Motility-organizing vesicles (MOVs) Elliptical fuzzy-coated vesicles, apparently unique to granuloreticuloseans. Thought to help organize microtubule patterns and movements in reticulopods.

Myosin Actin-associated mechanochemical ATPase.

Myosin subfragment 1 Globular proteolytic fragment of myosin that retains the ATPase and actin-binding sites. Used as a cytochemical marker for actin microfilaments.

Phalloidin Toxic alkaloid from the *Amanita phalloides* mushroom. Binds tenaciously to, and prevents disassembly of, F-actin. Fluorochromes can be linked to phalloidin, or the related compound phallacidin, and used as a fluorescent probe for actin filaments in cells.

Phase contrast microscopy Optical method (originally described by F. Zernike) that generates contrast due to phase differences between different areas of the specimen and/or the background due to refractive index or optical path differences.

Reticulopod Repeatedly branched and anastomosed thread-like pseudopod.

Saltatory transport Discontinuous translocation of organelles and cytoplasm. Moving particles undergo an instantaneous acceleration and travel in rectilinear or curvilinear paths, may stop and start, or spontaneously reverse direction.

Stress fibres Contractile fibrils of animal tissue cells composed of actin microfilaments, myosin and a host of accessory proteins, generally anchored at each end in an adhesion plaque.

Tubulin Globular protein subunits that self-assemble to form the microtubule lattice.

Z-Lines Filamentous electron-dense component of muscle contractile units (i.e. the sarcomeres) into which the F-actin filaments are anchored.

6
Symbiosis in foraminifera

JOHN J. LEE and O. ROGER ANDERSON

1.0	Introduction	157
2.0	Breadth of modern hosts	161
3.0	Overall morphology, cellular organization and regionalization	167
4.0	Identity of endosymbiotic algae	174
	4.1 Chlorophytes	175
	4.2 Dinoflagellates	177
	4.3 Rhodophytes	182
	4.4 Diatoms and chrysophytes	182
5.0	Functional aspects of symbiosis	189
6.0	Behavioural studies	193
7.0	The significance of light on the vitality and growth of larger benthic and planktonic foraminifera	195
8.0	Primary production and calcification	198
9.0	The roles of feeding in the metabolism of symbiont-bearing foraminifera	199
10.0	Mechanisms of carbon flow and other host–endosymbiont interactions	203
11.0	Chloroplast husbandry	207
12.0	Conclusions and outlook	212
	References	213

1.0 INTRODUCTION

Perhaps stimulated by Hedley's (1964) review and challenge for demonstration by modern techniques, and perhaps by heightened general interest of biologists in endocytobiology, the last quarter of a century has seen a flurry of interest in the algal endosymbionts of larger and planktonic foraminifera. At the same time, we have also recognized the phenomenon of chloroplast husbandry in a number of families (Nonionidae, Elphidiidae, Rotaliellidae). Considering the extent and diversity of algal endosymbionts in foraminifera, our previous neglect of the phenomenon is difficult to put into perspective. The appreciation by cell biologists that endosymbiosis was a major pathway of cellular evolution raises many interesting cellular and molecular questions about the algal-bearing and chloroplast-husbanding foraminifera. Contrary to the commonly held paradigm of the origin of major cellular organelles in eukaryotic cellular evolution, we recognize that foraminifera represent a very interesting line of cellular evolution, predisposed to endosymbiosis, which has not led to more tight genetic integration of the partners and loss of possible independence by the endosymbionts.

Since the Pennsylvanian there have been "evolutionary bursts" in which ordinary-sized benthic foraminifera (60–500 μm) gave rise to lineages which were $10-10^2$ times larger. Their non-random appearances in the fossil record correspond to episodes of global warming, relative drought, raised sea levels, expansion of tropical and semitropical habitats, and reduced oceanic circulation (Lee and Hallock, 1987). During such times, the rates of nutrient recycling to surface waters were reduced to such an extent that organic productivity in the sea dropped by as much as one or two orders of magnitude (Bralower and Theirstein, 1984). The conclusions drawn from the fossil record are that expanses of clear, oligiotrophic, shallow seas probably provided ideal conditions for diversifications of larger foraminifera. And that when the climate changed, and these habitats were lost, extinctions of larger foraminifera often occurred (Hallock et al., 1986). Their abundance in today's seas and their contribution to $CaCO_3$ shell production is not generally appreciated (Fig. 5). On Kudaka Jima (Okinana Préfecture, Japan) living "star sands" composed of calcarinids are so abundant they can be ladled from the floor of the back reef with large cooking spoons (production 6×10^2 g $CaCO_3/m^2$/year: Sakai and Nishihira, 1981). Deposition rates as high as kg/m/year in close proximity to coral reef margins have been reported (Hallock, 1981). Thus these "living sands" can play an important role in the world's CO_2/CO_3^{2-} budget.

Larger foraminifera are giants by protistan standards. Some Permian fusalinids reached impressive sizes (> 10 cm). All of the larger foraminifera and planktonic foraminifera (with the exception of *Hastigerina pelagica*) in the euphotic zones of tropical and semitropical seas are hosts for endosymbiotic algae, so it is reasonable to infer that this was also true in the past (reviewed in Lee, 1980, 1983; Hottinger, 1983; Lee and McEnery, 1983; Leutenegger, 1983, 1984). We have asserted that algal endosymbiosis was, in fact, a driving force in their evolution (Lee et al., 1979a; Lee and Hallock, 1987). It has been reasoned (Ross, 1974; Hottinger, 1982; Hallock and Schlager, 1986) and demonstrated (Röttger, 1972, 1976; Muller, 1977; Lee et al., 1979a, 1980b, 1988b, in press a, b; Röttger *et al.*, 1980; Hallock, 1981, 1985; ter Kuile and Erez, 1987) that algal endosymbiosis in larger foraminifera, with internal recycling of N, P. CO_2, O_2 and organic compounds, is an adaptation for survival and growth in the extremely oligotrophic tropical and semitropical seas where they are found; a trait which they share with many corals and other coelenterates, ascidians, sponges and molluscs (reviewed by Muscatine and Porter, 1977; Trench, 1979, 1987). Larger foraminifera are completely dependent on their algal endosymbionts for growth, because growth rates have light optima and are reduced, or absent, at lower light levels (Röttger and Berger, 1972; Röttger, 1976; Röttger *et al.*, 1980; Hallock, 1981; ter Kuile and Erez, 1984; Hallock *et al.*, 1986). Even when fed on an unlimited diet of algae, they are known to consume and assimilate (Lee *et al.*, 1988b); two species tested to date – *Amphistegina lobifera* and *Amphisorus hemprichii* – will not grow or survive very long (13 weeks) if they are incubated in the dark (Lee *et al.*, in press b).

The fact that today's larger foraminifera are hosts for an amazing diversity of algal types (e.g. dinoflagellates, chlorophytes, rhodophytes, chrysophytes and diatoms) suggests that they must have some fundamental properties which make them particularly good cellular habitats for the establishment and maintenance of algal endosymbionts. Some aspects of foraminiferal symbiosis set them apart from other groups of hosts in the sea. A number of highly complex families contain species specialized in husbanding only algal chloroplasts (Lee and Hallock, 1987; Lee *et al.*, 1988c; Lee and Lee, 1990). Another unusual aspect of foraminiferal endosymbiosis is the relative looseness of host/endosymbiont specificity among the diatom-bearing and dinoflagellate-bearing systems (Lee *et al.*, 1989; Lee and Lawrence, 1990). The same diatom-bearing host may have one, possibly two ($\sim 30\%$), of 20 small pennate species as endosymbiont(s). Foraminiferal populations at any station tend to harbour the same diatom species, but specimens of the same host collected elsewhere, or at the same station at different times of the year, may harbour other diatom species. The diversity is not endless – only 20 small, usually previously undescribed species, have been found in ~ 3000 isolations (Lee *et al.*, 1989, and work in progress).

The biological basis for the general adaptations of foraminifera to endosymbioses and the basis for the apparent looser adaptations in diatom- and dinoflagellate-bearing hosts are not yet fully understood. As a general conceptual framework to organize our thoughts, and perhaps stimulate further advancements in our knowledge of this phenomenon, several hypotheses and speculations have been tabulated (Table 1).

Table 1 Hypotheses and speculations on the general adaptations of foraminifera to endosymbioses

---

Hypothesis A: Foraminifera are particularly good habitats for the establishment and maintenance of algal symbionts

    Evidence: 1. The general cameral structure of foraminifera subdivides protoplasmic streams and promotes regionalization of cellular activities. Algal husbandry microhabitats are well separated from the host digestive activities (Müller-Merz and Lee, 1976; Lee and Hallock, 1987; Lee *et al.*, 1991; Faber and Lee, in press).
           2. Diversity of algal types found in foraminifera (reviewed in Lee and McEnery, 1983; Leutenegger, 1984; Lee, in press).
           3. Looseness of fit in some of the associations:
- Diatom-bearing hosts (Lee *et al.*, 1980a,b, 1983, 1986, 1989; Lee and Reimer, 1983; Reimer and Lee, 1988).
- Dinoflagellate-bearing soritids (Lee and Lawrence, 1990).

           4. The presence of rare or minor symbionts in some of the associations (Lee and Reimer, 1983; Reimer and Lee, 1984, 1988; Lee *et al.*, 1989).
           5. Breadth of the range of chloroplast scavengers from the tiny *Metarotaliella* to the relatively large and complex elphidids (reviewed by Lee, 1983; Leutenegger, 1984; Lee *et al.*, 1988c; Lee and Lee, 1990).
           6. The survivability of symbiotic algae in a non-symbiont-bearing foraminifera (one experiment *Chlamydomonas hedleyi* in *Rosalina leei*, reported in Lee and Zucker, 1969).
           7. The experimental replacement of some species of symbiotic diatoms with others after rendering host nearly aposymbiotic (Lee *et al.*, 1983, 1986).
           8. Asexual reproduction of foraminifera makes the transmission of symbionts to offspring automatic.
           9. Many, but not all, foraminifera have a gamontogamic sexual phase in their life-cycle in which the parents pair umbilicus to umbilicus and exchange gametes. Zygotes in these kinds of foraminifera could easily capture their parent's symbiotic algae.

Hypothesis B: Larger foraminifera have acquired some new common properties which promote maintenance of symbiosis

        1. Hosts are phototaxic and move to optimal lighting for their endosymbionts (Zmiri *et al.*, 1974; Lee *et al.*, 1980a).
        2. Larger foraminifera are somewhat plastic in their ontogeny and can change shape in response to environmental factors important to light gathering by their symbionts (e.g. Larsen, 1976; Hallock, 1979).
        3. Cell envelopes of algal symbionts are repressed when in hosts but they develop when symbionts are released and cultured in suitable media (Lee *et al.*, 1974, 1980a,b; Müller-Merz and Lee, 1976; Leutenegger, 1977a, 1984; McEnery and Lee, 1981; Lee and Reimer, 1983; Reimer and Lee, 1984, 1988; Lee and Hallock, 1987).

Table 1 *continued*

	4. On an evolutionary time-scale, many groups of larger foraminifera have developed spines, canal systems, complex chamberlets, pore-rim cups and other structures interpreted as morphological adaptations which promote symbiosis (Haynes, 1964; Hottinger, 1982; Lee and Hallock, 1987).
Speculations:	1. Larger foraminifera, or perhaps all foraminifera, have a food vacuole system which is responsive to signals given by potential algal symbionts.
	2. The small pennate diatoms which form symbioses with diatom-bearing larger foraminifera have surface antigens in common which they do not share with the free-living species on which their hosts feed and digest.
	3. The chloroplast envelopes of certain diatoms may have the same or some of the same antigens which could lead to their retention by some species of foraminifera.
	4. Initial fusion of lysosomes with phagosomes occurs in the unique granuloreticulopodal network outside the foraminifera. Thus any alga which escapes initial lysosomal fusion may completely escape digestion if the phagosomal vacuole (converted to symbiont vacuole) is taken inside.

2.0 BREADTH OF MODERN HOSTS

The range of algal-bearing foraminifera is quite broad. To a greater or lesser degree, the phenomenon has been documented by modern investigators in many recent families: Soritidae, Archaiadae, Peneroplidae, Alveolinidae, Nummulitidae, Amphisteginidae, Calcarinidae, Globigerinidae, Candeinidae, Pulleniatinidae and Globorotaliidae (Table 2). If symbiont diversity is used as the sole criterion, then the Miliolida seem to have been the most fertile group for the development of symbiosis (Figs 1–4). The fusiform to spherical alveolinids are the hosts for endosymbiotic diatoms. Each of the families of the disc-shaped and flabelliform superfamily Soritacea, is host to a different algal group respectively: Peneroplidae, rhodophytes; Archaiadae, chlorophytes; Soritidae, dinoflagellates (Table 1). The larger Rotaliid families, which are quite varied in form (annular discoid, globular, trochospiral to planispiral with spines) are hosts to pennate diatoms (Table 2, Figs 6–9). The four (arguably five) families of planktonic foraminifera are hosts to dinoflagellates or chrysophytes.

Table 2 Breadth of symbiont-bearing foraminifera

Family	Host foraminifera	Techniques used		Symbiont type	Identified symbiont	References
		TEM	Isolation			
Peneroplidae	Peneroplis pertusus	✓	✓	R	Porphyridium purpureum	Leutenegger (1984), Lee (1990)
	Peneroplis planatus	✓	✓	R	Porphyridium purpureum	Leutenegger (1984), Lee (1990)
	Peneroplis arietina	✓	✓	R	Porphyridium purpureum	Leutenegger (1984), Lee (1990)
Archaiadae	Archaias angulatus	✓	✓	C	Chlamydomonas hedleyi	Lee et al. (1974)
	Cyclorbiculina compressa	✓	✓	C	Chlamydomonas provasolii	Lee et al. (1979a)
	Androsina lucasi		✓	C	Chlamydomonas sp.	Lee et al. (in progress)
	"Peneroplis proteus"	✓		C	Chlamydomonas hedleyi	Leutenegger (1984)
Soritidae	Sorites marginalis	✓	✓	DF	Symbiodinium sp.	Müller-Merz and Lee (1976)
	Amphisorus hemprichii	✓	✓	DF	Symbiodinium sp.	McEnery and Lee (1981), Lee and Lawrence (1990), Ross (1974)
	Marginopora vertebralis	✓	✓	DF	Gymnodinium sp.	Lee and Lawrence (1990), work in progress
	Marginopora kudakajimensis		✓	DF	Symbiodinium sp.	
	Sorites orbiculus	✓		DF	Symbiodinium sp.	Leutenegger (1977a)
	Sorites "orbitolitoides"	✓		DF	Symbiodinium sp.	Leutenegger (1977a)
Alveolinidae	Borelis schlumbergeri	✓	✓	B	Fragliaria shiloi and other species	Leutenegger (1984), Lee et al. (1989)
	Aveolinelea quoii	✓		B		
Amphisteginidae	Amphistegina lobifera	✓	✓	B	Nitzschia frustulum var. symbiotica and other species	Leutenegger (1977a, 1984), Lee et al. (1989)
	Amphistegina lessonii	✓	✓	B		Leutenegger (1984), Lee et al. (1989)
	Amphistegina bicirculata	✓		B		Leutenegger (1984)
	Amphistegina radiata	✓		B		Leutenegger (1984), Lee et al. (1989)
	Amphistegina papilosa	✓		B		Leutenegger (1984), Lee et al. (1989)

Family	Species		Symbiont	Reference
Calcarinidae	Calcarina calcar	✓	Nitzschia frustulum var. symbiotica and other species	Hottinger and Leutenegger (1980)
	Calcarina gaudichaudii	✓	B	Lee et al. (in prep.)
	Calcarina delfranchii	✓	B	Lee et al. (in prep.)
	Calcarina hispida	✓	B	Lee et al. (in prep.)
	Calcarina spenglieri	✓	B	Hottinger and Leutenegger (1980)
	Baculogypsina sphaerulata	✓	B	Hottinger and Leutenegger (1980), Lee et al. (in prep.)
Nummulitidae	Cycloclypeus carpenteri	✓	B	Leutenegger (1984)
	Heterostegina depressa	✓	Nitzschia panduriformis var. continua and other species	Leutenegger (1984), Lee et al. (1989)
	Heterocyclina tuberculata	✓	B	Leutenegger (1984)
	Operculina ammonoides	✓	Entomoneis paldosa var. densistriata and other species	Leutenegger (1984), Lee et al. (1989)
	Nummulities cumingii	✓	B	Leutenegger (1984)
Globigerinidae	Globigerinella aequilateralis	✓	Ch	Gastrich (1988), Spindler and Hemleben (1980)
	Globigerinoides ruber	✓	Df Gymnodinium beii	Faber et al. (1989), Hemleben and Spindler (1983)
	Globigerinoides sacculifer	✓	Df Gymnodinium beii	Spindler and Hemleben (1980), Bé et al. (1977)
	Globigerina conglobatus	✓	Df Gymnodinium beii	Hemleben and Spindler (1983)
	Globigerina bulloides	✓	Ch	As above
	Globigerina cristata	✓	Ch	As above
	Turborotalita humilis	✓	Ch	Hemleben and Spindler (1983)
	Globigerina falconensis	✓	Ch	As above
	Orbulina universa	✓	Df Gymnodinium beii	Gastrich and Bartha (1988), Spero and Parker (1985), Spindler and Hemleben (1980), Bé et al. (1977)

Table 2 continued

Family	Host foraminifera	Techniques used		Symbiont type	Identified symbiont	References
		TEM	Isolation			
Candeinidae	Tinophodella glutinata	✓		Ch		Gastrich (1988)
	Candeina nitida	✓		Ch		Gastrich (1988)
Pulleniatinidae	Pulleniatina obliquiloculata	✓		Ch		Gastrich (1988)
Hastigerinidae	Hastigerina pelagica	✓		Ch	External associated algae	Bé et al. (1977)
Globorotaliidae	Neogloboquadrina dutertria	✓		Ch		Gastrich (1988)
	Globorotalia menardii	✓		Ch		Gastrich (1988)
	Globoratalia inflata	✓		Ch		Gastrich (1988)

R, unicellular red alga; C, unicellular chlorophyte; Df, dinoflagellate; B, bacilliarophyte = diatom; Ch, chrysophyte.
See Table 3. Diatom-bearing larger foraminifera host a variety of pennate species.

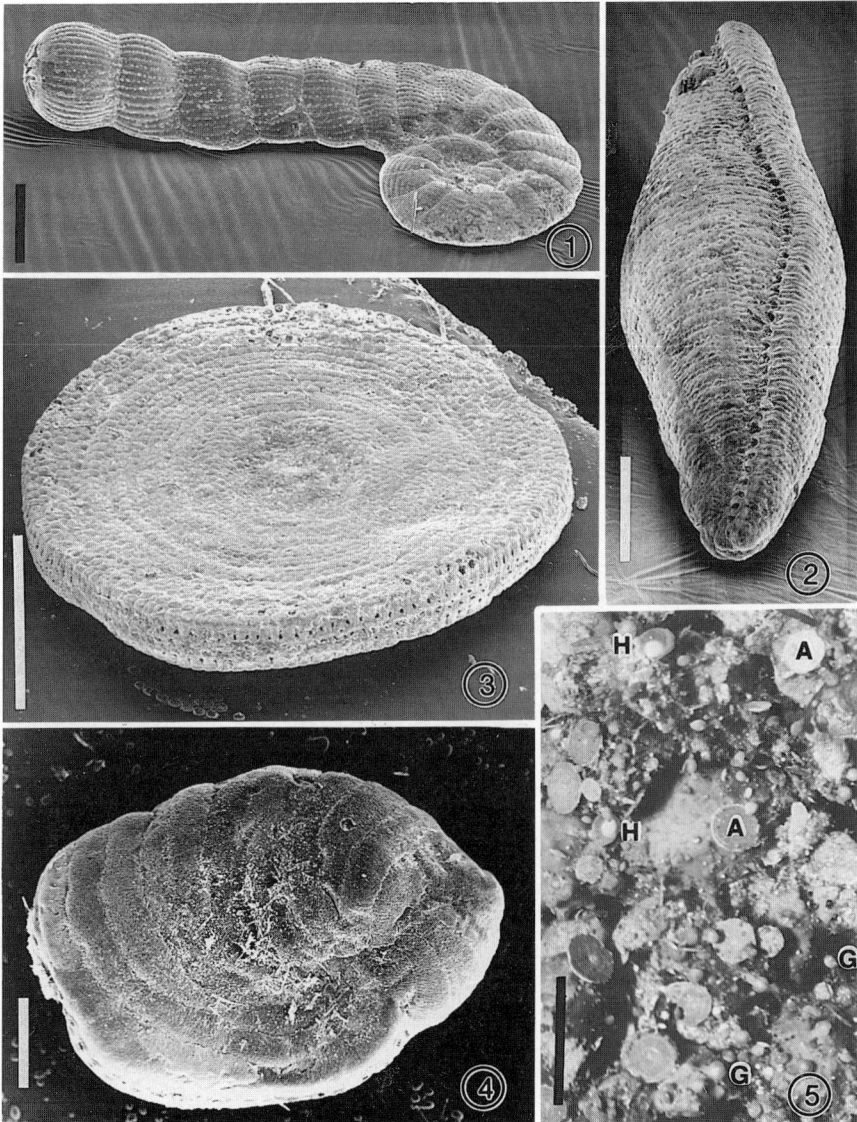

Figs 1–4. Scanning electron micrographs. Fig. 1. *Peneroplis arietina* (scale bar = 200 μm). Fig. 2. *Borelis schlumbergerii* (scale bar = 200 μm). Fig. 3. *Marginopora vertebralis* (scale bar = 1 mm). Fig. 4. *Androsina lucasi* (scale bar = 200 μm). **Fig. 5.** Underwater photograph of coral rubble substrate taken 30 m seaward of the H. Steinitz Marine Biological Laboratory in Elat, Israel. Photograph shows abundance of three species of larger foraminifera at that site: *Amphisorus hemprichii* (A); *Heterostegina depressa* (H); *Amphistegina* spp (G) (scale bar = 1 cm).

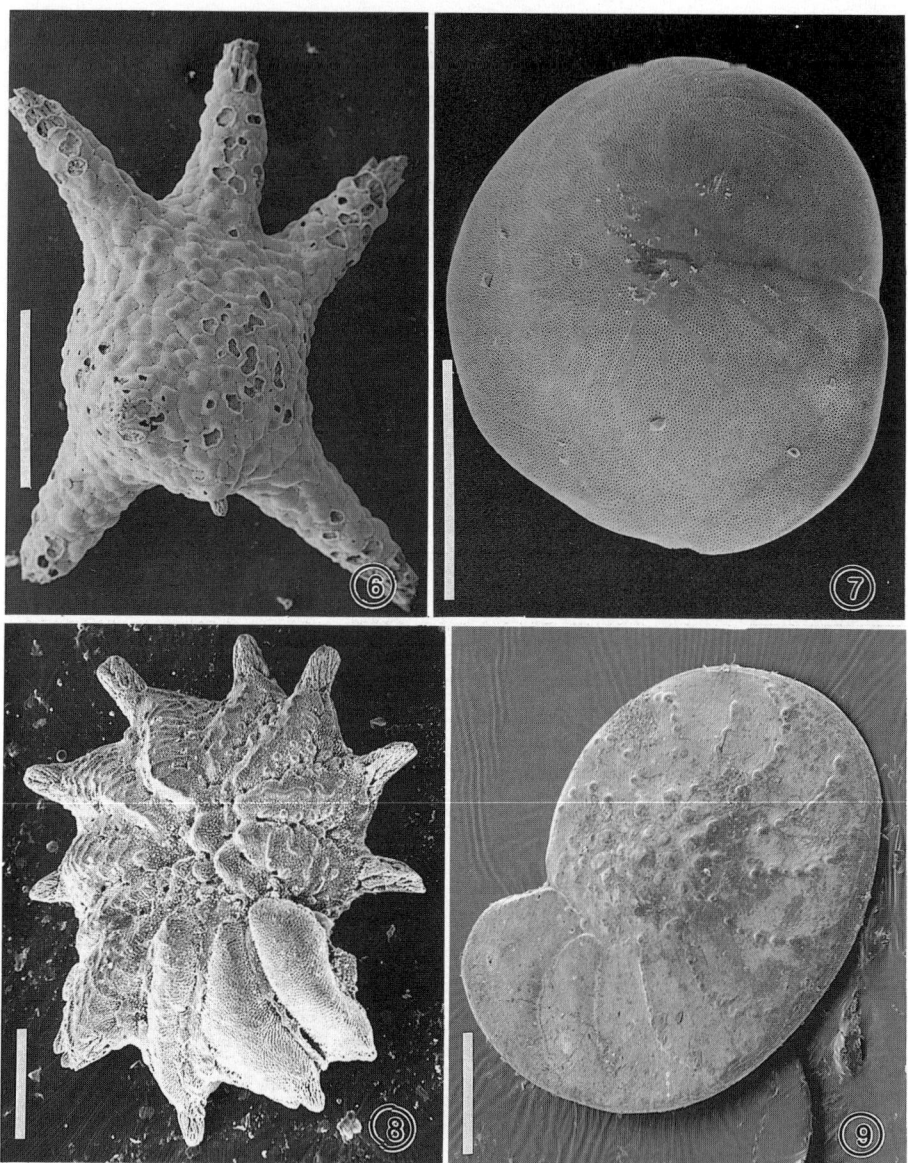

Figs 6–9. Scanning electron micrographs. Fig. 6. *Baculogypsina sphaerulata* (scale bar = 500 μm). Fig. 7. *Amphistegina lobifera* (scale bar = 400 μm). Fig. 8. *Calcarina calcar* (scale bar = 200 μm). Fig. 9. *Operculina ammonoides* (scale bar = 400 μm).

3.0 OVERALL MORPHOLOGY, CELLULAR ORGANIZATION AND REGIONALIZATION

Although many investigators have made subjective interpretations of the morphological complexities of larger foraminifera as indicators of possible functional adaptations to algal endosymbiosis, there is really a dearth of objective evidence. Yet some behavioural aspects of host–symbiont interaction are so remarkable as to offer almost *prima facie* evidence of complex co-adaptations. The elegantly simple, yet effective behaviour of planktonic foraminifera that move their symbionts peripherally into the rhizopodial network during the day to enhance photosynthesis and withdraw them into the shell at night (Bé et al., 1977) gives clear evidence of their remarkable adaptive plasticity. In contrast to larger benthic species, there appears to be little evidence in planktonic foraminifera of adaptations in shell architecture to enhance symbiotic associations; alternatively, behavioural plasticity may have been an evolutionary route exploited by these planktonic species.

Even if we cannot yet prove that morphological characters we see in larger foraminiferal shells are actually adaptations for symbioses, there is realistic hope that we may find objective underlying evidence. Cladistic analysis of only the morphological characteristics of the superfamily Soritacea clearly separated into three family clades: Archaiadae, Soritidae and Peneroplidae (Gudmundsson, 1990). As mentioned earlier, each of these families is host to a different algal group. Surely it is reasonable to look closely at this superfamily, to examine questions of morphological adaptation particular to major groups of algae.

Several different aspects of the architecture and organization of larger foraminifera may contribute to their functional abilities to husband algal endosymbionts with respect to (1) exposure of algal cells to appropriate light levels, (2) provide algal cells with physiologically suitable microhabitats and (3) protect symbionts from host digestive enzymes or other host cellular activities which may be deleterious to them.

A number of workers have suggested that much of the larger foram shell architecture should be interpreted as a compromise between hydrodynamic factors and expanding surface area for exposure of symbionts to light (Haynes, 1964; Hottinger, 1977; Hallock, 1979). Evidence for this is based on shape changes observed in larger foraminifera as a function of depth in the clear waters of the Gulf of Elat and elsewhere (Hottinger and Dreher, 1974; Larsen, 1976; Hottinger,

1977; Larsen and Drooger, 1977; Hallock, 1979). Further evidence was found by analysing more carefully the shell structure of *Amphistegina* collected at different depths. Hallock and Hansen (1979) found that the secondary lamellae were thinned with depth. In the laboratory, clones of *A. lessonii* and *A. lobifera* produced thicker tests when grown under strong light than they did when grown in reduced light (Hallock, 1979). Later studies confirmed and expanded these observations. Hallock and co-workers (1986) found that there were differences in shell thickness in individuals of *A. gibbosa* grown at different light levels (1.6–40.0 $\mu E/m^2/s$). Although the increase in diameter of individuals grown at 6.4 $\mu E/m^2/s$ was comparable to that at higher levels, the tests were thinner and lighter than those which received more light. Individuals grown at only 3.9 $\mu E/m^2/s$ were significantly flatter than those of the other trials. These same individuals had 30% thinner umbilical lamellae. The results with *A. lessonii* were quite similar. The flattest, lightest individuals grew at an intensity level of 3.9 $\mu E/m^2/s$.

In water-motion experiments, the same investigators (Hallock *et al.*, 1986) showed that individuals subjected to water motion grew significantly thicker tests. Both species of *Amphistegina* grew more slowly in diameter when culture vessels were placed on a shaker. A third species, *A. lobifera*, increased in diameter and mass more quickly in moving water. Water motion increased lamellar thickness on both the spiral and umbilical sides of the tests of *A. gibbosa* and *A. lessonii*.

The largest surface-to-volume ratios are found in the lineages of very flat, thin trochospiral, planispiral or discoidal tests. Flattening is accompanied by subdivision of the chamber into chamberlets to strengthen the tests and allow the outer walls to be thinner and more transparent. In the miliolines with their calcite crystal structure tending to reflect and diffuse light, the thinning of the lateral walls is very important (Ross and Ross, 1978). Pits and ribs found in many miliolines (Figs 10–12) may serve the same purpose and may in addition serve to increase diffusion of nutrients to the interior of the host–symbiont complex (Hansen and Dalberg, 1979).

Figs 10–14. Scanning electron micrographs. Figs 10 and 11. *Peneroplis planatus* from Kudaka Jima. Fig. 10. View of pits from a plane perpendicular to the test. Fig. 11. Same test viewed at an angle nearly horizontal to the test. Illumination chosen to emphasize the contour of the ribs and the abutment of the test plates (scale bars for both figs = 20 μm). Fig. 12. *Androsina lucasi* from a Caribbean collection by Dr Pamela Hallock. Shows deep pits in the surface of the test. Same specimen as in Fig. 4 (scale bar = 1 μm). Fig. 13. *Calcarina calcar* from Kudaka Jima. Fragment of the inside of a test wall showing vaulted pore rims (scale bar = 20 μm). Fig. 14. Portion of test of *Amphistegina lobifera* showing pores. Same specimen as in Fig. 7 (scale bar = 40 μm).

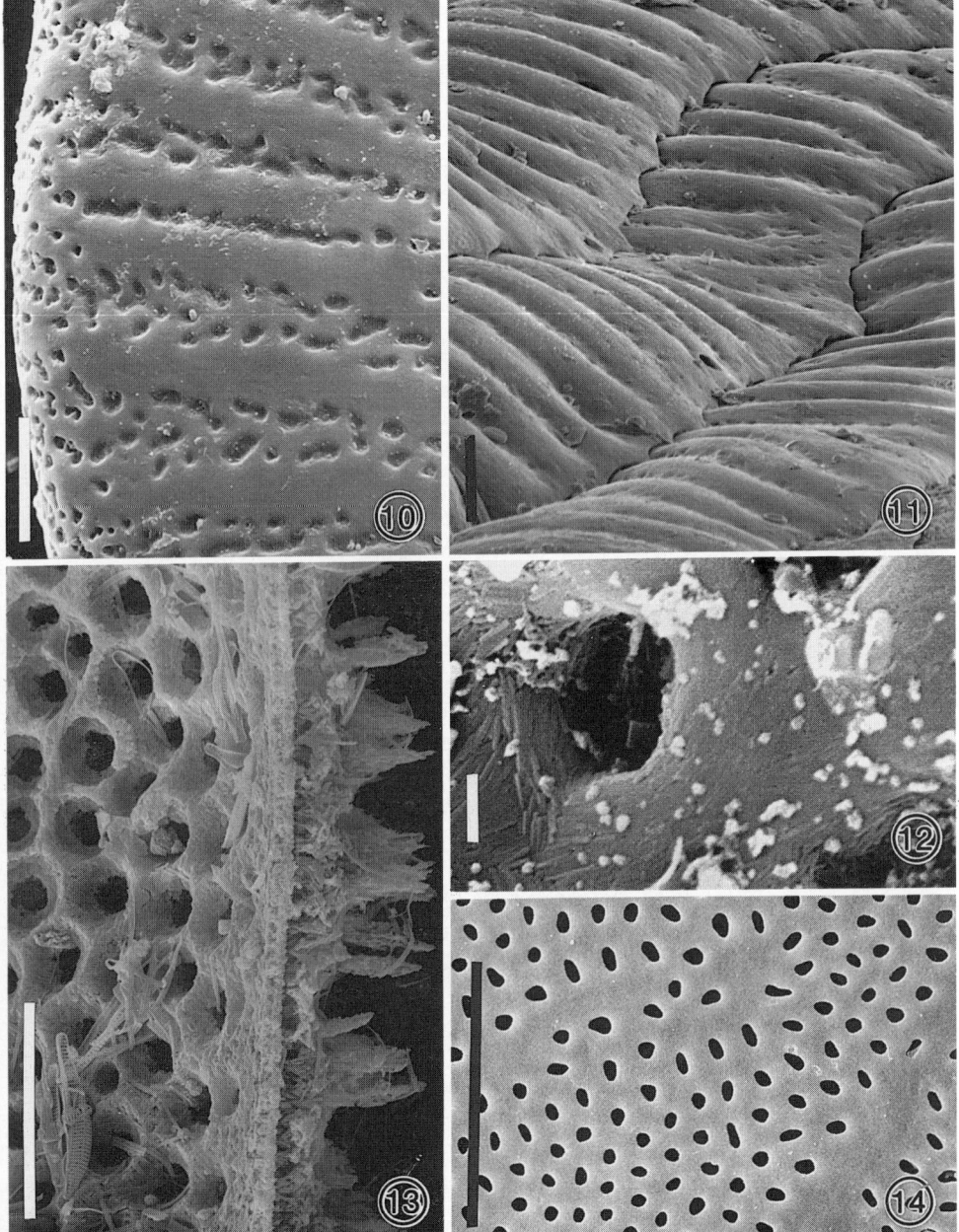

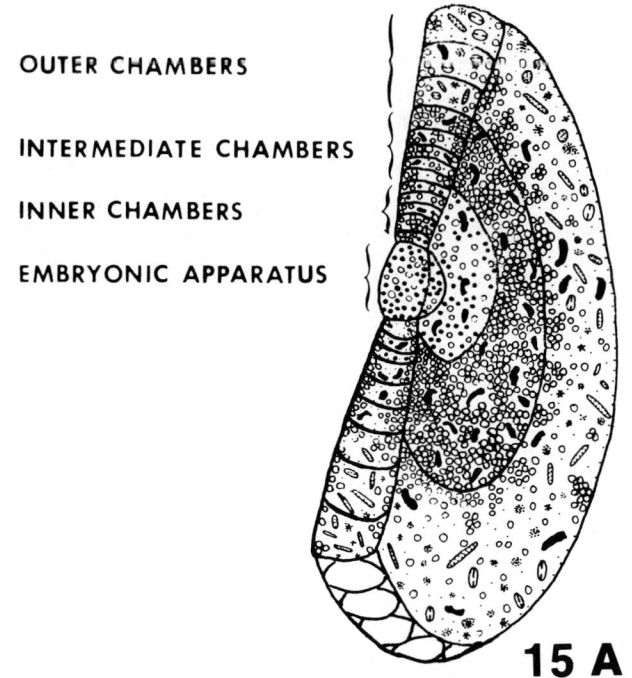

Fig. 15A

An interesting and quite obvious adaptation to symbiosis was noted by Hansen and Buchardt (1977). In the amphisteginids and nummulites, the inner surface of the test around the pores is excavated into cup-like pore-rims (Figs 13, 14). The symbiotic diatoms are concentrated along the surfaces of the cytoplasm in cytoplasmic bulges which fit into the pore-rim vaults (Figs 25, 26). Several experiments, one using the vital dye neutral red, and the other $^{14}CO_2$ have indicated that the pores are physiologically active even though they have an organic lining (Berthold, 1976; Leutenegger and Hansen, 1979). Becase CO_2/CO_3^{2-}

Figs 15–18. Light micrographs of histochemical preparations from a study of *Sorites marginalis* by Müller-Merz and Lee (1976). Fig. 15. Large section showing general distribution of symbionts (dark spheres) in this species. Larger and irregular dark bodies in outer whorls are food vacuoles. Proloculum (P) is near bottom of figure in centre. Compare with Fig. 15A (scale bar = 450 μm). Fig. 16. Section through an intermediate chamberlet showing the endosymbionts and stored starch (scale bar = 10 μm). Fig. 17. Section through an outer chamber showing food vacuoles, residua including a recognizable diatom frustule (*Cocconeis*; labelled C) (scale bar = 10 μm). Fig. 18. Section through an outer intermediate chamber showing two diatoms in a food vacuole (FV) in the early stages of digestion (scale bar = 30 μm).

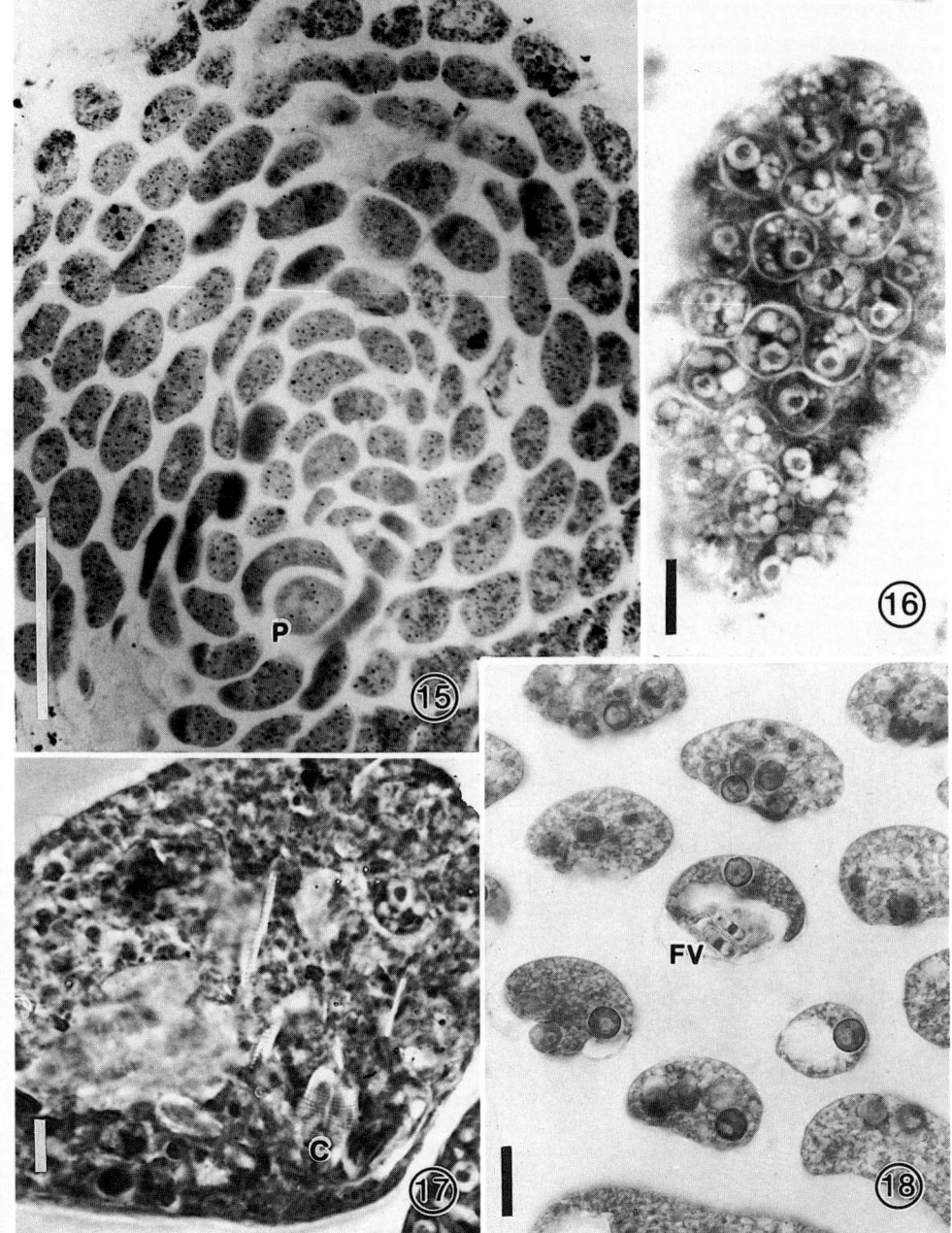

concentration can limit photosynthetic and calcification rates, the inference that the pores can facilitate nutrient flux is quite important.

Cytological, cytochemical and fine structural evidence tends to support the hypothesis that many, if not all, larger foraminifera are quite regionalized and that the endosymbiotic algae are usually distributed in parts of the cell away from digestive activities (Müller-Merz and Lee, 1976; McEnery and Lee, 1981; Leutenegger, 1984; Koestler et al., 1985; Faber and Lee, 1991; Lee et al., 1991). In *Sorites marginalis* and *Amphisorus hemprichii* (Fig. 3), two disc-shaped larger foraminifera, the endosymbiotic dinoflagellates are in highest concentration in a ring which extends from about one-quarter the diameter in from the apertural face to a ring in the centre which includes the inner chambers and embryonic apparatus (Figs 15–18). The symbiont ring is slightly larger in *Amphisorus* than it is in *Sorites*. Digestive vacuoles and less dense cytoplasm are found in the outer chambers (Fig. 17). The hundreds of generative nuclei of these heterokaryotic foraminifera are found in the embryonic apparatus and inner chambers, while the scores of somatic nuclei are distributed in the intermediate and outer chambers (Müller-Merz and Lee, 1976; McEnery and Lee, 1981).

The results of recent studies of the distribution of acid phosphatase in foraminifera indicate that digestion begins outside the test in the extended rhizopodia (Faber and Lee, 1991; Lee et al., 1991). It was reasoned that those algal species which avoid initial digestion by foraminifera have the potential to establish themselves as endosymbionts. Specimens of 20 different species of foraminifera were assayed using naphthol AS-BI phosphate for the presence of acid phosphatase (an indication of digestion). Ten of the species investigated harbour endosymbionts, five husband chloroplasts and five are not known to have symbionts. Although there was some variability in the concentration of acid phosphatase among the species studied, the

---

Figs 19–24. Green-filtered light micrographs of cytochemical preparations for the demonstration of acid phosphatase. Figs 19, 21 and 22 are aqueous preparations of whole specimens just after completing the reactions. Figs 20, 23 and 24 are paraffin-embedded sections. All figures are from Lee et al., 1991, and Faber and Lee, 1991. Fig. 19. *Marginopora kudakajimensis*. Note dark reaction product at the periphery of the cell where the apertures (A) are located (scale bar = 500 μm). Fig. 20. *Peneroplis planatus*. Note dark reaction product located near the apertures (scale bar = 200 μm). Fig. 21. *Baculogypsina sphaerulata*. Note the reaction product in the contracted pseuodopia on the spine in focus near the right edge of the figure (scale bar = 500 μm). Fig. 22. *Elphidium* sp. Note the reaction product in the fossettes in focus (scale bar = 200 μm). Fig. 23. A food vacuole stained to demonstrate acid phosphatase (scale bar = 5 μm). Fig. 24. Pseudopodia showing acid phosphatase reaction (scale bar = 1 μm).

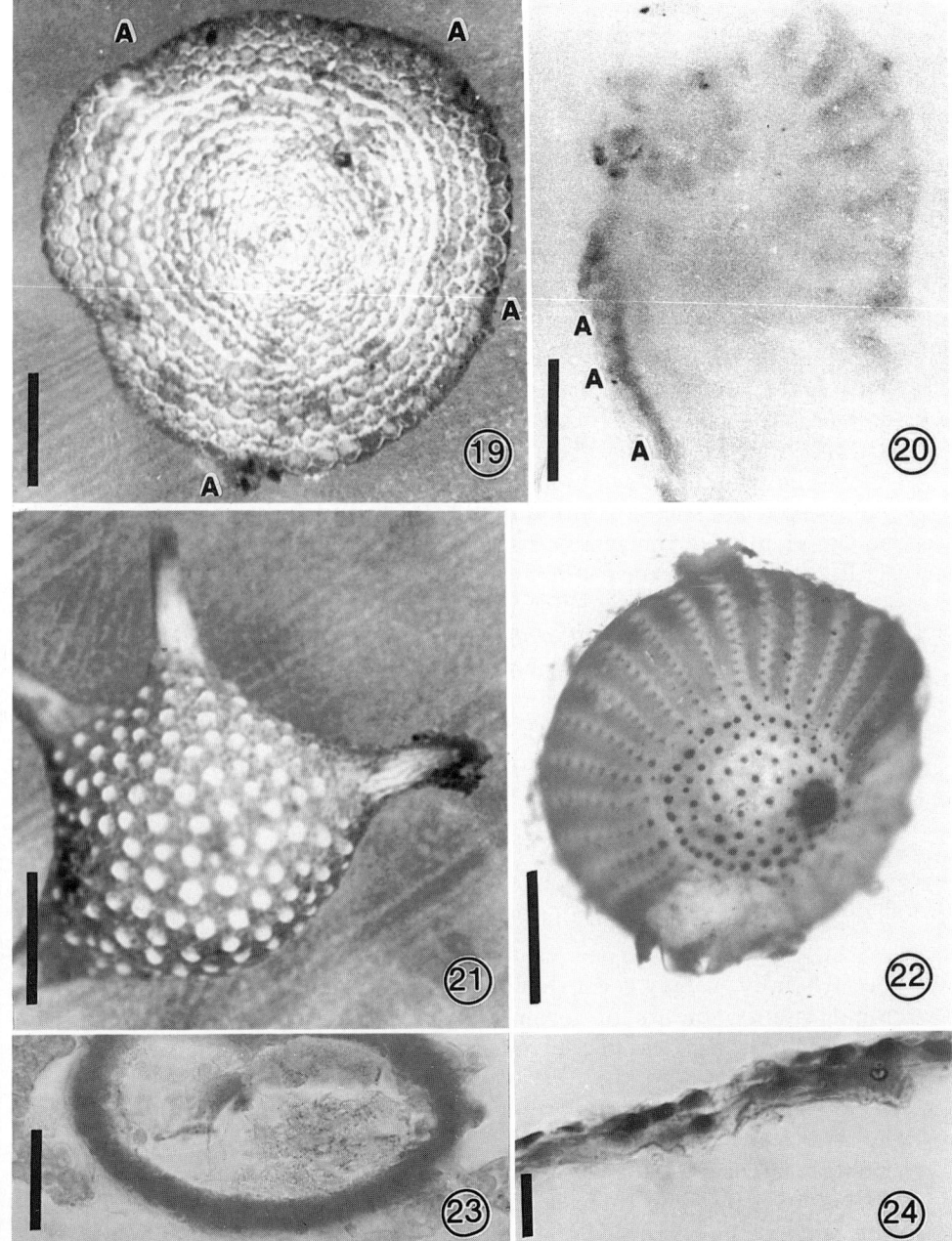

enzyme was found in the extended rhizopodia in regions of food capture, in masses of food gathered by foraminifera near their apertures, in outer chambers, in the fossettes of elphidids, in the marginal cords and canals of *Heterostegina* and *Operculina* and the spines of the calcarinids tested (Figs 19-24). This digestive enzyme, on the other hand, was never found near the location of the endosymbionts in the species which bear them.

Another line of evidence suggesting the same separation of symbionts and digestive activity was obtained in studies of symbiont specificity in *Amphistegina* (Lee et al., 1983; Koestler et al., 1985). The aim of the experiment was to bleach and starve *A. lessonii* until they were nearly aposymbiotic so that they could be used to test algal specificity and possible pecking order to persistence in that species. As might be anticipated, there were progressive and dramatic changes over time in the cytoplasmic organization as the symbionts were stressed with DCMU and the hosts were starved (Fig. 26). When the hosts got to the stage where the symbionts were being reduced in number, many micro-vesicles, some of which could have been lysosomes, were seen in the vicinity of the symbionts.

Recent fine structural studies of *Peneroplis pertusus* and *P. acicularis* seem to support the cytochemical observations (Lee, 1991) reported above. The rhodophytes, *Porphyridium purpureum*, are quite unique among foraminiferal endosymbionts. They are not surrounded by a second host vacuolar membrane (phagosome, symbiosome), but lie free in the host cytoplasm (Fig. 32). No vesicles resembling lysosomes were observed near the endosymbionts (Lee, 1990a). Digestive enzymes in these two species of *Peneroplis* were demonsttrated near the apertures (Fig. 24; Lee et al., 1991).

4.0 IDENTITY OF THE ENDOSYMBIOTIC ALGAE

Superficially, it may seem that few questions on the identities of the algal endosymbionts of foraminifera remain, but, in truth, many taxonomic issues are unresolved. Although the underlying causes are not yet evident, the algal endosymbionts within larger foraminifera have reduced cellular structure. Typical cell envelopes and flagellar apparatus are reduced or absent. Thus it is necessary to liberate them from their hosts and grow them in axenic (pure) culture. Fortunately, it has been possible to do so relatively easily, and in culture all of the isolates have formed the key morphological features which are necessary for proper diagnosis.

4.1 Chlorophytes

Chlorophytes are the only recognized algal endosymbionts of the family Archaiadae. Occasionally, chlorophytes have been isolated as minor endosymbionts or intracytoplasmic survivors from *Amphistegina* spp. and *Amphisorus hemprichii* from the Red Sea.

The endosymbiont *Chlamydomonas hedleyi* isolated from *Archaias angulatus* was the first of the larger foraminiferal algae to be grown in axenic (pure) culture (Lee et al., 1974; Figs 27, 28). Spherical non-motile cells (10–24 μm in diameter) are the dominant phase in the host (Lee and Zucker, 1969) and in culture (Lee et al., 1974). The chloroplast is large, lobed and parietal. The nucleus and a large embedded pyrenoid lie between the chloroplast lobes. We used the structure of the pyrenoid, which is surrounded by a starch sheath and penetrated by a single thylakoid, as a diagnostic feature to separate this species from *Chlamydomonas provasolii* (Lee et al., 1979a; Figs 29–31). Motile, biflagellated cells appear in culture 3–5 days after transfer to fresh media. The motile forms of *C. hedleyi* are relatively long and slender (12–18 × 6–8.5 μm). The anterior end of the cell is slightly furrowed and truncated, the chloroplast is displaced anteriorly, and the nucleus is spherical and located anterior-centrally. A prominent stigma is observed in cultured forms. Asexual reproduction was observed in both the motile and non-motile stages. Fine structural evidence from ultrathin sections of *Peneroplis proteus* also suggests that this host harbours *C. hedleyi* (Leutenegger, 1984).

A very similar species of *Chlamydomonas*, *C. provasolii*, was isolated from *Cyclorbiculina compressa*. *C. provasolii* is easily distinguished from *C. hedleyi in vitro* and *in situ* if observations are made with a TEM. The pyrenoids of *C. provasolii* are penetrated by many tubules, each with a single thylakoid stack in the centre (Figs 29–31). Small starch grains surround the pyrenoid (Lee et al., 1979a; Müller-Merz and Lee, 1976; Lee and McEnery, 1983). *C. provasolii* has also been identified in low numbers in TEM sections of *Sorites marginalis*, a dinoflagellate-bearing foraminifer (Müller-Merz and Lee, 1976; Fig. 29).

Although the fine structure of these species is known well enough to characterize them, their separation from other species of marine *Chlamydomonas* remains to be clarified. The fine structure of many species assigned to the genus, but only briefly described by light microscopy, requires additional attention. Newer molecular methods might also be useful in clarifying the taxonomy of this large genus.

Recently, Pamela Hallock found some living *Androsina lucasi* in the field, a relatively rare green species, and sent them to one of us

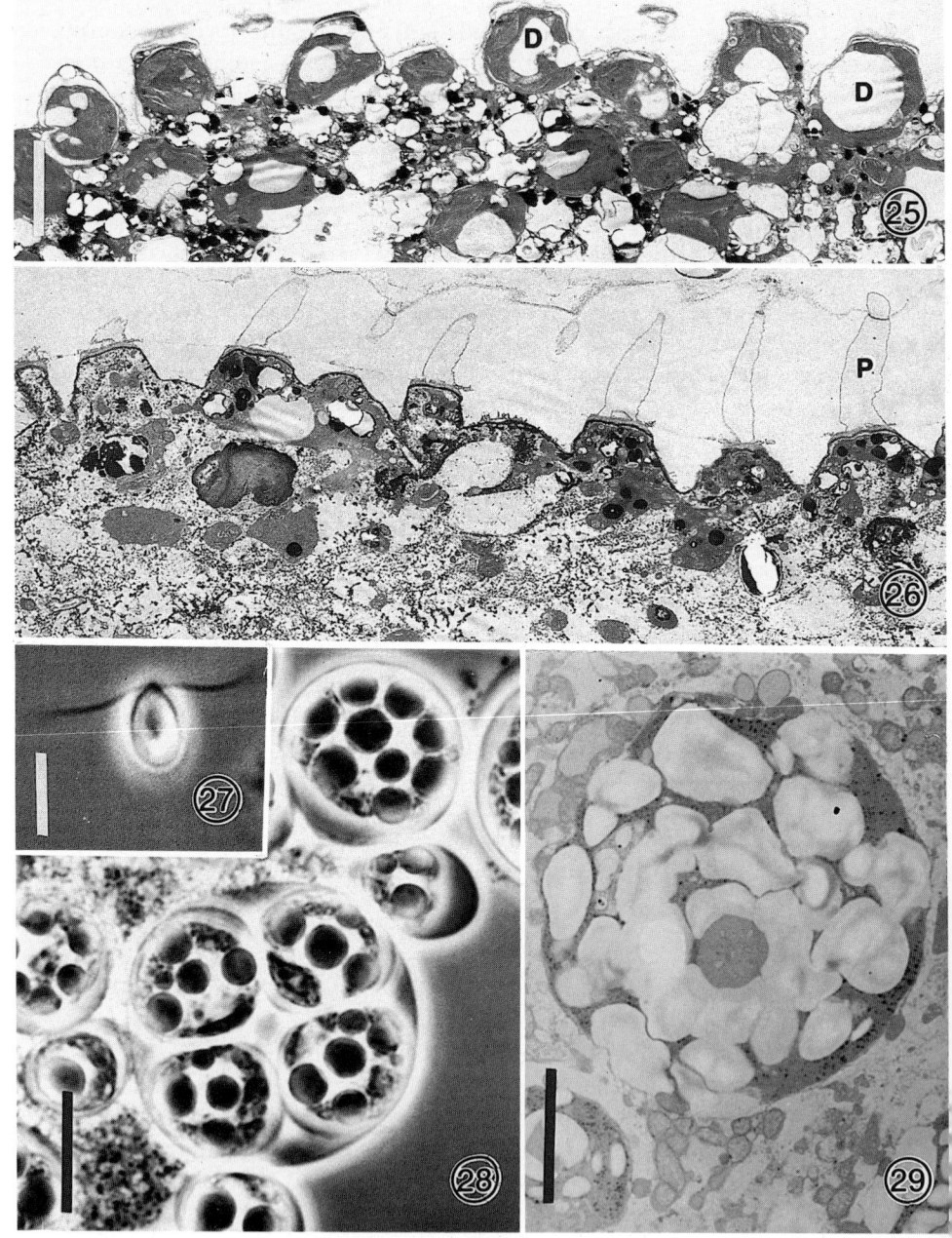

(J.J.L.). Fortunately, the host foraminifera were easy to clean and their endosymbionts grew in the same media as *C. hedleyi* and *C. provasolii*. At the light microscopical level, the new endosymbiont also seems to be a species of *Chlamydomonas*. It is smaller and more globuse than either of the aforementioned species. Its fine structure and physiology are presently being studied.

The "minor endosymbiotic chlorophytes" from *Amphistegina*, *Amphisorus* and *Heterostegina* were more easily characterized due to the work of Kessler and co-workers (Kessler and Czygan, 1970; Kessler and Zweier, 1971; Kessler, 1974, 1976, 1978). The *Chlorella* we isolated belong to species which differ from each other in cultural, fine structural and physiological characteristics and autospore production (Lee et al., 1980a, 1982). Biochemical and physiological tests using Kessler's methodology suggested that the isolate from *H. depressa* is a halotolerant strain of *Chlorella saccharophila* Kruger. The other species from *Amphistegina lessonii* and *Amphisorus hemprichii* did not grow in the thiamine-supplemented ammonium test medium or nitrate test media and could not be characterized further by physiological tests. Its fine structure was distinct from *Chlorella nana*, *C. spaercki*, *C. marina* and *C. stigmatophora*. It was closer in structure to *C. ovalis* and *C. salina*, but could be distinguished from both by the structure of its cell wall (Lee et al., 1982).

4.2 Dinoflagellates

The endosymbionts of the soritids *Amphisorus hemprichii*, *Sorites marginalis*, *Marginopora vertebralis* and *M. kudakajimensis* Gudmundson 1990, and a number of planktonic species, *Globigerinoides ruber*, *G.*

Figs 25, 26. Transmission electron micrographs. Sections through the periphery of a chamber of *Amphistegina* showing endosymbiotic diatoms in the expanded pore-rim cups at the base of the pores in the shell. Compare Figs 25 and 26. Figure 26 has been treated with DCMU and incubated in the light. Note the granulated and vacuolated cytoplasm in Fig. 26. Also note pore-rims not occupied by symbionts in Fig. 26. Some diatoms in Fig. 26 undergoing autolysis or digestion; none appear normal in comparison with Fig. 25. Organic lining of pores (P) visible in many pores in section shown in Fig. 26 (scale bar = 10 μm). **Fig. 27.** Phase micrograph of *Chlamydomonas hedleyi* motile stage (scale bar = 10 μm). **Fig. 28.** Phase micrograph of asexually produced vegetative cells within parental cell wall of *C. hedleyi* filled with starch (scale bar = 10 μm). **Fig. 29.** Transmission electron micrograph of *Chlamydomonas provasolii* within host. Note starch grains and distinctive pyrenoid (see Fig. 31) (scale bar = 4 μm). Figures 25 and 26 from Koestler et al. (1985); Figs 27 and 28 from Lee and Zucker (1969); Fig. 29 from Müller-Merz and Lee (1976).

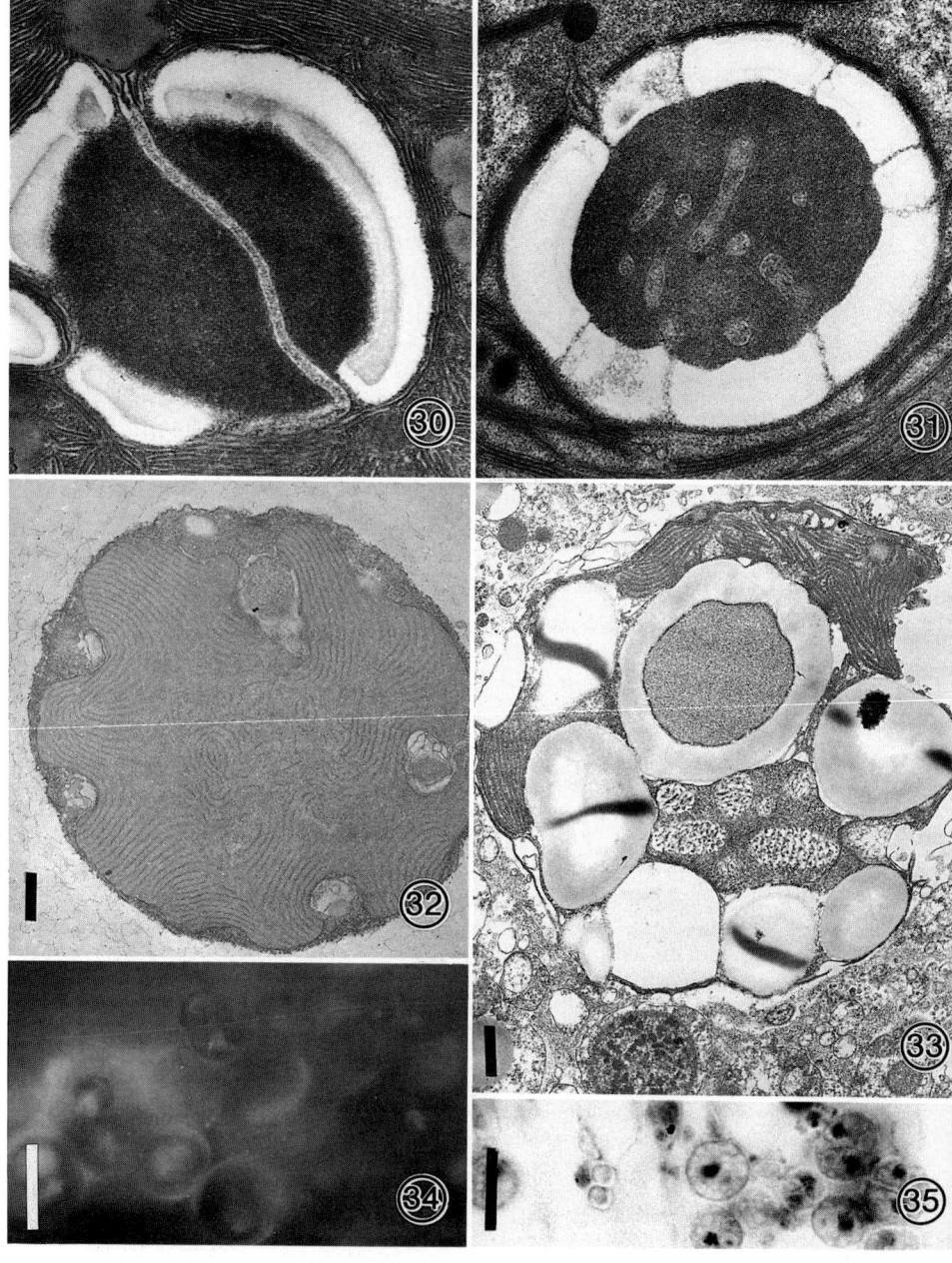

conglobatus, *G. sacculifer* and *Orbulina universa*, are dinoflagellates (Lee *et al*., 1965; Ross, 1974; Müller-Merz and Lee, 1976; Leutenegger, 1977a; Spindler and Hemleben, 1980; Figs 33, 35–37, 40, 41). In addition, several species of dinoflagellates (*Pyrocystis noctiluca* and *P. robusta*) have been described as commensals in the bubble capsule of *Hastigerina pelagica*.

The dinoflagellate from *O. universa* has been isolated, cultured and characterized as a new species, *Gymnodinium beii* (Spero, 1987; Fig. 37). The dinoflagellates found in *Globigerinoides conglobatus*, *G. ruber* and *G. sacculifer* seem identical in fine structure to those in *O. universa*. In their hosts, *G. beii* are small (5–10 μm) and coccoid; in culture, they are flagellated gymnodinoids with an equal-sized epicone and hypocone. It has been estimated that the populations of *G. beii* reach 3×10^3 symbionts in mature *O. universa*; one very large specimen (~ 900 μm) harboured even more (2.3×10^4).

Though light microscopical and fine structural studies have clearly shown that the endosymbiotic algae in *Sorites marginalis*, *S. orbiculus*, *S. orbitolites*, *Amphisorus hemprichii*, *Marginopora vertebralis* and *M. kudakajimensis* are dinoflagellates (Müller-Merz and Lee, 1976; Leutenegger, 1977a,b, 1983, 1984; Lee *et al*., 1979b; McEnery and Lee, 1981), the precise identities of the endosymbionts are in doubt. This uncertainty grows out of the perceptive and tireless work of Trench and his co-workers, which overthrew the dogma that there was a single pandemic dinoflagellate species (Freundenthal, 1962), *Symbiodinium microadriaticum*, which was endosymbiotic in a broad range of corals,

---

Figs 30–33. Transmission electron micrographs. Pyrenoids from *Chlamydomonas hedleyi* (Fig. 30) and *Chlamydomonas provasolii* (Fig. 31). Note the single thylakoid stack piercing the pyrenoid in *C. hedleyi* in contrast with *C. provasolii*. Fig. 30 × 24 300; Fig. 31 × 27 000. Fig. 32. *Porphyridium purpureum* within *Peneroplis pertusus*. Note invaginations of cell surface which seem to be the source of cell sheath fibres (scale bar = 1 μm). Fig. 33. *Symbiodinium* sp. within its host *Sorites marginalis* (scale bar = 1 μm). **Fig. 34.** Epifluorescent light micrograph of a fluorescent antibody-reacted preparations. Endosymbiotic diatoms freshly released from *Amphistegina lessonii*. The cells were incubated with polyvalent antisera made in a rabbit against the cell envelope fraction of *Fragilaria shiloi*. The serum was first reacted with *Nitzschia panduriformis* before reacting it with the freshly released endosymbionts. A second antibody, a goat antibody against rabbit antibody which was conjugated with the fluorescent dye FITC, was used to visualize the results (scale bar = 10 μm). **Fig. 35.** Fuelgen-stained preparation of endosymbiotic diatoms from *Amphistegina lessonii*. Several cells near the figure number are in division (scale bar = 10 μm). Figures 30 and 31 from Lee *et al*. (1979a) (with permission from Micropaleontology); Fig. 32 from Lee (1990); Fig. 33 from Müller-Merz and Lee (1976); Fig. 34 from Lee *et al*. (1988b) (with permission from the Society of Protozoology); Fig. 35 from Lee *et al*. (1980c).

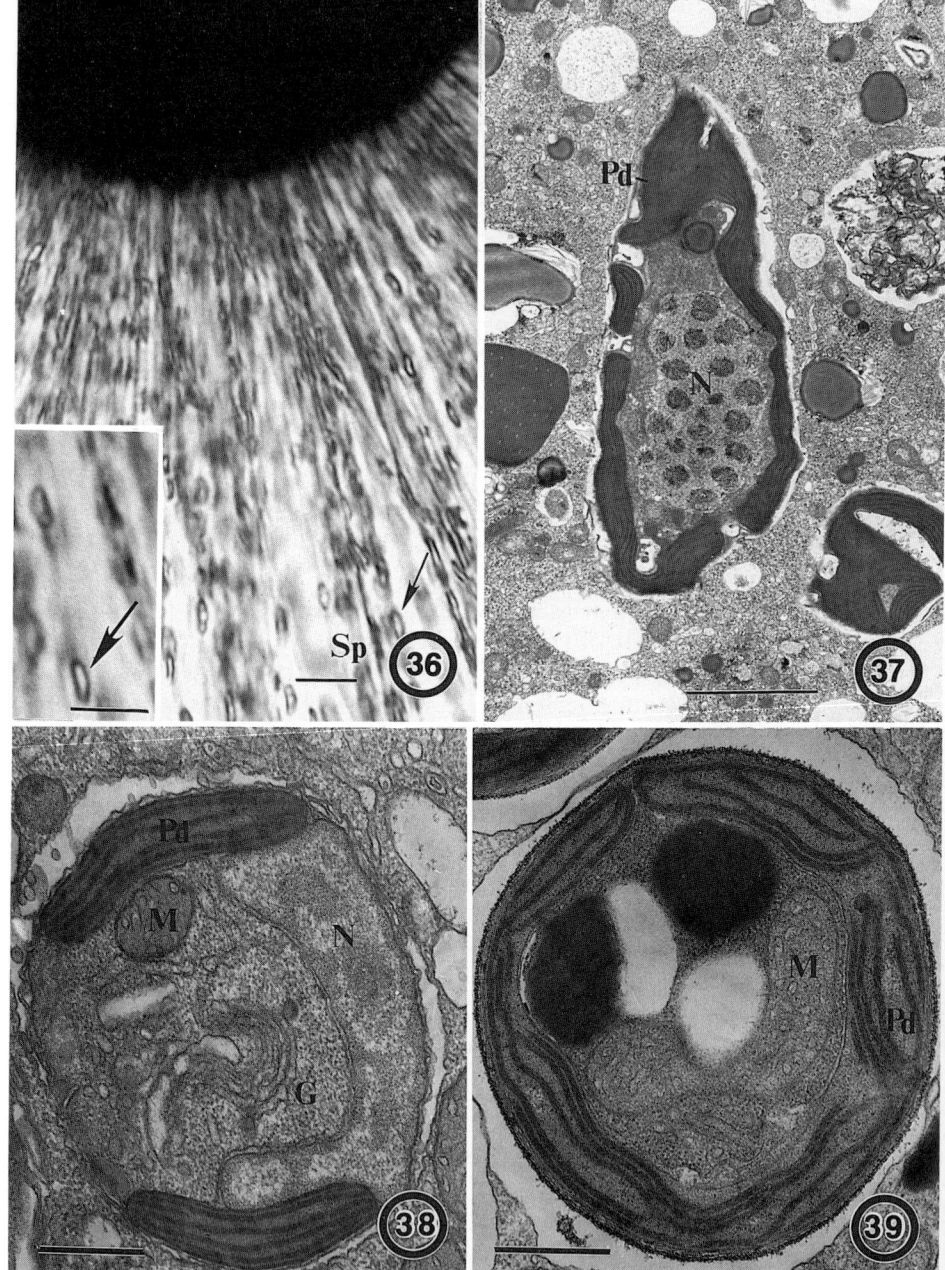

jellyfish, anemones and giant clams (reviewed in Trench, 1987, 1988). Though similar in gross morphology at the light microscopical level, various new species of *Symbiodinium* have been established on the basis of cell envelope characteristics, karyotypes, physiology and isoenzyme patterns (McLaughlin and Zahl, 1959; Schoenberg and Trench, 1980a,b,c; Blank and Trench, 1985a,b; Blank, 1987; Trench and Blank, 1987). None of the foraminiferal endosymbiotic species has yet been characterized in such detail, although we can report a small amount of progress and can pick up a little detail from already published fine structural micrographs.

The endosymbionts in *Amphisorus hemprichii* in *in situ* micrographs seem to have more chromosomes than those of *Sorites marginalis* (Müller-Merz and Lee, 1976; Leutenegger, 1977a,b, 1983; McEnery and Lee, 1981). There is also evidence that some of the dinoflagellates in *Amphisorus* have elements of flagellar apparatus *in situ* (Leutenegger, 1983).

The task of isolating and studying the endosymbiotic dinoflagellates from soritids is quite formidable because of the architecture of the host tests. Bacteria and diatoms grow epizootically on the test and are particularly difficult to remove from the crevices formed where the calcareous plates, which cover the surfaces of the disc-shaped test, are joined (Figs 48, 49). After a number of false starts, we developed a method to dissolve the test and inhibit the bacteria and diatoms without killing the dinoflagellates (Lee and Lawrence, 1990).

Much to our surprise, we isolated two different genera of dinoflagellates from *Amphisorus hemprichii*. The most numerous was a *Symbiodinium* sp., but an *Amphidinium* sp. was also frequently (16%) isolated (Figs 40 and 41). Neither will be easy to fully characterize. The *Amphidinium* from *A. hemprichii* resembles *Amphidinium rhychocephalum* in microscopic morphology when observed in the SEM (Roberts *et al.*, 1988). It does not seem to have the numerous polygonal amphaesmal vesicles that have recently been described for *A. carterae* (Klut *et al.*, 1989). The comparative structure of the ventral ridge

---

Figs 36–39. Algal symbionts of planktonic foraminifera. Fig. 36. Dinoflagellate symbionts, *Gymnodinium beii* (Spero, 1987), of *Orbulina universa* (arrows) within the rhizopodia along the spines (Sp) (scale bars = 20 μm). Fig. 37. Transmission electron microscopic view of a section through the cytoplasm of *O. universa* showing a dinoflagellate symbiont within a perialgal vacuole. The plastids (Pd) occur at the periphery of the cytoplasm. A nucleus (N) contains puffy chromosomes characteristic of mesokaryotic dinoflagellates (scale bar = 3 μm). Type I (Fig. 38) and type II (Fig. 39) chrysophyte algal symbionts, associated with the planktonic foraminifer *G. aequilateralis*, containing a nucleus (N), plastids (Pd), Golgi body (G) and mitochondria (M) (scale bars = 1 μm). From Faber *et al.* (1988).

microtubules and the vesicular complex which extends from the flagellar openings to various components of the flagellar apparatus need to be described before the *Amphidinium* from *A. hemprichii* can be assigned to a taxon.

We have carried out a comparative gross morphological study of the envelopes of the motile and vegetative stages of the *Symbiodinium* spp. we isolated from *A. hemprichii* and *M. kudakajimensis* (Lee and Lawrence, 1990). We have compared our new isolate in the SEM against strains of *Symbiodinium microadriaticum* and *S. pilosum* kindly supplied by Dr R. Trench. The foraminiferal isolates are different from both of Trench's strains. Both have very thin theca with amphaesmal vesicles that are extremely thin and which tear easily (Fig. 41).

4.3 Rhodophytes

Fine structural studies of *Peneroplis* spp. clearly showed that the endosymbiotic rhodophyte was a species of *Porphyridium* (Leutenegger, 1977a, 1984; Hottinger, 1982). Because the cells in their hosts had reduced cell envelopes, specific identification awaited the isolation and cultivation of this symbiont. In culture, the symbionts form a thick cell sheath (Fig. 53). The fine structure of isolates from *Peneroplis pertusus* and *P. acicularis* was compared with a reference strain of *Porphyridium purpureum* (UTEX 161) grown in the same medium and under the same conditions. The strains were morphologically identical and on this basis it was concluded that they are conspecific (Lee, 1990). If in the future new criteria are applied to the taxonomy of the group, the question of identity should be re-examined.

4.4 Diatoms and chrysophytes

Because they have not yet been isolated in culture, the golden and yellow-green endosymbiotic algae, which have been described from the cytoplasm of *Globigerinella aequilateralis* cannot be properly identified (Faber *et al.*, 1988). Both are very small (3.5 μm in diameter) and have eccentrically located nuclei enveloped by double perinuclear-plastidal membranes. A distinction was made between symbiont types I and II on the basis of chloroplast location, thylakoid spacing and position of a large vacuole (Figs 38, 39). The chloroplasts of alga type I are parietal and scattered at the periphery of the cell, whereas those of type II are more continuous, often encircling the cell periphery. Both types have three thylakoids per lamella but those of type I are more closely spaced. The vacuole in type I alga is eccentric, whereas the one in type II is centrally located (Faber *et al.*, 1988). While Faber and co-workers concluded that the symbionts were probably chrysophytes, Gastrich (1988) concluded that there was only one species of endosymbiont and that it was more properly categorized as a prymnesiophyte. She also used fine structural data to suggest that a number of other planktonic species (e.g. *Globigerinita glutinata*) also harbour endosymbiotic chrysophytes (Gastrich, 1988).

It has been thought for quite some time that the endosymbiotic algae from some species of larger foraminifera are probably diatoms (Deitz-Elbraechter, 1971). In retrospect, if we examine some of the earliest micrographs of the endosymbionts *in situ*, it should have been clear to us that more than one species is involved in the phenomenon (Hansen and Buchardt, 1977; Leutenegger, 1977b; Berthold, 1978; Schmaljohann and Röttger, 1978). Later fine structural studies of the diatom endosymbionts clearly showed that this diversity could be demonstrated *in situ* (Leutenegger, 1983, 1984) even though frustules were not formed. Fortunately, it was possible to isolate the endosymbionts in culture and they formed diagnostic frustules. In the course of the last decade, 20 species or varieties of diatoms have been isolated from over 3000 specimens of larger foraminifera (*Amphistegina lessonii, A. lobifera, Heterostegina depressa, Borellis schlumbergerii, Operculina ammonoides* and *Calcarina calcar*) from the Red Sea, Indian Ocean, Hawaii, Palau and the Great Barrier Reef (Lee *et al.*, 1980a,b,c, 1989; Lee and Reimer, 1983; Reimer and Lee, 1984, 1988, work in progress; Figs 6–9). The diatoms are all very small (10 μm) pennate species belonging to the genera *Fragilaria, Navicula, Nitzschia, Amphora, Achnanthes, Cocconeis* and *Protokeelia* (Figs 42–47). All of the endosymbiotic species are rare in natural communities or have never been found other than as symbionts (Lee *et al.*, 1989). Half the endosymbionts found were new species or varieties. One, *Navicula muscatinei*, has a very unusual life-cycle for a diatom (Lee and Xenophontes, 1989). Several other isolates undergo very unusual patterns of size rejuvenation.

The most commonly isolated diatom was *Nitzschia frustulum* var. *symbiotica*. This species was found in 33.7% of the hosts examined (Lee *et al.*, 1989). The next most common were *Nitzschia panduriformis* var. *continua, Fragilaria shiloi, Nitzschia laevis* and *Amphora roettgerii* (14.5, 9.9, 8.7 and 6% respectively of the total isolations; Table 3). More than half the isolations were done in the vicinity of the H. Steinitz Marine Biological Laboratory on the Red Sea, so that these results must be interpreted with that bias in mind.

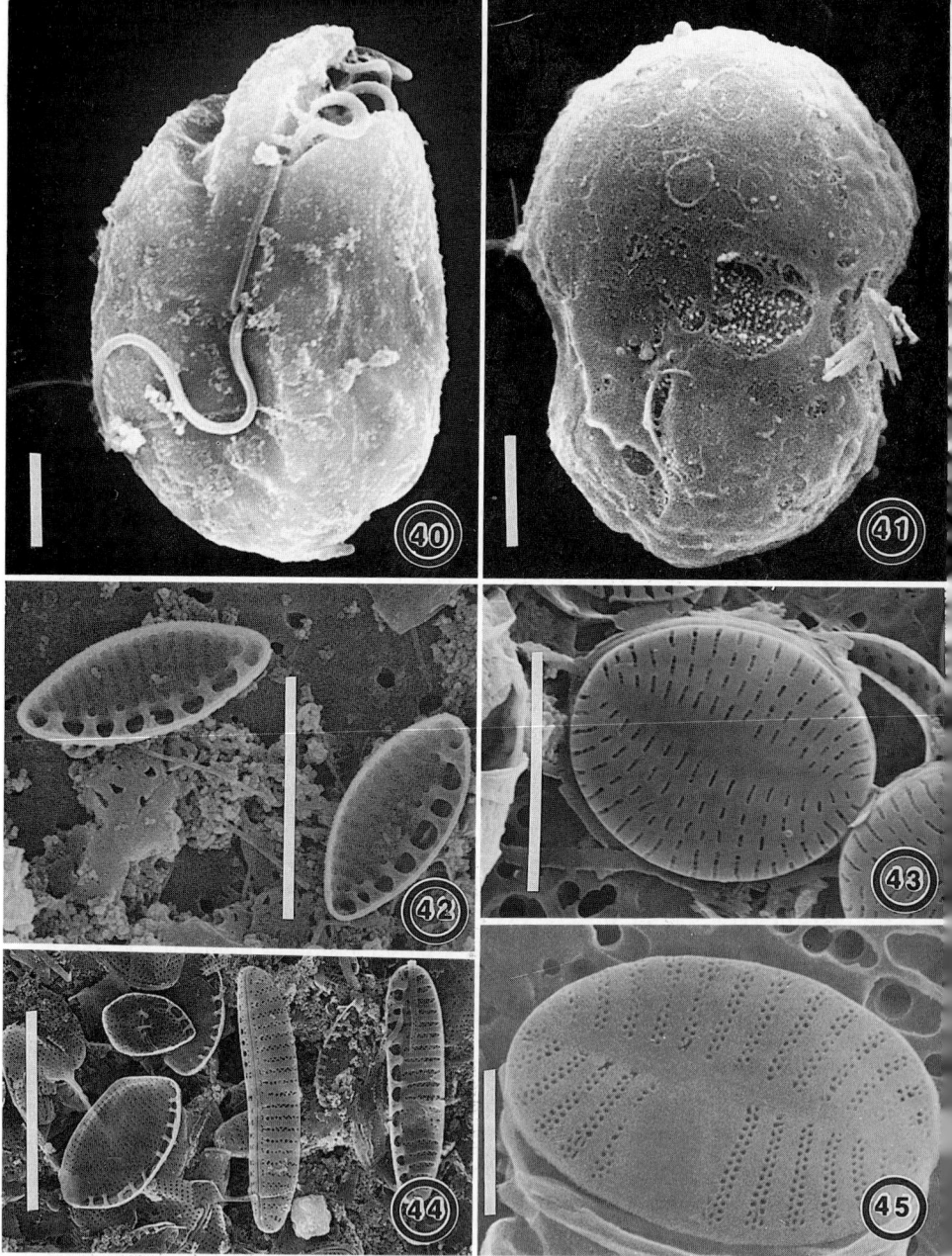

Significant numbers of *N. frustulum* var. *symbiotica*, *N. laevis* and *N. panduriformis* var. *continua* were isolated from hosts collected at every depth. In contrast, *Fragilaria shiloi* was rarely isolated from hosts collected at depths greater than 25 m and *Achnanthes maceneryae*, *Protokeelia hottingeri* and *Amphora* sp. were only isolated from hosts harvested from depths greater than 25 m. Although host species often harboured more than one species of endosymbiont, most (71%) hosted only one species at a time. In this study, it was reported that several hosts tended to harbour particular species of endosymbiont. *N. laevis* and *F. shiloi* were the most common endosymbionts in *Borellis schlumbergerii*. *N. laevis* and *A. maceneryae* were most common in *Operculina ammonoides*, and *N. frustulum* var. *symbiotica* was most commonly isolated (93%) from *Calcarina calcar*. The latter, however, was represented by the fewest specimens in the samples studied. Work on Indopacific collections with large numbers of calcarinnids is in progress.

Two experimental studies of symbiont persistence in *Amphistegina lessonii* have been reported (Lee et al., 1983, 1986; Koestler et al., 1985). These experiments arose out of questions raised about the possibilities that particular diatom species might be adaptive for hosts at different depths in the water column or at different seasons. The hosts were rendered nearly aposymbiotic by incubating them in flasks with 1×10^{-5} M DCMU (3-3,4-dichlorophenyl)-1,1-dimethyl urea) in the light at a depth of 5 m in the Gulf of Elat. At the end of 5 days, the hosts had bleached and fine structural studies and ^{14}C primary production experiments indicated that they were nearly aposymbiotic (Fig. 26). The hosts were then fed mixtures of different algae, each mixture containing two symbiotic diatom species and one non-symbiotic species, and they were then reincubated at 10 and 20 m in flasks anchored to the sea bed. Ten different combinations or permutations of species were used in the experiment. The results suggested that some algal species were selected over, or were more competitive than, others during the rebrowning of the host, *A. lessonii*. By comparing the results of different combinations of algae in the mixtures, *Nitzschia*

Figs 40–45. Scanning electron micrographs. Fig. 40. *Amphidinium* sp. from *Amphisorus hemprichii* (scale bar = 2 μm). Fig. 41. *Symbiodinium* sp. from *Amphisorus hemprichii*. Note thin amphiesmal vesicles on surface of cell. The outer surface of some of those in the sulcus and cingulum are torn (scale bar = 2 μm). Fig. 42. *Nitzschia frustulum* var. *symbiotica* (scale bar = 4 μm). Fig. 43. *Cocconeis andersonii* (scale bar = 4 μm). Fig. 44. A mixture of *Nitzschia laevis* (at left) and *N. valdestriata* (right) (scale bar = 10 μm). Fig. 45. *Achnanthes maceneryae* (scale bar = 2 μm).

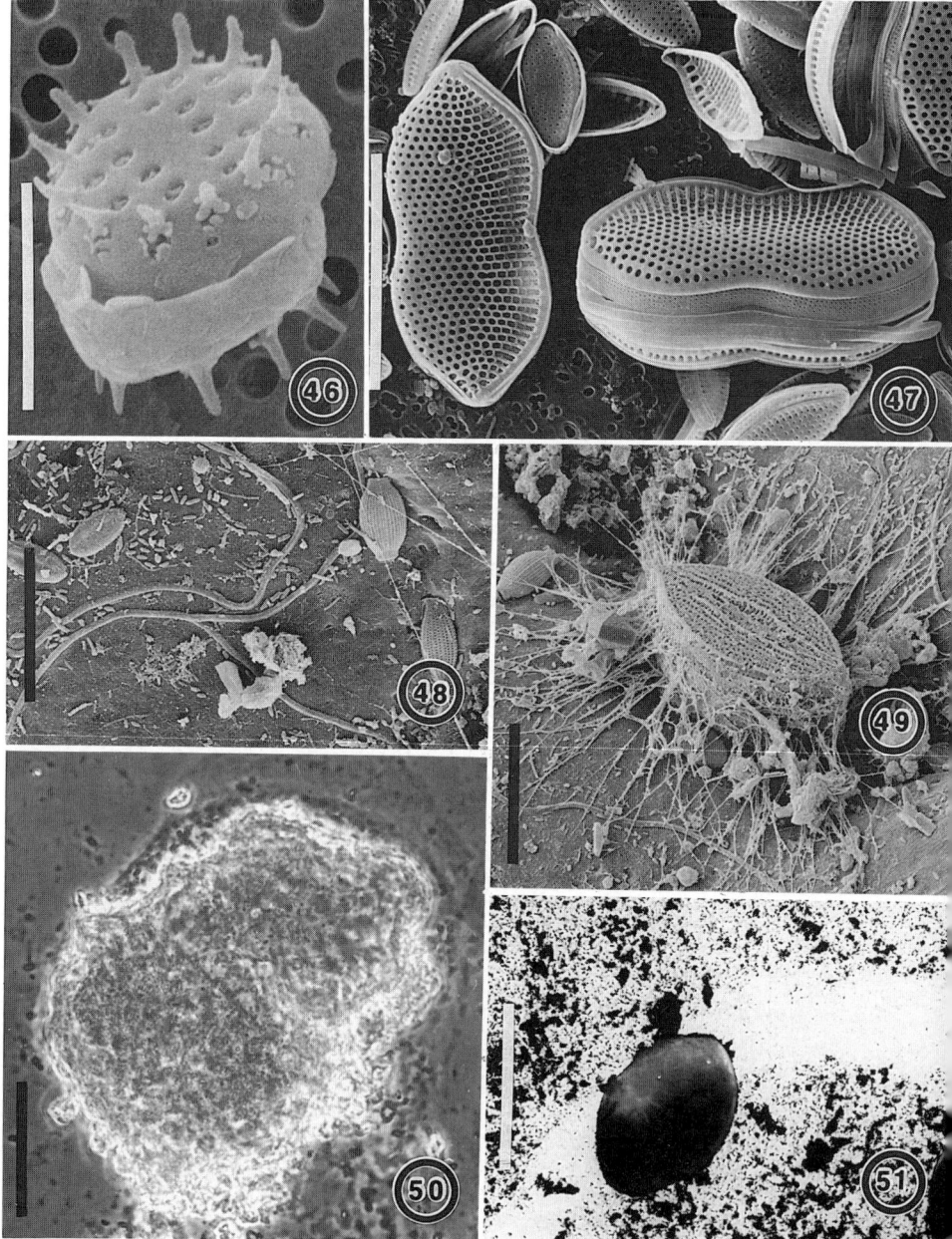

valdestriata and *N. laevis* were the most successful endosymbionts (Fig. 44) and *Fragilaria shiloi* the least. It should be noted that none of the free-living diatoms fed in the mixtures was isolated from rebrowned hosts. Presumably, they were all digested. With the exception that at 20 m *N. valdestriata* was preferred over *N. laevis*, no other differences were noted in the incubations at both depths.

In a second experiment, the population of *A. lessonii* originally hosted mixtures of symbionts: *Nitzschia laevis, N. panduriformis* var. *continua, N. frustulum* var. *symbiotica* and *Amphora roettgerii*. As before, they were rendered nearly aposymbiotic by incubation with DCMU and then they were incubated with mixtures of endosymbiotic diatoms. In this experiment, *N. laevis*, which was preferred over other symbionts in the first experiment, was also recovered in many "rebrowned" foraminifera (Lee *et al.*, 1986). The simplest interpretation of the data would be that little or no replcement of the original symbionts took place during the algal repopulation of the hosts after DCMU treatment was stopped. Changes involving a loss of diversity of endosymbionts (e.g. reduction of the number of foraminifera hosting *N. panduriformis* and *N. frustulum* var. *symbiotica*) may suggest that *N. laevis* had a competitive edge in the repopulation dynamics.

Because very few small pennate diatoms, which belong to diverse taxonomic families, are involved in endosymbiosis with larger foraminifera, it is quite reasonable to ask what are the properties of these species which permit them to establish themselves and persist in their hosts? One of the more logical places to look for cell recognition properties is the cell envelope. A first step along these lines was recently reported (Lee *et al.*, 1988a). Antisera against the cell envelope fraction (frustules) of several endosymbiotic diatom species (*Nitzschia panduriformis, Fragilaria shiloi, Amphora tennerima*) were raised in rabbits.

Figs 46–49. Scanning electron micrographs. Fig. 46. *Fragilaria shiloi* (scale bar = 2 μm). Fig. 47. Mixed isolation with two exceptionally large *Nitzschia panduriformis* (left and middle) and *N. frustulum* var. *symbiotica* (top and bottom) (scale bar = 10 μm). Figs 48, 49. Diatoms and bacteria on the surface of *Amphistorus hemprichii*. Fig. 48 shows some *Amphora* sp. in the crevice formed by the junction of test plates (at night) and some *Cocconeis* sp. (at left) on the broad surface of the test. Fig. 49 shows mucilagenous threads produced by *Mastogloea*, which anchor it in a crevice formed by the junction of test plates (scale bars = 20 μm). **Fig. 50.** Phase contrast micrograph of an egested food residium freshly deposited by *Amphistegina lobifera* (scale bar = 20 μm). **Fig. 51.** Taken with the aid of a dissection microscope with a transillumination stage. A feeding track made by a specimen of *A. lobifera* which has grazed through a field of diatoms broadcast on the bottom of a petri dish (scale bar = 1.5 μm). Figures 46 and 47 from Lee *et al.* (1989); Figs 48 and 49 from Lee and Lawrence (1990); Figs 50 and 51 from Lee *et al.* (1988b).

Table 3 Percentage of occurrence of the diatom species isolated as endosymbionts from different host species, regardless of the location, depth or season of the collection. From Lee et al. (1989). Reproduced with permission of *Micropaleontology*

	Host species					
	Amphistegina lessonii	Amphistegina lobifera	Heterostegina depressa	Borellis schlumbergerii	Operculina ammonoides	Calcarina calcar
Fragilaria shiloi	10.2	12.6	0	16.0	8.5	0
Fragilaria sp. (K)	0	1.6	0	0	0	0
Achnanthes maceneryae	2.0	0	17.7	2.1	29.8	0
Achnanthes sp. (L)	0	2.6	0	0	0	0
Cocconeis andersonii	2.0	1.3	0	0	6.4	0
Amphora roettgerii	5.4	5.0	12.1	1.1	10.6	0
Amphora tennerrima	0.8	2.0	2.1	0	0	0
Amphora erezii	1.4	2.6	6.4	5.3	0	0
Amphora sp. (J)	0	0	0.4	31.9	0	0
Entomoneis paludosa var. densistriata	0	1.0	0	0	0	0
Navicula hanseniana	3.7	4.8	0.7	1.1	0	0
Navicula reissi, N. muscatinei	4.8	0.4	0	5.3	0	0
Navicula sp. (W)	2.3	2.6	0	5.3	0	3.6
Protokeelia hottingeri	0.8	1.5	2.5	0	0	0
Nitzschia frustulum var. symbiotica	34.9	39.0	20.2	7.4	12.8	92.9
Nitzschia frustulum variety	2.3	3.0	0.7	0	0	0
Nitzschia laevis	10.0	7.8	5.0	13.8	25.5	0
Nitzschia panduriformis var. continua	18.2	10.3	25.2	6.4	6.4	3.6
Nitzschia valdestriata	1.4	1.8	7.1	4.3	0	0
Number of isolates	650	913	282	94	47	47
Percentage of total	32.3	45.3	14.0	4.7	2.3	2.3

The antisera were then incubated in various experiments with other species of endosymbiotic diatoms and with diatoms liberated from freshly crushed diatom-bearing hosts (Fig. 34). After washing away unreacted antiserum, the cells were incubated with a second antibody, a goat antibody against rabbit antibody which was conjugated with FITC (a fluorescent stain). At the conclusion of the incubation, the cells were washed free of unreacted antibody, transferred to microscope slides, and then observed in an epifluorescent microscope. By first reacting the antisera with one diatom species before reacting the same antiserum with a second species, it was possible to show that the polyvalent antisera contained different groups of antibodies some of which (1) reacted with all the pennate diatoms tested, (2) reacted only with the endosymbiotic diatoms tested and (3) reacted only with the particular species which was used as the antigen given to the rabbit (Lee et al., 1988a, and unpublished student project of D. Cooke in J.J. L.'s lab). Thus there is preliminary qualitative evidence that the endosymbiotic diatoms share some common surface antigens not found in free-living diatoms. This question begs deeper molecular probing. Tracer labelled immunoprecipitation or western blotting are reasonable quantitative next steps.

5.0 FUNCTIONAL ASPECTS OF SYMBIOSIS

A number of studies have been directed towards testing the nutrient requirements of axenic cultures of endosymbionts to see if they have any unusual requirements which might be interpretable in terms of endosymbiont needs and regulation. One of the earliest studies was on the physiology of *Chlamydomonas hedleyi* (Lee et al., 1974). The symbiont was isolated from *Archaias angulatus*, which is extremely abundant in the shallow water (<5 m) of the Florida Keys. Monosaccharides and vitamins were tested as potential growth factors or stimulants. A mixture of purines, nucleotides, nucleosides, various amino acids, urea, NH_4^+ and NO_3^- were tested as nitrogen sources. Inorganic PO_4^{3-}, sodium glycerol phosphate and nucleotides were tested as P sources. It was reasoned that some of the organic metabolites and vitamins might be derived from organisms eaten and digested by the hosts (bacteria and non-symbiotic algae) and then passed to the symbionts. Except for thiamine, none of the vitamins or monosaccharides tested stimulated the growth of *C. hedleyi*. An absolute requirement for thiamine could not be demonstrated after six serial transfers without the vitamin, but growth was more than tripled in its presence.

In some experiments, biotin also stimulated growth when the medium was supplemented with glutamic acid (1 mM), histidine (1–10 mM) and methionine (1–10 mM). Although growth was boosted by as much as 50% by these nutrients, none of them was an absolute requirement. Urea (20 μM) was the best nitrogen source. Growth was only 50% as good on NO_3^- (20 μM); similar growth was obtained at the optimum level of NH_4^+ (2 μM). Purines and pyrimidines did not serve as nitrogen sources. The optimum PO_4^{3-} concentration was 0.1 μM when urea and NH_4^- were the nitrogen sources and 1 μM when NO_3^- was used. Although we do not know anything about the nutrient levels in the microhabitat of *Thalassia testudinum* stems where *Archaias angulatus* lives, the results of the experiments suggest that the growth of the endosymbiont alga *C. hedleyi* could be stimulated by nutrients derived secondarily from its host's food.

A similar study of *Chlamydomonas provasolii* from *Cyclorbiculina compressa* and *Symbiodinium* sp. from *Sorites marginalis* from the same Florida habitat as *Archaias angulatus* also showed that vitamins could stimulate the growth of the endosymbiotic algae (Lee et al., 1979b). Growth of *C. provasolii* was doubled in the presence of 1.0 μM biotin and tripled in the presence of 0.1 μM vitamin B_{12}. Thiamine was not stimulatory. *Symbiodinium* sp. from *Sorites* had absolute requirements for vitamin B_{12}, biotin (0.01 μM) and thiamine ($> \sim 10$ μM). The optimum phosphate concentration (as $Na_2HPO_4.7H_2O$) for both isolates was ~ 0.1 mM, a very high level indeed. This suggests that available phosphate is always probably limiting the growth of these two symbionts (Lee et al., 1979b). Both NO_3^- and NH_4^+ ($< \sim 0.2$ mM) served as nitrogen sources.

Nutritional studies of the endosymbiotic diatom clones *Nitzschia panduriformis* (isolates 24 and 17), *N. laevis*, *N. frustulum*, *N. valdestriata*, *Amphora tenerrima*, *Navicula reissii* and *Fragilaria shiloi*, suggested that all of the isolates required thiamine because the clones failed to grow on the second or third transfer to media without it (Lee et al., 1980a). Biotin (0.01 μM) stimulated the growth of six of the eight clones tested. Only the clone of *Nitzschia frustulum* was stimulated by vitamin B_{12} (0.1 μM). The optimum concentration of NO_3^- varied greatly among the clones tested. One of the clones of *N. panduriformis* (#24) had the lowest optimal value (2 μM). The optimum for four other clones tested – *N. laevis*, *N. frustulum*, *N. valdestriata* and *F. shiloi* – was three orders of magnitude higher (2 mM). At the depths in the Gulf of Elat where most of the foraminifera were captured, the concentration of NO_2^- and NO_3^- rarely exceeds 1 μg/l (Levanon-Spanier et al., 1979), a condition which suggests that either the foraminifera/algal symbiotic systems with the lowest N requirements must sequester and tightly recycle bound nitrogen or that the systems are constantly nitrogen-limited. The profile for phosphate (as sodium glycerol phosphate) was similar. *Nitzschia valdestriata* and *Amphora tenerrima* had the lowest (1 μM) requirements for optimal growth, while *N. panduriformis* (#17), *N. laevis* and *Navicula reissi* had the highest (100 μM). At the same depths, the level of phosphate in the Gulf of Elat rarely exceeds 0.3 μg/l (Levanon-Spanier et al., 1979).

Recent related studies aimed at understanding how phytoplankton respond and adapt to periodic nitrate supplies in oligotrophic waters have relevance to foraminiferal/symbiont systems. It was shown that a N-limited population of the diatom *Phaeodactylum tricornutum* growing at a steady state has a tremendous affinity for nitrate. This diatom seems able to take up very low nitrate concentrations at rates greater than the steady-state growth rate (Raimbault and Gentilhomme, 1990). Thus it is sequestering scarce fixed nitrogen to satisfy future needs, and it seems to do so not only during a short initial period but also during the entire period of nitrate consumption, if the cell density is low. Initial transient uptake rates obtained after nitrate additions of 0.1 μM were quite near the V_{max} rate estimated for concentrations of nitrate >5 μM. The authors interpreted their results as a mechanism by which phytoplankton growing in nutrient-depleted areas could adjust to ephemeral patches or other pulse supply processes. By extension, one can imagine larger foraminiferal/algal systems responding to the nutrient pulses brought into the benthos of the same waters.

Field observations, field and laboratory experiments (Zmiri et al., 1974; Lee et al., 1980c, d), and growth experiments (Röttger, 1972, 1976; Hallock, 1981) of intact host–symbiont systems suggest that there are optimal ranges of light intensity and spectral quality for system function (reviewed below). The responses of four species of diatom endosymbionts – *Nitzschia valdestriata, N. laevis, N. panduriformis* and *Fragilaria shiloi* – were studied in the laboratory (M.J. Lee et al., 1982). Two of the diatoms, *F. shiloi* and *N. laevis*, isolated from *Amphistegina lessonii*, grew fastest at high light intensities (312 μW/cm^2). The other two species, *N. valdestriata* and *N. panduriformis*, isolated from *Heterostegina depressa*, grew best at low light intensities (19 μW/cm^2). Photosynthetic rates measured by respirometry suggested that the diatoms were photoinhibited at high light intensities and did well at moderately low light intensities (175 μW/cm^2). The photocompensation points (the light level at which primary production balances respiration) of all four diatoms tested were at a light level which approximated 2% of the light measured in the spring in the Red

Sea at Elat at a depth of 1 m. If the algae were free, the photocompensation depth would be reached between 40 and 50 m at Elat. The shells of the host must attenuate the light to a certain degree, and therefore the photocompensation depth must be reached at depths of 40 m or less in the waters of Elat. Changes in chlorophyll a and c concentrations were also noted when the test diatoms were grown at different light levels.

This type of laboratory data fits in well with observed distributions of larger foraminifera in the field (Leutenegger, 1983). Species with chlorophyte endosymbionts (e.g. *Archaias angulatus* and *Cyclorbiculina compressa*) are found only in relatively shallow waters (reviewed in Leutenegger, 1983; Reiss and Hottinger, 1984). Zoochlorellal pigments are more effective in harvesting light at long wavelengths, and because these are selectively attenuated in the upper levels of the water column, chlorophyte-bearing species would be poorly adapted for greater depths. Peneroplids with red algal symbionts with pigments that are effective in harvesting light energy at short wavelengths have a much wider depth range. The depth ranges of diatom-bearing species are quite interesting. Some species (e.g. *Heterostegina depressa* and *Operculina ammonoides*) seem to occupy wide ranges of depth, but when examined carefully this statement needs qualification. *H. depressa* is found on the shady sides of tide pools in Hawaii (Röttger, 1972). Near Elat on the Red Sea, it and *O. ammonoides* are often found a few millimetres below the surface of the sediments between *Halophila stipulacea* patches and on the surface of shaded sediments beneath the patches (Reiss and Hottinger, 1984; J.J.L., personal observation). Thus for these species the light range is narrower than what might be presumed by considering only the light-filtering properties of the water column. *Amphistegina* spp., which also seem to have a wide depth range distribution, do not seem to seek shade at depths of 10–15 m or more. (One of the authors, J.J.L., has not found them in great numbers at shallower depths). These living sands are found on the illuminated surfaces of algae, macrophytes and sediments. Other diatom-bearing larger foraminifera seem to have more restricted depth ranges (reviewed in Leutenegger, 1983; Reiss and Hottinger, 1984). Some species (e.g. *Baculogypsina sphaerulata* and *Calcarina calcar*) are found only in shallow waters, whereas others (*Heterocyclina tuberculata* and *Cyclopeus carpenteri*) are only found in deeper (>70 m) waters.

It is not yet clear whether the specificities of the diatoms are the primary factors affecting the distributions of their hosts. On the basis of fine structure of the chloroplasts and pyrenoids, Leutenegger (1983) recognized three diatom groups which correlated with depth distribution patterns. The actual species involved or their photophysiology could not be inferred by this approach. Some additional information on this question emerged from the isolation programme reported earlier (Lee et al., 1989). Significant numbers of some diatom species (e.g. *Nitzschia frustulum* var. *symbiotica*, *N. laevis* and *N. panduriformis* var. *continua*) were isolated from hosts collected from every depth sampled in the study. On the other hand, *Fragilaria shiloi* was rarely isolated from hosts collected at depths greater than 25 m, and three other species – *Achnanthes maceneryae*, *Protokeelia hottingeri* and *Amphora* sp. (J) – were only collcted from hosts at depths greater than 25 m. The field data on the distribution of *F. shiloi* correlated well with the laboratory growth experiments and photosynthetic rate study mentioned earlier (M.J. Lee et al., 1982). Growth and photophysiological studies on additional diatom species could significantly contribute to our greater understanding of the distribution patterns of diatom-bearing larger foraminifera.

6.0 BEHAVIOURAL STUDIES

Several simple experimental studies have shown phototaxis in symbiont-bearing larger foraminifera and photoresponse in the symbiont distribution of planktonic species. *Amphistegina lessonii* was phototaxic at photonic fluxes between 10^{11} and 10^{12} photons/cm^2/s and unresponsive at lower light levels (Zmiri et al., 1974). The action spectrum for this response peaked near 500 nm. *A. lobifera* was positively phototaxic at an incident illumination between 0.1 and 1.0 klx and negatively phototaxic at higher light levels: ~ 4 klx (<6 klx >1.7 klx) to 20 klx (Lee et al., 1980a). *Amphisorus hemprichii* was positively phototaxic at some point between 11 and 22 klx and <6 klx. The investigators noted that the phototaxic response in *A. hemprichii* was stronger than its feeding response.

Light is a triggering response for algal farming in a number of planktonic species: *Orbulina universa*, *Globigerinoides ruber*, *G. sacculifer* and *G. conglobatus*. In nature, the symbiotic algae, which are non-motile dinoflagellates or chrysophytes, emerge from the shell in the pseudopods at dawn and are moved by cytoplasmic flow to the distal parts of the spines and the rhizopodial net. In the evening, they are withdrawn (Anderson and Bé, 1976; Bé and Hudson, 1977; Bé et al., 1977). This adaptation maximizes the exposure of the algae to light and nutrients. Bé and his associates (1977) demonstrated that light was the trigger for this phenomenon in a very simple experiment. The foraminifera were kept in the dark for several hours and then illuminated under a microscope. Upon exposure to the light, the algae emerged from the shell and were withdrawn again when the light was switched off.

The algal symbionts of planktonic foraminifera are enclosed in perialgal vacuoles within the intratest cytoplasm and in the network of rhizopodia surrounding the test. Algal-containing vacuoles are moved by host cytoplasmic streaming and exhibit a predictable pattern of withdrawal in the evening and dispersal outwards into the peripheral halo of rhizopodia during daylight, usually before dawn. The symbionts are coccoid and lack flagella. They are moved by the host and not by their own motility. It is not known why the algal symbionts are withdrawn into the test at night. This may provide protection for the symbionts and/or enhance transfer of symbiont photosynthetic products to the host. There is fine structural evidence of dense deposits of organic matter near the surface of the symbiont, and in direct contact with the perialgal membrane, when the symbiont is in the intratest cytoplasm (Anderson and Bé, 1976), suggesting possible secretion of organics to the host. The control of symbiont movement is entrained to a diel light cycle. If the normal cycle is reversed by placing the foraminifera in darkness during the day and illuminating them at night, the pattern of symbiont ingress and egress becomes reversed relative to the natural cycle, and entrainment to the new light/dark cycle usually occurs within 24 h. Thus, the pattern of cytoplasmic streaming controlling the distribution of the symbionts appears to be triggered by environmental light/dark events rather than an innate physiological rhythm, and may be mediated by symbiont physiological rhythms entrained to the illumination cycle. There is no fine structural evidence for a light-receptive organelle in the host cytoplasm; however, dispersed light-harvesting molecules may be present that cannot be resolved by electron microscopy. On the whole, most evidence points towards the symbionts as triggering the light/dark distribution cycle. When the foraminifera are treated with a photosynthetic inhibitor (DCMU) that suppresses the light-harvesting capacity of the alga, symbionts remain sequestered within the test cytoplasm as though they were in continuous darkness (Bé et al., 1982). We do not know the mechanism of symbiont–host communication in the regulation of these cyclical events, and a significant contribution to cellular physiology undoubtedly could be made by elucidating this mechanism as a model of a more generalized control mechanism in cells.

This interesting aspect of their biology led Jørgensen and his co-workers (1985) to measure photosynthesis in the intact *Globigerinoides sacculifer* host–symbiont system. A micromanipulator was used to place a very fine needle microelectrode into various parts of the symbiont halo during shifts from light to dark. The microelectrode had a very high spatial resolution (10 μM), quick response times (0.2 s) and sensitivity to oxygen. The photocompensation level for the host symbiont system was 26–30 $\mu E/m^2/s$. Light saturation (I_k) was reached at 160–170 $\mu E/m^2/s$, at which level the gross photosynthetic rate of each foraminifer was 18.1 nM O_2/h. The action spectrum of the system peaked at 450 and 670 nm corresponding to chlorophyll a and 500–550 nm corresponding to peridinin. The authors calculated that the high rate of photosynthesis by the symbiotic algae could theoretically provide all the organic carbon needed by the system and that the hosts would be limited in the uptake of dissolved nitrogen and phosphorus. This condition made the capture of prey necessary as a source for the nutrients.

Freshly collected specimens of the planktonic foraminifer *Orbulina universa* had photosynthetic characteristics that suggested that their endosymbionts were sun-adapted: $I_k = 386\ \mu E/m^2/s$; a high light-saturated photosynthetic rate of $1.72 \times 10^6\ \mu$M C/symbiotic alga/h; no susceptibility to photoinhibition at intensities up to 800 $\mu E/m^2/s$. The light-saturated photosynthetic rates in the afternoon were 10 times higher than those in the evening and early morning (Spero and Parker, 1985).

7.0 THE SIGNIFICANCE OF LIGHT ON THE VITALITY AND GROWTH OF LARGER BENTHIC AND PLANKTONIC FORAMINIFERA

Several different approaches have been used to demonstrate the different functional aspects of light on the host–symbiont relationships in larger foraminifera. The principal interrelated questions asked have been those related to: (1) the absolute need for endosymbionts by the foraminifera; (2) the relationship of light to growth and calcification; (3) the role of endosymbionts in the carbon budgets and nutrition of their hosts; and (4) the mechanisms or means of carbon transfer.

It has not been easy to separate the effects of light on vitality from that of light on growth, because nutritional needs of the host foraminifera have remained ill-defined. Within particular light ranges, the growth of each of the symbiont-bearing foraminiferal species brought into the laboratory for study, has been shown to be correlated with light (Röttger, 1972, 1974; Muller, 1978; Caron et al., 1981). *Heterostegina depressa* was the first larger foraminifera to be studied in detail for this effect (Röttger, 1972). This diatom-bearing species survives and grows well in the absence of any obvious concentration of food if it is incubated in the light. Steady, but slow growth was observed at an irradiance of 45 lx; optimal growth was observed at an irradiance of 300 lx. Above that level (300–2000 lx), the organisms grew more slowly. This work was extended and expanded to another species, *Amphistegina lessonii* (Röttger et al., 1980). Both *H. depressa* and *A. lessonii* grew best at 600–800 lx (12–16 $\mu E/m^2/s$). Most important was the demonstration that *A. lessonii* would not grow in the dark even in the presence of autoclaved "mashed *Cladophora socialis*", unspecified "detritus of unicellular algae, bacteria, protozoa and fungi" or "yeast". Starved dark controls also did not grow but had slightly higher levels of protein at the end of the experiment. Another research team (Hallock et al., 1986) confirmed that the maximum growth rate of clones of *A. lessonii* from Hawaii was attained at light levels between 14 and 40 $\mu E/m^2/s$. Individuals grown at the highest light intensity tested (40 $\mu E/m^2/s$) were pale or mottled at the end of the experiment, whereas those grown at lower light intensities were more normal in appearance. Very slow growth rates were obtained at 1.6 $\mu E/m^2/s$. The light saturation level for Caribbean clones of *Amphistegina gibbosa* was also between 14 and 40 $\mu E/m^2/s$. Populations grown at 6.4 and 3.9 $\mu E/m^2/s$ had lighter, thinner and flatter tests than those grown at higher light intensities.

In *Archaias angulatus*, carbon fixation by the symbionts is proportional to light flux over a range of 0–200 $\mu E/m^2/s$ (Duguay and Taylor, 1978). Saturation was reached at 380 $\mu E/m^2/s$.

In shorter, in situ experiments over 9 days, the growth rates of *Amphistegina lobifera* and *Amphisorus hemprichii* were an order of magnitude higher in the light than in dark plastic-wrapped cages (ter Kuile and Erez, 1984). The growth rate of *A. lobifera* increased with depth from 3.3% per day at 10 m to 6% per day at 20 and 30 m and declined to 4% per day at 35 m. Visual examination suggested that more algae were present in cages incubated at 20 m than at 10 m.

Refining the technique further, ^{14}C tracer feeding experiments were used to study the uptake and assimilation of different species of unicellular algae by the same two species (*Amphistegina lobifera* and *Amphisorus hemprichii*), and *Peneroplis planatus* (Lee et al., 1988b, in press a; Faber and Lee, 1991). Once the algal species eaten and assimilated by the foraminifera were identified, they were used in growth experiments. *P. planatus* ingested *Cocconeis placentula* and *Amphora* sp. (*Halamphora* group) at a rate which was five times greater than it did other species of algae tested (Faber and Lee, 1991). *P. planatus* would not grow in culture when it was starved, and grew very slowly when it was in darkness but fed. Thus this red algal symbiont-bearing foraminifer requires both light and food for growth and survival. This species grew well between 30 and 50 $\mu E/m^2/s$, but grew larger and formed more chambers when the light was increased to 200–400 $\mu E/m^2/s$ (Faber and Lee, 1991). Growth of both *A. lobifera* and *A. hemprichii* was enhanced on diets of particular mixtures of algae. Unfed controls of *A. lobifera* grew as well or better than those fed unialgal diets. However, even when fed weekly and supplied with nutrients, neither species survived in the dark. All fed individuals of *A. hemprichii* died after 8 weeks of incubation in the dark. Fed *A. lobifera* survived longer in the dark, but all were dead after 13 weeks (Lee et al., in press b). In the same experiments, *A. hemprichii* grew faster when the medium was changed weekly. Both species withdraw nitrate and phosphate (*A. lobifera*: 0.78 nM NO_3^-, 0.099 nM PO_4^{3-}/foraminifer/day; *A. hemprichii*: 1.2 nM NO_3^-, 0.817 nM PO_4^{3-}/foraminifer/day) from the media when grown in chemostats.

Light also affects the growth and reproduction of planktonic foraminifera and endosymbionts. When specimens of *Globigerinoides sacculifer* were cultured at 400–500 $\mu E/m^2/s$, they formed more chambers and lived longer than specimens cultured at a second, much lower light intensity (20–50 $\mu E/m^2/s$: Caron et al., 1981). Shell growth rates were unaffected by increasing light intensity, but gametogenesis was delayed. When individual specimens were larger than 250 μm, continuous darkness induced gametogenesis. Feeding frequency had a greater effect than light intensity on growth and reproduction. When specimens of *G. sacculifer* were rendered aposymbiotic by treating them for 72 h with a photosynthetic inhibitor DCMU, they behaved as though they had been incubated in the dark (Bé et al., 1982). Symbiont elimination resulted in earlier gametogenesis (shortened survival time) and smaller shell sizes.

There were differences in the growth and longevity of *Globigerinella aequilateralis* when specimens with the two different, mutually exclusive symbiont types were compared in the laboratory (Faber et al., 1989). Overall growth rates were greater in specimens harbouring chrysophyte type 1. Both the light and feeding regimens affected growth and reproduction.

Specimens bearing type 1 symbionts, which were fed brine shrimp, lived longer but formed chambers more slowly than those bearing type II at the highest light intensity tested (100–200 $\mu E/m^2/s$). These differences were less dramatic in specimens cultured at a lower light intensity (50 $\mu E/m^2/s$). Type II-bearing specimens appeared to be more capable of growth and reproduction without being fed than those with type I symbionts (Faber et al., 1989). Feeding seemed more important in *G. aequilateralis* than it did in *G. sacculifer* because all the treatments which were fed, either in the light or dark, grew and reproduced. Darkness as a factor did not induce gametogenesis in the former species.

8.0 PRIMARY PRODUCTION AND CALCIFICATION

The photosynthetic rates of algae associated with foraminifera are comparable (within the same order of magnitude) to the rates found in algal associations in radiolaria and anemones (Spero and Parker, 1985; Gastrich and Bartha, 1988). A large adult *Orbulina universa* has approximately 2×10^4 symbionts and would be expected to fix 3.4×10^{-2} μM C/h at the light-saturated photosynthetic rate found by Spero and Parker (1985) for this species. Using estimates of surface oceanic abundances of planktonic foraminifera obtained by others, and assuming that the data for *O. universa* is representative of symbiont-bearing planktonic foraminifera as a whole, Spero and Parker (1985) estimated that a single planktonic foraminifera would contribute ~0.2% of the fixed carbon in a cubic metre volume of sea water. This would mean that in most oligotrophic waters with foraminiferal densities between 1 and 5 specimens/m^3 and individuals >200 μm, symbiont productivity would approach 1%. In habitats where foraminiferal patches are denser (e.g. Gulf of Elat, 130 specimens/m^3) the planktonic foraminifera may account for as much as 25% of the primary production (Spero and Parker, 1985). Using a different incubation approach, Gastrich and Bartha (1988) obtained data on the primary productivity of *Globigerinoides ruber* that was in agreement with the data obtained on *O. universa* (Spero and Parker, 1985). The new data on radiocarbon incorporation of *G. ruber* was highly variable and showed little statistical correlation with cell diameter or chlorophyll a content. The authors interpreted this variance as a reflection of the unknown age, health status and feeding history of the specimens they collected.

Primary production, as measured by ^{14}C incorporation, in two large benthic species of *Amphistegina* (*A. lobifera* and *A. lessonii*) was 2.9×10^{-5} and 1.95×10^{-5} mg ^{14}C/h/foraminifer respectively (Muller, 1978; Hallock, 1981). Similar studies of *Marginopora vertebralis* measured an incorporation rate of 0.05 ng C/min at Lizard Island, Queensland (Smith and Wiebe, 1977). At a light saturation of 380 μE/m^2/s, the rates of total carbon fixed into the organic and shell fractions of *Archaias angulatus* was 170 ng C/mg dry wt/h as measured in 2- and 4-h incubations (Duguay and Taylor, 1978). The weight specific amount of radiocarbon incorporated by *Sorites marginalis* at Bimini Harbor was approximately three times that obtained for *A. angulatus* (Duguay,

1983). The rates of total carbon incorporated by *Cyclorbiculina compressa* collected at Bimini approximated the rate of *A. angulatus* (Duguay, 1983).

A number of authors have observed that light enhances calcification rates. In three Caribbean species – *Archaias angulatus, Sorites marginalis* and *Cyclorbiculina compressa* – calcification rates in the light were approximately 2–3 times those in specimens incubated in the dark (Duguay, 1983). In the laboratory, the calcification rates of *A. angulatus* and *S. marginalis* were proportional to light intensity in the range 0–200 $\mu E/m^2/s$ and saturated at higher light intensities.

In a series of experiments, ter Kuile, Erez and collaborators (ter Kuile and Erez, 1987; ter Kuile *et al.*, 1987, 1989a, b; see also Chapter 4 in this book) have found evidence that there are differences in the uptake, kinetics and internal carbon cycling of *Amphistegina lobifera* and *Amphisorus hemprichii*. By extension, they suggest their evidence is fundamentally applicable to differences between perforate and imperforate larger foraminifera. Using ^{14}C tracer techniques, they found that diffusion limits the total uptake of inorganic carbon in *A. lobifera*. Radionuclide pulse-chase experiments enabled them to demonstrate the transfer of photosynthetically acquired ^{14}C into the skeleton of *A. lobifera* but not in *A. hemprichii* (ter Kuile and Erez, 1987). Their work suggests that perforate species have a large internal inorganic carbon pool which serves mainly for calcification and possibly also for photosynthesis, while imperforate species take up carbon for calcification indirectly from sea water or have a very small inorganic carbon pool (ter Kuile and Erez, 1987). Kinetic experiments on the uptake of inorganic carbon in the presence and absence of DCMU (which inhibits photosystem II), and in the presence of carbonic anhydrase (the enzyme which catalyses $HCO^{3-} \rightarrow CO_2 + H_2O$), suggested that in *A. lobifera* there was competition between photosynthesis and calcification for inorganic carbon (ter Kuile *et al.*, 1989a). Because there is little or no internal inorganic carbon pool in *A. hemprichii*, diffusion of inorganic carbon to the sites of calcification is the rate-limiting step for this process in this larger foraminifer (ter Kuile *et al.*, 1989b).

9.0 THE ROLES OF FEEDING IN THE METABOLISM OF SYMBIONT-BEARING FORAMINIFERA

Although all foraminifera seem to have the enzymatic capacity to feed and digest food in their rhizopodial nets (Lee *et al.*, 1991), the relative roles of feeding in the nutritional budgets of symbiont-bearing forms varies widely. As mentioned earlier in this chapter, feeding may serve only a minor role as a source of energy and organic carbon compounds in some of the larger foraminifera (e.g. *Heterostegina*: Röttger, 1972; Röttger *et al.*, 1980). In these species, food may supply needed metabolites which are not synthesized within the association (e.g. vitamins) and may serve as a source of fixed N and P. At the other extreme, it is clear that feeding is a major contributor of organic compounds and energy to the planktonic foraminiferal host–symbiont systems (Anderson and Bé, 1976; Anderson *et al.*, 1979; Bé *et al.*, 1981; Spindler *et al.*, 1984).

Light and transmission electron microscopy were used to examine freshly collected specimens of eight species of spinose planktonic foraminifera (*Hastigerina pelagica, Globigerinella aequilateralis, Globigerinoides conglobatus, G. ruber, G. sacculifer, Globigerina falconensis, Globigerinita glutinata*) and six non-spinose species (*Globoquadrina dutertrei, Globorotalia hirsuta, G. inflata, G. menardii, G. truncatulinoides* and *Pulleniatina obliquiloculata*). All of them had recognizable organisms within their food vesicles and all, with the possible exception of *H. pelagica*, are omnivorous (Anderson *et al.*, 1979). *H. pelagica* may be exclusively carnivorous or feed very infrequently on phytoplankton. Diatoms and dinoflagellates were the most commonly consumed phytoplankton but other types of algae were noted in the food vacuoles (Anderson *et al.*, 1979). In laboratory studies, *Globorotalia truncatulinoides* survived longest (21.6 days) on a diet of *Emiliani huxleyi* and did less well on monoalgal diets of either *Thalassiosira pseudonana* (a diatom) (11.4 days) or *Gymnodinium* sp. (a dinoflagellate) (11.4 days). Starved, illuminated controls lived 16.9 days (Anderson *et al.*, 1979). When *Hastigerina pelagica* was fed on *Artemia salina* nauplii in the laboratory, median survival time (26.8 days) was longest when the foraminifera were fed every 6 days; but this datum seems misleading because when the foraminifera were fed *Artemia* nauplii daily, many began gametogenesis, thus ending the life of the original organism (Anderson *et al.*, 1979). Building on these data, Spindler and co-workers (1984) concluded that spinose planktonic foraminifera feed mainly on copepods and other zooplankton. In contrast, they concluded that the non-spinose species feed mainly on diatoms and other phytoplankton. Spinose species accepted most of the calanoid copepod species offered to the foraminifera as food and digested them within 7–9 h. The acceptance rates varied with the species tested. *Globigineroides sacculifer* had a calanoid copepod acceptance rate of 80%, whereas *G. ruber* accepted only 6%. The cyclopoid copepods tested were less well accepted. *Orbulina universa* accepted 33% of the cyclopoid copepods offered but none were accepted by *G. ruber* (Spindler et al., 1984). The digestion times were longer (9–20+ h). Harpaticoid copepods were accepted and digested by only two species (*O. universa* and *G. sacculifer*) of the five that were tested. These researchers concluded that in nature spinose planktonic foraminifera capture a zooplankter (copepods, amphipods, ostracods, crustacean larvae, tintinnids, radiolarians, polychaete and gastropod larvae, heteropods, pteropods or tunicates) about every 24 h.

Feeding and food quality are also important in the nutrition and carbon budgets of some of the species of larger foraminifera which have been studied in this respect. Respirometric and tracer feeding techniques were used to study the importance of feeding in two Caribbean species: *Archaias angulatus* and *Sorites marginalis* (Lee and Bock, 1976). Fine structural studies of freshly collected *S. marginalis* suggested at the onset that this species was an active algal eater (Müller-Merz and Lee, 1976). The algae used in tracer feeding studies were either organisms which had previously been found to be excellent food organisms for a variety of littoral benthic foraminifera (*Phaeodactylum tricornutum*, *Nitzschia acicularis*, *Dunaliella salina*, *Chlamydomonas hedleyi*), or diatoms co-occurring in the habitat (*Achnanthes haukiana*, *Cocconeis placentula*, *Mastogloea* sp., *Amphora coffeaeformis*) isolated from the blades of *Thalassia testudinum*, which had heavy populations of the two species of foraminifera. Tracer feeding studies clearly indicated that feeding was the more important process at midday. The ratio of carbon gain in both foraminiferal/symbiont systems by feeding exceeded primary production by $\geqslant 10:1$. the rate of primary production was greater in *A. angulatus* than it was in *S. marginalis* (Lee and Bock, 1976). With the exception of one alga (*Mastogloea* sp.), which was eaten by both, the diets of the two species of foraminifera hardly overlapped. *Archaias angulatus* consumed twice the amount of *C. hedleyi* and *C. placentula* and ten times the amount of *A. coffeaeformis* than did *S. marginalis*. On the other hand, *S. marginalis* ate seven times the amount of *D. salina* than did *A. angulatus*. Feeding seemed to enhance carbon fixation in both species, but no direct correlation was found. The pulse-chase tracer feeding methodology was used recently to study the red algal-bearing species *Peneroplis planatus* from the Gulf of Elat (Faber and Lee, 1991). Again the algae used in the experiments were either those previously found to be good food for benthic foraminifera or isolated locally (in this case they were epiphytes on *Halophila stipulacea*). Again selectivity in feeding was noted. *Cocconeis placentula* were ingested at a rate five times faster than that of other species tested. After 24 h, almost 100% of the carbon ingested by eating 6 of the 10 species was assimilated. *P. planatus* did not grow when starved, but incubated in the light. It was concluded that this species, in common with the other imperforate species studied (*A. angulatus, S. marginalis* and *A. hemprichii*), acquires much of its energy for growth from food and cannot grow solely on organic compounds produced by its endosymbiotic algae, even though light is required by the host–symbiont system (Faber and Lee, 1991).

Radionuclide tracer methodology and microscopic observations were also used to make a detailed appraisal of feeding, and factors which might affect feeding, egestion and aspects of assimilated carbon flow in two species of larger foraminifera, *Amphistegina lobifera* and *Amphisorus hemprichii*, from the Red Sea (Lee et al., 1988b; Figs 50–52). Locally isolated algae were present in excess (Fig. 51). Although both species of foraminifera consumed some of all the species present in the experimental dishes, again selectivity was observed. The number of algae eaten per milligram by *A. hemprichii* during a 3-h exposure varied from a low of 354 per mg of *Amphora bigibba* to a high of 13 028 per mg of *Amphora* sp. Both species of *Amphora* have approximately the same biomass and are weakly silicified. A similar differential uptake (× 8) was noticed among *Chlorella* clones. Three times more *Entomoneis* was taken up than *Cocconeis placentula* or *Navicula* sp. *Amphistegina lobifera* also ate fewer *Amphora bigibba* (533 per mg) than it did other algal species (*Amphora* sp., 6057 mg per foraminifer; *Chlorella* sp., 6967 mg per foraminifer). This foraminifer also ingested eight times less of one clone of *Chlorella* and less of *C. placentula* and *Navicula* sp. than it did *Entomoneis* sp. The feeding rates of both species of foraminifera were approximately the same in the light and in the dark. In both species, feeding and egestion were episodic. Significant fractions of the carbon ingested by both foraminiferal species is egested after 24 h (Fig. 51). The characteristics of the food are one factor affecting egestion. Neither species of foraminifera seemed able to digest the cell envelope of *Chlorella* sp.; its remains appeared as empty ghosts in the egesta. Overall, *A. hemprichii* fed more than *A. lobifera*. It consumed approximately 5% of its organic weight in a 3-h feeding episode compared to 2.5% by *A. lobifera*. Approximately 4% of the labelled carbon in food was eventually recovered in the skeletal fraction. By contrast, only 1.4% of the carbon ingested by *A. lobifera* was recovered in the skeletal fraction (Lee et al., 1988b). In related experiments, the addition of inorganic phosphate and nitrate stimulated *A. lobifera* to grow five times faster than starved specimens, but fed specimens grew slightly faster (ter Kuile et al., 1987). At the same time, specimens of A. hemprichii which were in media enriched with nitrate and phosphate grew only at twice the rate of starved specimens. Fed specimens grew at a rate four times faster than starved specimens in this short (2-week) experiment. These observations of A. hemprichii agree with those reported above. Each of the imperforate species which has been studied in this respect acquires a significant fraction of its energy and organic compounds from food. In A. lobifera, ingested food serves mainly as a source of nutrients (ter Kuile et al., 1987).

10.0 MECHANISMS OF CARBON FLOW AND OTHER HOST–ENDOSYMBIONT INTERACTIONS

A number of cytological and fine structural studies have provided static circumstantial evidence that there are several possible pathways for the transfer of algal metabolites to the foraminifera. The cytoplasm of both Sorites marginalis and Cyclorbiculina compressa, harvested from Key Largo Sound, Florida, contained large numbers of starch grains (see Müller-Merz and Lee, 1976, and figures from the same study in Lee and McEnery, 1983). The endosymbionts in both hosts were filled with starch grains which resembled those found in the host (Fig. 54). Starch grains were also found in the cytoplasm of Peneroplis planatus collected in the Mediterranean Sea (Leutenegger, 1977a). It is reasonable to assume that they were synthesized and accumulated by the symbionts and then transferred to the host cytoplasm. How? One could postulate that the starch grains are either exocytosed by the symbionts, or that the symbionts autolyse, or that they are digested by the host which releases the starch grains into its cytoplasm. On the other hand, the starch grains may be residua of digested algae remaining because amylase is not present in the host.

Recent fine structural studies of Peneroplis planatus and its red algal endosymbiont, Porphyridium purpureum, provided static circumstantial evidence for still another potential pathway for organic carbon to flow from endosymbionts to their host. The symbiosis in Peneroplis is unusual because the endosymbionts are not separated from the host cytoplasm by a symbiosome membrane, as are all other foraminiferal endosymbionts, but lie free in the cytoplasm (Hawkins and Lee, 1990; Lee, 1990). In culture, the endosymbionts are surrounded by a heavy fibrous polysaccharide sheath, but when the symbionts are in situ the sheath is reduced or absent (Figs 32, 53). Fine structural studies suggest

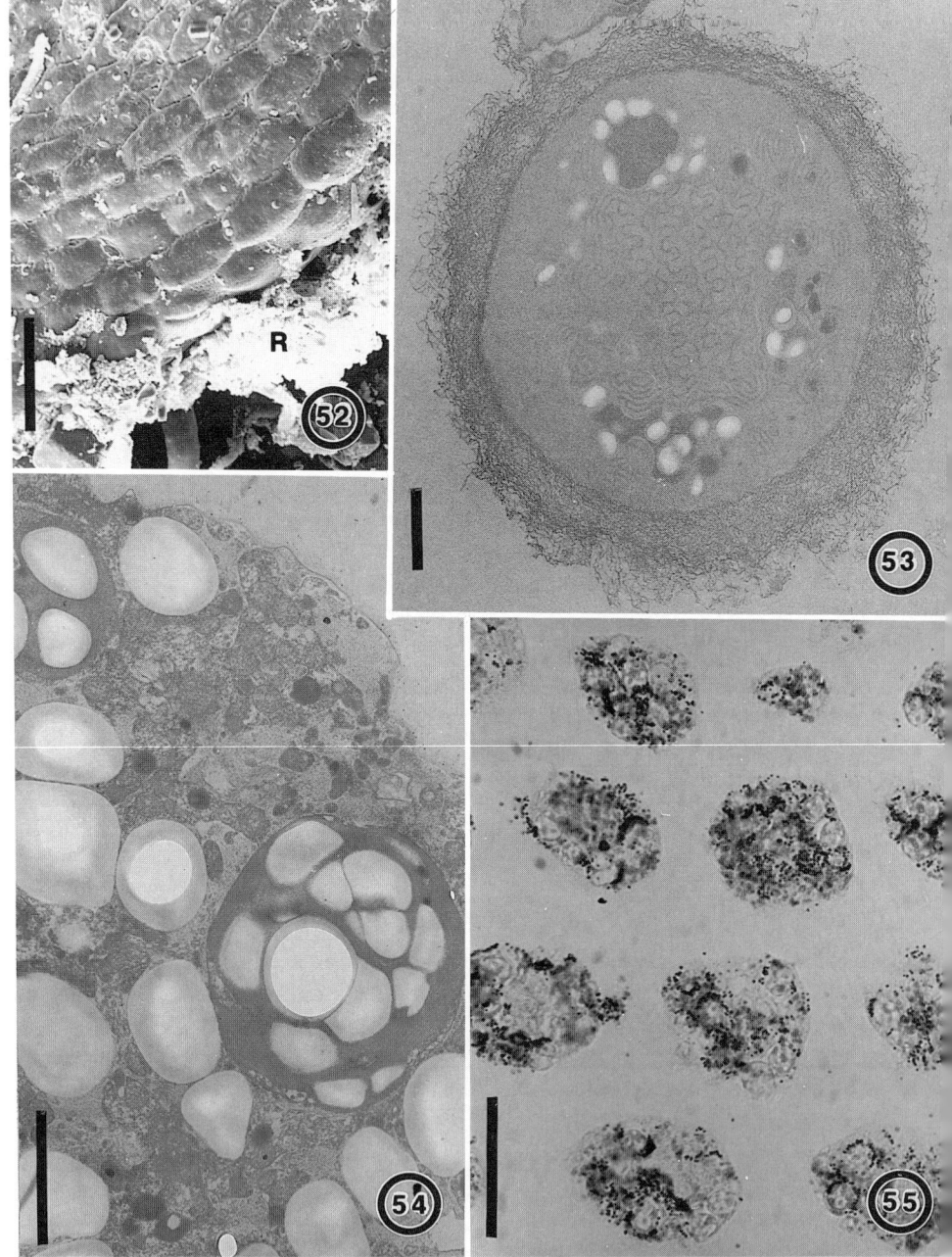

that the sheath material may be produced by the symbionts in their hosts and that sheath fibre digestion might be a pathway of energy transfer from symbionts to their hosts. The presence of starch grains in the symbionts, but not in the host cytoplasm, might also be construed as evidence supporting this hypothesis.

Fine structural evidence is at best static and circumstantial. Continued and more detailed fine studies could determine how starch grains get from the endosymbionts to the host, but they will not give us a dynamic picture of rates of transfer unless they are part of a pulse-label study.

Radioautographs of *Amphistegina lobifera* and *Amphisorus hemprichii* fed ^{14}C-labelled food in pulse-labelled experiments provided evidence of carbon transfer from the host to its symbionts (Fig. 55). After feeding the foraminifera with tracer-labelled food, they were then incubated for an additional 24 h with unlabelled food. The foraminifera were then fixed, prepared for cytological sectioning, sectioned and incubated with photographic emulsion. The symbionts were heavily labelled after 24 h, showing passage of the ^{14}C label from the food through the host to the symbionts (Lee et al., 1988b). This type of experiment does not demonstrate the nature of the carbon transferred; it could be a metabolized and respired label (CO_2, HCO_3^-) or it could be organic metabolites released from the digestive vacuoles.

It is possible that algal symbionts release dissolved metabolites to their hosts. *Chlamydomonas hedleyi* from *Archaias angulatus* releases large quantities of a metabolite when it is grown in axenic culture (Lee et al., 1974). At the end of log-phase growth (10 days), more of the ^{14}C tracer (57%) that had been bound into organic compounds was found in the medium, as mannitol, than was found in the cells. *Chlamydomonas provasolii*, from *Cyclorbiculina compressa*, under the same conditions released only 7% of its photosynthate (Saks, unpublished).

Fig. 52. Scanning electron micrograph of a portion of the test of *Amphisorus hemprichii* showing mass of food residua (R) and debris at periphery of shell (scale bar = 100 μm).
Fig. 53. Transmission electron micrograph of *Porphyridium purpureum* from *Peneroplis pertusus* grown in liquid culture. Note the thick fibrous sheath surrounding the cell (scale bar = 1 μm). **Fig. 54.** Transmission electron micrograph of *Chlamydomonas provasolii* and numerous starch grains in the cytoplasm of *Cyclorbiculina compressa* (scale bar = 10 μm).
Fig. 55. Autoradiograph of a histochemical preparation. Developed silver grains in the photographic emulsion above histological sections of *Amphisorus hemprichii* which were fed ^{14}C-labelled food. The silver grains above the endosymbionts indicates that some of the ^{14}C-labelled carbon which entered the foram as food transferred to the endosymbionts during the experimental incubation period (scale bar = 40 μm). Fig. 53 from Hawkins and Lee (1990); Fig. 54 from Müller-Merz and Lee (1976); Fig. 55 from Lee et al. (1988b).

Kremer and co-workers (1980) used a $H^{14}CO_3^-$ tracer to study the major photosynthates in six larger foraminiferal–algal associations. Chromatographic techniques were used to separate and identify the photosynthates in algal–host extracts. They identified floridoside (2-O-D glycerol-D-galacto pyranoside) and polyglucan in extracts of *Peneroplis arietina* and *P. pertusus*. In *Amphisorus hemprichii*, 74% of the label was in unspecified lipids and 3.5% was in glycerol. A large percentage of the label in *Amphistegina lessonii* (31%), *A. lobifera* (51%) and *Heterostegina depressa* (33%) was also found in lipids and glycerol (5, 6 and 11% respectively). Further work along these lines, with careful separation of the host cytoplasm from the algal cells by differential centrifugation or another methodology, could be quite useful in elucidating the pathways of transfer, if they are present.

Although there is a dearth of evidence (Kremer *et al.*, 1980; Koestler *et al.*, 1985; Lee, 1990), it is possible that autolysis or digestion are major pathways of carbon energy transfer from symbionts to hosts. This is a question which begs further careful examination.

It has been shown in a variety of associations between algae and invertebrates that the release of metabolites is stimulated by host tissues (reviewed by Trench, 1980). Sterile homogenates of freshly collected and crushed *Amphistegina* also affected the release of photosynthetates from the diatoms tested (endosymbiotic diatoms: *Nitzschia valdestriata, N. laevis, N. panduriformis, N. frustulum* var. *symbiotica, Amphora tenerrima, Fragilaria shiloi, Navicula hanseniana*; free-living isolate: *Amphora* sp.). In the presence of host homogenate, all the diatoms tested increased their levels of release of photosynthetate dramatically (Lee *et al.*, 1984). The increase of release ranged from 190 to 9000%. In terms of the percentage of total carbon fixed (not corrected for respiration), the most highly affected species *N. valdestriata* was stimulated to release 76% of its total photosynthetate, whereas the other species tested released from 25 to 54% of their photosynthetates. The homogenate also stimulated the release of photosynthetate from the free-living isolate tested, leading the investigators to be very cautious in interpreting their results (Lee *et al.*, 1984). Quite a number of physiological conditions and surface active molecules could induce rapid leakiness by alteration of cell membrane properties. Even more subtle factors might be operating in the host milieu. The "milking factor" in foraminiferal/endosymbiotic diatom relationships is not proved to be one which is specific between the partners. Separation and general categorization of the active factor(s) in the host homogenates, and scaling up the cultures so that the molecules released can be characterized, are the next logical steps.

Another interesting demonstration came from the same experiments. To various degrees, the host homogenate affected the formation of new frustules of the growing and dividing cells (Lee et al., 1984; Figs 56, 57, 60). *Fragilaria shiloi* was the most affected species (Figs 56, 60). New cells were spheroids and had only vestiges of frustules or none at all. It was inferred that "host substances" are probably responsible for the maintenance of the frustule-less state *in vivo*.

11.0 CHLOROPLAST HUSBANDRY

Some members (perhaps many) of three families of foraminifera (Nonionidae, Rotaliellidae and Elphiidae) are known to retain and husband chloroplasts of some of the algae they partially consume (Lopez, 1979; Lanners, in Lee, 1983; Lee et al., 1988c; Fig. 62). Perhaps other groups of foraminifera are also involved in the phenomenon, but they have not yet been shown to do so. A wider search would certainly be worthwhile. The number of chloroplasts husbanded by the nonionids and the elphiidids is quite high. At shallow stations (0–0.5 mm, and 6 m) in a Danish brackish water embayment (Limfjörden), three freshly collected species, *Elphidium williamsoni*, *E. excavatum* and *Haynesina germanica* husbanded $9.7 \pm 4.9 \times 10^3$, $1.2 \pm 0.7 \times 10^3$ and $5.2 \pm 1.6 \times 10^3$ chloroplasts per individual respectively (Lopez, 1979). The number of chloroplasts in *E. williamsoni* and *H. germanica* was the same in individuals collected in the summer or in the winter, and it was concluded that these species husband a steady-state population of chloroplasts, independent of season and population changes which may take place in the algal assemblages upon which the foraminifera are feeding. *H. germanica* (Fig. 59) collected in the summer at the Great Sippewissett salt marsh, Falmouth, Massachusetts, had fewer chloroplasts ($1.8 \pm 0.5 \times 10^2$) per individual (Lee and Lee, 1990). *Elphidium crispum* collected in the summer near Drake's Island in Plymouth Harbour, England, husbanded approximately $1.5 \pm 0.4 \times 10^2$ chloroplasts per individual (Lee and Lee, 1990). The population of *Elphidium translucens* (Fig. 58) from Towd Point marsh husbanded approximately $2.5 \pm 2.0 \times 10^2$ chloroplasts per individual. *Elphidium incertum* from Great Sippewissett salt marsh husbanded $9.2 \pm 2.0 \times 10^2$ chloroplasts per individual (Lee and Lee, 1990).

Primary production studies have shown that the chloroplasts are quite functional. In the Danish study, ^{14}C-labelled HCO_3^- was used to measure primary production. At saturating light intensity, *E. williamsoni* fixed 2.3×10^{-3} mg C/mg ash-free dry wt/h. In *H. germanica*, primary

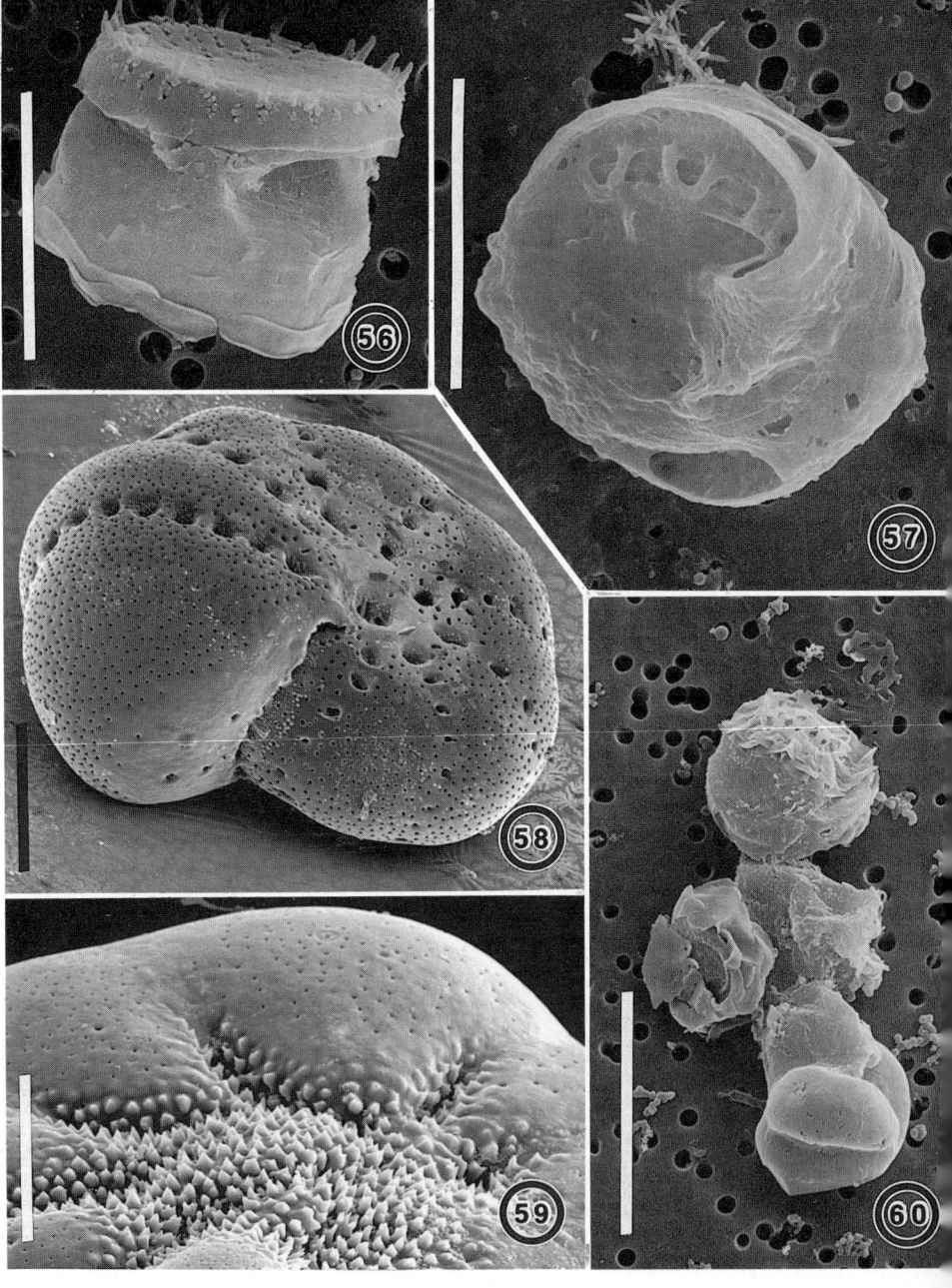

production was measured as 5.0×10^{-4} mg C/mg ash-free dry wt/h (Lopez, 1979). When *E. excavatum* was the experimental organism, there was no significant differences in the tracer uptake between specimens incubated in the light or the dark. These data suggest that the chloroplasts were not functional in the experiment. Specimens of *E. crispum* harvested at a depth of 38 m in the Red Sea fixed 1.5 µg C/mg dry wt/48 h (Lee *et al.*, 1988c). This was slightly more than specimens of *Heterostegina depressa* (1.28 µg C/mg dry wt/48 h) (Lee *et al.*, 1988c) collected at the same depth and site.

Fine structural studies and photosynthetic pigment analyses have been used to give additional information on the nature of the chloroplast donors and the relative preservation state of the chloroplasts and their pigments (Lopez, 1979; Knight and Mantoura, 1985; Lee *et al.*, 1988c). Two-dimensional thin-layer chromatography was used to separate the pigments of the Danish foraminifera. Six spots were found in the pigments extracted from *E. williamsoni* and *H. germanica* and four were found in *E. excavatum*. The spots corresponded to: (1) chlorophyll c and phaeophytin c; (2) fucoxanthin; (3) chlorophyll a; (4) xanthophyll; (5) phaeophytin a; and (6) carotenes (Lopez, 1979). *Elphidium excavatum* had only pigments 1, 4, 5 and 6. Phaeophytins are partially degraded and non-functional chlorophyll derivatives. Chlorophylls a and c are characteristic of chromophyte algae (e.g. diatoms, chrysophytes). The xanthophyll with maximum absorption at 451 nm and a shoulder at 470–475 nm was diatoxanthin or a mixture of diatoxanthin and diadinoxanthin. This suggested that the chloroplasts were of diatom origin, a finding which was in consonance with the fine structural evidence. Knight and Mantoura (1985) used HPLC to demonstrate and characterize the chloroplast retention phenomenon in additional species of foraminifera.

Scanning electron microscopy combined with the araldite embedding technique introduced to the field by Hottinger (1978, 1979) have given us new insights into the complexities of the canal systems of modern

---

Figs 56–60. Scanning electron micrographs. Fig. 56. A specimen of *Fragilaria shiloi* from a log growth phase of an axenic culture. The frustule at the top of the photograph appears normal. The rest of the cell envelope seems frustuleless. The latter was presumably formed during growth in the presence of host homogenate (scale bar = 4 µm). Fig. 57. *Nitzschia frustulum* var. *symbiotica*. A partially malformed frustule produced under the same conditions as Fig. 56 (scale bar = 4 µm). Fig. 58. *Elphidium translucens* (scale bar = 100 µm). Fig. 59. Umbilical region of *Haynesina germanica* (scale bar = 40 µm). Fig. 60. *Fragilaria shiloi* from the same study as Fig. 56. Note uppermost specimen has only fragments of a malformed frustule (scale bar = 5 µm). Figures 56, 57 and 60 from Lee *et al.* (1986). Figs 58 and 59 from Lee and Lee (1990).

elphidiids and nonionids (Hottinger, 1978; Billman et al., 1980; Hottinger and Leutenegger, 1980; Gudina and Levtchuk, 1989). If the various canals (e.g. spiral, septal, umbilical), lacunae and other morphological features have functions in terms of chloroplast husbandry, they have not yet been worked out. In view of the fact that the tiny and very simple foraminifer, *Metarotaliella parva*, also husbands chloroplasts, it is reasonable to suggest that the canal systems *per se* are not an absolute requirement for chloroplast husbandry in foraminifera. Recent cytochemical studies have shown that digestion in elphidiids takes place in the pseudopodial network in the fossettes (Lee et al., 1991; Fig. 22). The calcareous teeth or denticles lining the fossettes appear to act as combs or sieves to trap the captured diatom frustules as diatoms are being digested (Fig. 61).

Several additional questions about chloroplast husbandry immediately came to mind as soon as the phenomenon was known. How long do individual chloroplasts last within foraminifera? Is there any selectivity in the chloroplast husbandry phenomenon? Are there any factors which could affect the retention time of sequestered chloroplasts? Several studies began to address these questions. The persistence of chloroplasts in *Elphidium williamsoni* and *Haynesina germanica* from the Limfjörden was measured in starved individuals which were incubated either in total darkness or in a day/night (18 h light/6 h dark) cycle (Lopez, 1979). Every second day, a $H^{14}CO_3^-$ tracer was used to measure the photosynthetic capacity of an aliquot of cells. The chlorophyll content and numbers of chloroplasts were measured in other aliquots of the population. The chloroplasts in both species survived for a longer period of time in individuals adapted to continuous darkness than did those kept in alternating light/dark cycles. To keep a steady-state population of chloroplasts, Lopez (1979) calculated that under normal light/dark conditions, *E. williamsoni* must eat at least 65 chloroplasts/individual/h, but *H. germanica* needed to consume only 20/individual/h.

Using a slightly different approach in their studies, Lee and Lee (1990) calculated the turnover time of chloroplasts, the survival time of starved hosts and did some food quality and sequestration studies on the *Elphidium* spp. and *H. germanica* they collected. *H. germanica*, which were starved and incubated in the dark, survived only 9 weeks,

---

Fig. 61. Scanning electron micrograph of two fossettes showing diatom (D) trapped in denticles at opening (scale bar = 10 μm). **Fig. 62.** Transmission electron micrograph of a chamber showing nearly normal looking chloroplasts. Vacuolar membrane (arrow) surrounding chloroplast is visible in one specimen even at the low magnification used for this micrograph (scale bar = 2 μm). Figures 61 and 62 from Lee et al. (1988c).

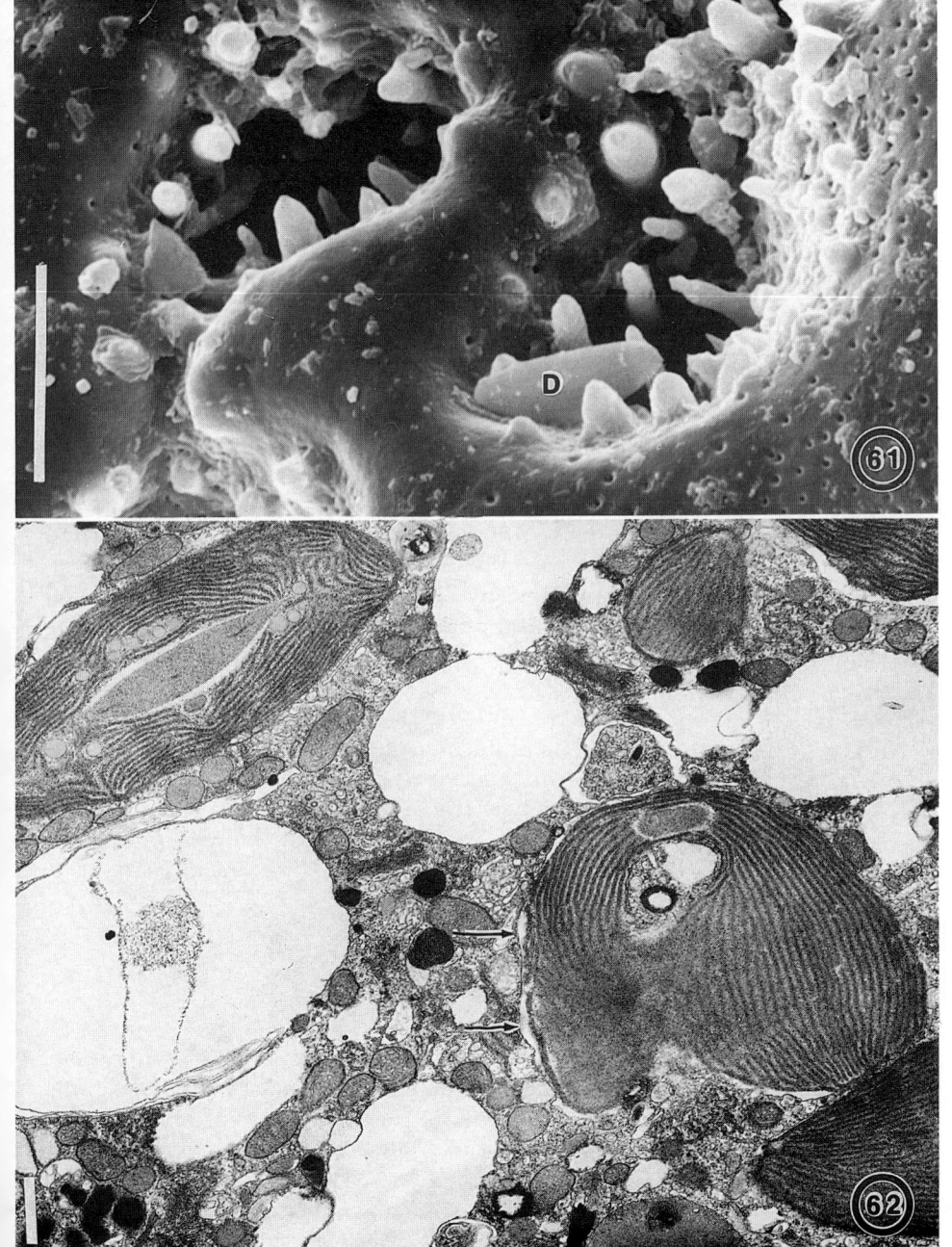

whereas those starved and incubated in the light survived 13.5 weeks. There was a steady loss of chloroplasts in the dark starved population; the chloroplast half-life was estimated as 2 weeks. The loss of chloroplasts of *H. germanica* in the light was more gradual. There was an initial drop to half the number after 6 weeks, after which the chloroplast loss became more gradual ($T_{1/2} > 10$ weeks). When *H. germanica* was fed a mixture of *Navicula menisculus* and *Amphora tenerrima*, two small pennate diatoms, it sequestered a steady-state population of 80 chloroplasts in the light and half as many when it was incubated in the dark. *Elphidium crispum* from Drake's Island rapidly lost chloroplasts in the dark ($T_{1/2} \sim 3$ weeks) and all perished after 10 weeks. In the light, there was a biphasic curve with an initial 2-week rapid loss in chloroplasts ($T_{1/2} \sim 3.5$ weeks), followed by a second phase ($T_{1/2} \sim 10$ weeks) which lasted through the sixth week when the experiment was terminated. Some quality aspects of chloroplast retention were noted in feeding experiments which were incubated in the light. Although after the first week approximately equal numbers (150 per individual) of chloroplasts were found in populations of *E. crispum* fed on monoalgal diets of either *Cocconeis placentula*, *Amphora* sp., *Entomoneis* sp., *Nitzschia subcommunis* and *Navicula* sp., the numbers gradually changed during the sixth week of incubation. By the second week, the number of chloroplasts sequestered in the population declined to 30 per individual, and by the sixth week it was down to only 10 per individual. On the other hand, the number of chloroplasts sequestered by populations fed only *Cocconeis* or *Amphora* rose to 160 and 180 per individual respectively over the same interval. The chloroplasts sequestered by populations fed only *Navicula* or *Nitzschia* were lower (100 and 140 per individual respectively) after 6 weeks. Very little difference was detected between light and dark incubation in the retention time of the chloroplasts of *Elphidium translucens* ($T_{1/2} < 2.5$ weeks) or *E. insertum* ($T_{1/2} < 1.3$ weeks).

12.0 CONCLUSIONS AND OUTLOOK

Considering the morphological complexities developed by the elphidiids and nonioniids, perhaps as adaptations to chloroplast husbandry, one wonders if there have been corresponding molecular/genetic adaptations for the phenomenon. Perhaps, as in higher plants, the genes for the synthesis of the proteins needed for chloroplast maintenance and function may not all be encoded in the chloroplast genome, but might be found in the nucleus. Is it possible that some of the nuclear–chloroplast function genes have been transferred to the nucleii of the foraminifera in these two families? The long half-life of some of the chloroplasts, and the differential qualitative aspects of the sequestration phenomenon, make this a question well worth probing. Along the same lines of thought, one wonders why the chloroplast-sequestering foraminifera don't continue to digest the chloroplasts after they have partially digested the algae in which they are organelles? What are the molecular signals and responses given by both parties involved in the phenomenon?

It is quite clear that the association of foraminifera with unicellular algae has been a long and very successful one. The phenomenon is widely spread in the photic zones of the tropics and semitropics. Underappreciated as it was some 25 years ago, the phenomenon has received a great flurry of attention since then. Yet as we look over the various topics covered in this chapter, it is quite clear that there are more unanswered questions than there were before the recent work began. It is with great excitement that we look forward to advances in this aspect of the biology of foraminifera.

REFERENCES

Anderson, O.R. and Bé, A.W.H. (1976). The ultrastructure of a planktonic foraminifer, *Globigerinoides sacculifer* (Brady), and its symbiotic dinoflagellates. *J. Foram. Res.* **6**:1–21.

Anderson, O.R., Spindler, M., Bé, A.W.H. and Hemleben, Ch. (1979). Trophic activity of planktonic foraminifera. *J. Mar. Biol. Ass. U.K.* **59**:791–799.

Bé, A.W.H. and Hutson, W.H. (1977). Ecology of planktonic foraminifera and biogeographic patterns of life and fossil assemblages in the Indian Ocean. *Micropaleontol.* **23**:369–414.

Bé, A.W.H., Hemleben, Ch., Anderson, O.R., Spindler, M., Hacunda, J. and Tuntivate-Choy, S. (1977). Laboratory and field observations of living planktonic foraminifera. *Micropaleontol.* **23**:155–179.

Bé, A.W.H., Caron, D.A. and Anderson, O.R. (1981). Effects of feeding frequency on life processes of the planktonic foraminifer *Globigerinoides sacculifer* in laboratory culture. *J. Mar. Biol. Ass. U.K.* **61**:257–277.

Bé, A.W.H., Spero, H.J. and Anderson, O.R. (1982). Effects of symbiont elimination and reinfection on the life processes of the planktonic foraminifer *Globigerinoides sacculifer. Mar. Biol.* **70**:73–86.

Berthold, W.U. (1976). Ultrastructure and function wall perforations in *Patellina corrugata* Williamson, Foraminifera. *J. Foram. Res.* **6**:22–29.

Berthold, W.U. (1978). Ultrastrukturanalyse der endoplasmatischen Alegn von *Amphistegina lessonii* d'Orbigny, Foraminifera (Protozoa) und ihre systematische Stellung. *Arch. Protistenk.* **120**:16–62.

Billman, H., Hottinger, L. and Oesterle, H. (1980). Neogene to recent rotaliid foraminifera from the Indopacific Ocean: Their canal system, their classification and their stratigraphic use. *Schweiz. Paläntol. Abh.* **101**:71–113.

Blank, R.J. (1987). Evolutionary differentiation in gymnodinoid zooxanthellae. *Ann. N.Y. Acad. Sci.* **503**:530–533.

Blank, R.J. and Trench, R.K. (1985a). Specification and symbiotic dinoflagellates. *Science* **229**:656–658.

Blank, R.J. and Trench, R.K. (1985b). *Symbiodinium microadriaticum*: A single species? In *Proceedings of the Fifth International Coral Reef Congress*, Vol. 6, pp. 113–117. New York Academy of Sciences, New York.

Brawlower, T.J. and Theirstein, H.R. (1984). Low productivity and slow deep water circulation in mid-Cretaceous oceans. *Geol.* **12**:614–618.

Caron, D.A., Bé, A.W.H. and Anderson, O.R. (1981). Effects of variations in light intensity on life processes of the planktonic foraminifer *Globigerinoides sacculifer* in laboratory culture. *J. Mar. Biol. Ass. U.K.* **62**:435–452.

Dietz-Elbraechter, G. (1971). Untersuchungen über die Zooxanthellen der Foraminifere *Heterostegina depressa* d'Orbigny 1826. *Meteor Forsch. Ergebr.* **6**:41–47.

Duguay, L.E. (1983). Comparative laboratory and field studies on calcification and carbon fixation in foraminiferal–algal associations. *J. Foram. Res.* **13**:252–261.

Duguay, L.E. and Taylor, D.L. (1978). Primary production and calcification by the soritiid foraminifera *Archaias angulatus* (Fichtel & Moll). *J. Protozool.* **25**:356–361.

Faber, W.W., Jr and Lee, J.J. (1991). Feeding and growth of the foraminifer *Peneroplis planatus* (Fichtel & Moll) Montfort. *Symbiosis* **10**:63–82.

Faber, W.W., Jr, Anderson, O.R., Lindsey, J.L. and Caron, D.A. (1988). Algal–foraminiferal symbiosis in the planktonic foraminifer *Globigerinella aequilateralis*: I. Occurrence and stability of two mutually exclusive chrysophyte endosymbionts and their ultrastructure. *J. Foram. Res.* **18**:334–343.

Faber, W.W., Jr, Anderson, O.R. and Caron, D.A. (1989). Algal–foraminiferal symbiosis in the planktonic foraminifer *Globigerinella aequilateralis*: II. Effects of two symbiont species on foraminiferal growth and longevity. *J. Foram. Res.* **19**:185–193.

Freudenthal, H.D. (1962). *Symbiodinium* gen. nov. and *S. microadriaticum* sp. nov., a zooxanthella: Taxonomy, life cycle, morphology. *J. Protozool.* **9**:45–52.

Gastrich, M.D. (1988). Ultrastructure of a new intracellular symbiotic alga found within planktonic foraminifera. *J. Phycol.* **23**:623–632.

Gastrich, M.D. and Bartha, R. (1988). Primary productivity in the planktonic foraminifer *Globigerinoides ruber*. *J. Foram. Res.* **18**:137–142.

Gudina, V.I. and Levtchuk, L.K. (1989). Fossil and modern elphidids of arctic and boreal regions: Morphology and taxonomic classification. *J. Foram. Res.* **19**:20–37.

Gudmundsson, G. (1990). Systematics of recent species of the superfamily Soritidae Ehrenberg 1939 with notes on systematic principles. Ph.D. dissertation, CUNY, University Microfilms.

Hallock, P. (1979). Trends in test shape with depth in large, symbiont-bearing foraminifera. *J. Foram. Res.* **9**:61–69.

Hallock, P. (1981). Light dependence in *Amphistegina*. *J. Foram. Res.* **11**:40–46.

Hallock, P. (1985). Why are larger foraminifera large? *Paleobiol.* **11**:195–208.

Hallock, P. and Hansen, H.J. (1979). Depth distribution in *Amphistegina*: Change in lamellar thickness. *Geol. Soc. Denmark Bull.* **27**:99–104.

Hallock, P. and Schlager, W. (1986). Nutrient excess and the demise of coral reefs and carbonate platforms. *Palaios* **1**:389–398.
Hallock, P., Forward, L.B. and Hansen, H.J. (1986). Influence of environment on the test shape of *Amphistegina*. *J. Foram. Res.* **16**:224–231.
Hansen, H.J. and Buchardt, B. (1977). Depth distribution of *Amphistegina* in the Gulf of Elat. *Utrecht Microplaeontol. Bull.* **1**:225–239.
Hansen, H.J. and Dalberg, P. (1979). Symbiotic algae in milioline foraminifera: CO_2 uptake and shell adaptations. *Bull. Geol. Soc. Denmark* **28**:47–55.
Hawkins, E.K. and Lee, J.J. (1990). Fine structure of the cell surface of a cultured endosymbiont strain of *Porphyridium* sp. *Trans. Amer. Micro. Soc.* **109**:352–360.
Haynes, J.R. (1964). Symbiosis, wall structure and habitat in foraminifera. *Cushman Found. Foram. Res. Contrib.* **16**:40–43.
Hedley, R.H. (1964). The biology of foraminifera. *Int. Rev. gen. esp. Zoo.* **1**:1–45.
Hemleben, C. and Spindler, M. (1983). Recent advances in research on living planktonic foraminifera. *Utrecht Micropaleontol. Bull.* **30**:141–170.
Hottinger, L. (1977). Distribution of larger Peneroplidae, *Borelis* and Nummulitidae in the Gulf of Elat, Red Sea. *Utrecht Micropaleontol. Bull.* **15**:35–109.
Hottinger, L. (1978). Comparative anatomy of shell structures in selected foraminifera. In *Foraminifera* (eds R.H. Hedley and C.G. Adams), Vol. 3, pp. 203–266. Academic Press, London.
Hottinger, L. (1979). Araldit als Helfer in der Mikropaläontologie. *Ciba-Geigy Aspekte* **3**:1–10.
Hottinger, L. (1982). Larger foraminifera, giant cells with a historical background. *Naturwiss.* **69**:361–371.
Hottinger, L. (1983). Processes determining the distribution of larger foraminifera in space and time. *Utrecht Micropaleontol. Bull.* **30**:239–253.
Hottinger, L. and Dreher, D. (1974). Differentiation of protoplasm in Nummulitidae (Foraminifera) from Elat, Red Sea. *Mar. Biol.* **25**:41–61.
Hottinger, L. and Leutenegger, S. (1980). The structure of calcarinid foraminifera. *Schweiz. Paläont. Abh.* **101**:115–151.
Jørgensen, B.B., Erez, J., Revsbech, N.P. and Cohen, Y. (1985). Symbiotic photosynthesis in planktonic foraminifera, *Globigerinoides sacculifer* (Brady), studied with microelectrodes. *Limnol. Oceanogr.* **30**:1253–1267.
Kessler, E. (1974). Physiologishe und biochemische Beitrage zur Taxonomie der Gattung *Chlorella* (Chlorophyceae). IX. Salzresistenz als taxonomisches Merkmal. *Arch. Microbiol.* **100**:51–56.
Kessler, E. (1976). Comparative physiology, biochemistry, and the taxonomy of *Chlorella* (Chlorophycea). *Plant Sys. Evol.* **125**:129–138.
Kessler, E. (1978). Physiological and biochemical contributions to the taxonomy of the genus *Chlorella*. XII. Starch hydrolysis and a key for the identification of 13 species. *Arch. Microbiol.* **19**:13–16.
Kessler, E. and Czygan, F.-C. (1970). Physiological and biochemical contributions to the taxonomy of the genus *Chlorella*. IV. Utilization of organic nitrogen compounds. *Arch. Microbiol.* **70**:211–216.
Kessler, E. and Zweier, I. (1971). Physiological and biochemical contributions to the taxonomy of the genus *Chlorella*. V. Auxotrophic and mesotrophic species. *Arch. Microbiol.* **79**:44–48.
Klut, M.E., Bisalputra, T. and Antia, N.J. (1989). Some details of the cell surface of two marine dinoflagellates. *Bot. Mar.* **32**:89–95.

Knight, R. and Mantours, R.F.C. (1985). Chlorophyll a and carotenoid pigments in foraminifera and their symbiotic algae: Analysis by high performance liquid chromatography. *Mar. Ecol. Prog. Ser.* **23**:241–249.

Koestler, R.J., Lee, J.J., Reidy, J., Sheryll, R.P. and Xenophotos, X. (1985). Cytological investigation of digestion and re-establishment of symbiosis in the larger benthic foraminifer *Amphistegina lessonii*. *Endocytol. C. Res.* **2**:21–54.

Kremer, B.P., Schaljohann, R. and Röttger, R. (1980). Features and nutritional significance of photosynthates produced by unicellular algae symbiotic with larger foraminifera. *Mar. Ecol. Prog. Ser.* **2**:225–228.

ter Kuile, B. and Erez, J. (1984). *In situ* growth rate of experiments on the symbiont-bearing foraminifera *Amphistegina lobifera* and *Amphisorus hemprichii*. *J. Foram. Res.* **14**:262–276.

ter Kuile, B, and Erez, J. (1987). Uptake of inorganic carbon and internal carbon cycling in symbiont-bearing benthonic foraminifera. *Mar. Biol.* **94**:499–509.

ter Kuile, B, Erez, J. and Lee, J.J. (1987). The role of feeding in the metabolism of larger symbiont-bearing foraminifera. *Symbiosis* **4**:335–350.

ter Kuile, B., Erez, J. and Padan, E. (1989a). Mechanisms for the uptake of inorganic carbon by two species of symbiont-bearing foraminifera. *Mar. Biol.* **103**:241–251.

ter Kuile, B., Erez, J. and Padan, E. (1989b). Competition for inorganic carbon between photosynthesis and calcification in the symbiont-bearing foraminifer *Amphistegina lobifera*. *Mar. Biol.* **103**:253–259.

Larsen, A.R. (1976). Studies of Recent *Amphistegina*: Taxonomy and some ecological aspects. *Israel J. Earth-Sci.* **25**:1–26.

Larsen, A.R. and Drooger, C.W. (1977). Relative thickness of the test in *Amphistegina* species of the Gulf of Elat. *Utrecht Micropaleontol. Bull.* **15**:225–239.

Lee, J.J. (1980). Nutrition and physiology of the Foraminifera. In *Biochemistry and Physiology of Protozoa*, 2nd edn, Vol. 3, pp. 43–46. Academic Press, London.

Lee, J.J. (1983). Perspective on algal endosymbionts in larger foraminifera. *Int. Rev. Cytol.* **11**:49–77 (suppl.).

Lee, J.J. (1990). Fine structure of the rhodophycean *Porphyridium purpureum in situ* in *Peneroplis pertusus* (Forskol) and *P. acicularis* (Batsch) and in axenic culture. *J. Foram. Res.* **20**:162–169.

Lee, J.J. and Bock, W.D. (1976). The importance of feeding in two species of soritid foraminifera with algal symbionts. *Bull. Mar. Sci.* **26**:530–537.

Lee, J.J. and Hallock, P. (1987). Algal symbiosis as the driving force in the evolution of larger foraminifera. *Ann. N.Y. Acad. Sci.* **503**:330–347.

Lee, J.J. and Lawrence, C. (1990). Endosymbiotic dinoflagellates from the larger foraminifera *Amphisorus hemprichii* and *Sorites marginalis*. In *Endocytobiology IV* (eds P. Nardon, V. Gianinazzi-Pearson, A.M. Grenier, L. Margulis and D.C. Smith), pp. 221–223. Institut National de la Recherche Agronomique, Paris.

Lee, J.J. and Lee, R.E. (1990). Chloroplast retention in elphidids (Foraminifera). In *Endocytobiology IV* (eds P. Nardon, V. Gianinazzi-Pearson, A.M. Grenier, L. Margulis and D.C. Smith), pp. 215–220. Institut National de la Recherche Agronomique, Paris.

Lee, J.J. and McEnery, M.E. (1983). Symbiosis in foraminifera. In *Algal Symbiosis* (ed. L.J. Goff), pp. 37–68. Cambridge University Press, Cambridge.

Lee, J.J. and Reimer, C.W. (1983). Isolation and identification of endosymbiotic diatoms from larger foraminifera of the Great Barrier Reef, Australia, Makapuu Tide Pool, Oahu, Hawaii, and the Gulf oof Elat, Israel with the description of three new species *Amphora roettgerii, Navicula hanseniana* and *Nitzschia frustulum* var. *symbiotica*. In *Proceedings of the 7th International Diatom Symposium* (ed. D.G. Mann), pp. 327–343, Philadelphia, August 1982. Otto Koeltz, Koenigstein, Germany.

Lee, J.J. and Xenophotes, X. (1989). The unusual life cycle of *Navicula muscatinei*. *Diatom Res.* **4**:69–77.

Lee, J.J. and Zucker, W. (1969). Algal flagellate symbiosis in the foraminifer *Archaias*. *J. Protozool.* **16**:71–81.

Lee, J.J., Freudenthal, H., Kossoy, V. and Bé, A. (1965). Cytological observations on two planktonic foraminifera, *Globigerina bulloides* d'Orbigny, 1826, and *Globigerinoides ruber* (d'Orbigny, 1839), Cushman, 1927. *J. Protozool.* **12**:531–542.

Lee, J.J., Crockett, L.J., Hagen, J. and Stone, R. (1974). The taxonomic identity and physiological ecology of *Chlamydomas hedleyi* sp. nov., algal flagellate symbiont from the foraminifer *Archaias angulatus*. *Br. J. Phycol.* **9**:407–422.

Lee, J.J., McEnery, M.E., Kahn, E. and Schuster, F. (1979a). Symbiosis and the evolution of larger foraminifera. *Micropaleontol.* **25**:118–140.

Lee, J.J., McEnery, M.E., Shilo, M. and Reiss, Z. (1979b). Isolation and cultivation of diatom endosymbionts from larger foraminifera (Protozoa). *Nature* **280**:57–58.

Lee, J.J., McEnery, M.E. and Garrison, J. (1980a). Experimental studies of larger foraminifera and their symbionts from the Gulf of Elat on the Red Sea. *J. Foram. Res.* **10**:31–47.

Lee, J.J., McEnery, M.E., Lee, M., Reidy, J., Garrison, J. and Röttger, R. (1980b). Algal symbionts in larger foraminifera. In *Endocytobiology I* (eds W. Schwemmler and H.E.A. Schenk), pp. 113–124. Walter de Gruyter, Berlin.

Lee, J.J., McEnery, M.E., Röttger, R. and Reimer, C.W. (1980c). The isolation, culture and identification of endosymbiotic diatoms from *Heterostegina depressa* d'Orbigny and *Amphistegina lessonii* d'Orbigny (larger foraminifera) from Hawaii. *Bot. Mar.* **23**:297–302.

Lee, J.J., Reimer, C.W. and McEnery, M.E. (1980d). The identification of diatoms isolated as endosymbionts from larger foraminifera from the Gulf of Elat (Red Sea) and the description of 2 new species, *Fragilaria shiloi*, sp. nov. and *Navicula reissii* sp. nov. *Bot. Mar.* **23**:41–48.

Lee, J.J., Reidy, J. and Kessler, E. (1982a). Symbiotic *Chlorella* species from larger foraminifera. *Bot. Mar.* **25**:171–176.

Lee, M.J., Ellis, R. and Lee, J.J. (1982b). A comparative study of photoadaptation in four diatoms isolated as endosymbionts from larger foraminifera. *Mar. Biol.* **68**:193–197.

Lee, J.J., McEnery, M.E., Koestler, R.L., Lee, M.J., Reidy, J. and Shilo, M. (1983). Experimental studies of symbiont persistence in *Amphistegina lessonii*, a diatom-bearing species of larger foraminifera from the Red Sea. In *Endocytobiology II* (eds H.E.A. Schenk and W. Schwemmler), pp. 487–514. Walter de Gruyter, Berlin.

Lee, J.J., Saks, N.M., Kapioutou, F., Wilen, S.H. and Shilo, M. (1984). Effects of host cell extracts on cultures of endosymbiotic diatoms from larger foraminifera. *Mar. Biol.* **82**:113–120.

Lee, J.J., Erez, J., McEnery, M.E., Lagziel, A. and Xenophotos, X. (1986). Experiments on persistence of endosymbiotic diatoms in the larger foraminifera *Amphistegina lessonii*. *Symbiosis* **1**:211–226.

Lee, J.J., Chan, Y. and Lagziel, A. (1988a). An immunofluorescence approach toward the identification of endosymbiotic diatoms in several species of larger foraminifera. In *Immunochemical Approaches to Coastal Estuarine and Oceanographic Questions* (eds C.M. Yentsch, F.C. Mague and P.K. Hogan), pp. 230–241. Springer-Verlag, New York.

Lee, J.J., Erez, J., ter Kuile, B., Lagziel, A. and Burgos, S. (1988b). Feeding rates of two species of larger foraminifera *Amphistegina lobifera* and *Amphisorus hemprichii*, from the Gulf of Elat (Red Sea). *Symbiosis* **5**:61–102.

Lee, J.J., Lanners, E. and ter Kuile, B. (1988c). Retention of chloroplasts by the foraminifer *Elphidium crispum*. *Symbiosis* **5**:45–60.

Lee, J.J., McEnery, M.E., ter Kuile, B., Erez, J., Röttger, R., Rockwell, R.F., Faber, W.W., Jr and Lagziel, A. (1989). Identification and distribution of endosymbiotic diatoms in larger foraminifera. *Micropaleontol.* **35**:353–366.

Lee, J.J., Faber, W.W., Jr and Lee, R.E. (1991). Granular reticulopodal digestion – a possible preadaptation to benthic foraminiferal symbiosis? *Symbiosis* **10**:47–61.

Lee, J.J., Sang, K., ter Kuile, B., Strauss, E., Lee, P.J. and Faber, W.W., Jr (in press b). Nutritional and related experiments aimed at the laboratory maintenance of three species of symbiont-bearing larger foraminifera. *Mar. Biol.*

Leutenegger, S. (1977a). Ultrastructure de foraminiferes perfores et imperfores ainsi que de leurs symbiotes. *Cahiers de Micropaleontogie* **3**:1–52.

Leutenegger, S. (1977b). Ultrastructure and motility of dinophyceans symbiotic with larger, imperforated foraminifera. *Mar. Biol.* **44**:157–164.

Leutenegger, S. (1983). Specific host–symbiont relationship in larger foraminifera. *Micropaleontol.* **29**:111–125.

Leutenegger, S. (1984). Symbiosis in benthic foraminifera: Specificity and host adaptations. *J. Foram. Res.* **14**:16–35.

Leutenegger, S. and Hansen, H.J. (1979). Ultrastructural and radiotracer studies of pore-function in foraminifera. *Mar. Biol.* **54**:11–16.

Levanson-Spanier, I., Padan, E. and Reiss, Z. (1979). Primary production in a desert enclosed sea – the Gulf of Elat (Aqaba), Red Sea. *Deep Sea Res.* **26**:673–685.

Lopez, R. (1979). Algal chloroplasts in the protoplasm of three species of benthic foraminifera: Taxonomic affinity, viability and persistence. *Mar. Biol.* **53**:201–211.

McEnery, M.E. and Lee, J.J. (1981). Cytological and fine structural studies of three species of symbiont-bearing larger foraminifera from the Red Sea. *Micropaleontol.* **27**:71–83.

McLaughlin, J.J.A. and Zahl, P.A. (1959). Axenic zooxanthellae from various invertebrate hosts. *Ann. N.Y. Acad. Sci.* **77**:55–72.

Muller, P.H. (1977). Some aspects of the ecology of several large, symbiont-bearing foraminifera and their contribution to warm, shallow-water biofacies. Ph.D. dissertation, University of Hawaii, University Microfilms.

Muller, P.H. (1978). ^{14}Carbon fixation and loss in a foraminiferal–algal symbiont system. *J. Foram. Res.* **8**:35–41.

Muscatine, L. and Porter, J.W. (1977). Reef corals: Mutualistic symbioses adapted to nutrient-poor environments. *BioSci.* **27**:454–460.

Müller-Merz, E. and Lee, J.J. (1976). Symbiosis in the larger foraminiferan *Sorites marginalis* (with notes on *Archaias* spp.). *J. Protozool.* **23**:390–396.

Raimbault, P. and Gentilhomme, V. (1990). Short- and long-term responses of a marine diatom *Phaeodactylum tricornutum* to spike additions of nitrate at nanomolar levels. *J. Exp. Mar. Biol. Ecol.* **135**:161–176.

Reimer, C.W. and Lee, J.J. (1984) A new pennate diatom: *Protokeelia hottingeri* gen. et sp. nov. *Proc. Acad. Nat. Sci. Phil.* **136**:194–199.

Reimer, C.W. and Lee, J..J. (1988). New species of endosymbiotic diatoms (Bacillariophyceae) inhabiting larger foraminifera in the Gulf of Elat (Red Sea), Israel. *Proc. Acad. Nat. Sci. Phil.* **140**:339–351.

Reiss, Z. and Hottinger, L. (1984). *The Gulf of Aqaba*. Springer-Verlag, New York.

Roberts, K.R., Farmer, M.A., Schneider, R.M. and Lemoine, J.E. (1988). Microtubular cytoskeleton of *Amphidinium rhynchocephalum* (Dinophycean). *J. Phycol.* **24**:544–553.

Ross, C.A. (1974). Evolutionary and ecological significance of large calcareous foraminiferida (Protozoa), Great Barrier Reef. In *Proceedings of the Second International Coral Reef Symposium*, Vol. 1, pp. 327–333. Great Barrier Reef Committee, Brisbane.

Ross, C.A. and Ross, J.R.P. (1978). Adaptive evolution in the soritids *Marginopora* and *Amphisorus* (Foraminiferida). *Scan. Electron Microscopy* **2**:53–60.

Röttger, R. (1972). Die Bedeutung der Symbiose von *Heterostegina depressa* (Foraminifera, Nummulitidae) fur hohe Siedlungsdichteund Karbonatproduktion. *Abh. dt. Zool. Ges.* **65**:43–47.

Röttger, R. (1974). Larger foraminifera: Reproduction and early stages of development in *Heterostegina depressa*. *Mar. Biol.* **26**:5–12.

Röttger, R. (1976). Ecological observations of *Heterostegina depressa* (Foraminifera, Nummulitidae) in the laboratory and in its natural habitat. *Maritime Sediments, Spec. Pub.* **1**:75–80.

Röttger, R. and Berger, W.H. (1972). Benthic foraminifera: Morphology and growth in clone culture of *Heterostegina depressa*. *Mar. Biol.* **15**:89–94.

Röttger, R., Irwan, A., Schmaljohann, R. and Franzisket, L. (1980). Growth of the symbiont-bearing foraminifera *Amphistegina lessonii* d'Orbigny and *Heterostegina depressa* d'Orbigny (Protozoa). In *Endocytobiology I* (eds W. Schwemmler and H.E.A. Schenk), pp. 125–132. Walter de Gruyter, Berlin.

Sakai, K. and Nishihira, M. (1981). Population study of the benthic foraminifera *Baculogypsina sphaerulata* on an Okinawan reef flat and preliminary estimation of its annual production. In *Proceedings of the Fourth International Coral Reef Symposium*, Vol. 2, pp. 763–766. Marine Science Centre, University of the Philippines, Quezan City.

Schmaljohann, R. and Röttger, R. (1978). The ultrastructure and taxonomic identity of the symbiotic algae of *Heterostegina depressa* (Foraminifera: Nummunlitidae). *J. Mar. Biol. Ass. U.K.* **58**:227–237.

Schoenberg, D.A. and Trench, R.K. (1980a). Genetic variation in *Symbiodinium* (= *Gymnodinium*) *microadriaticum* Freudenthal, and specificity in its symbiosis with marine invertebrates. I. Isoenzyme and soluble protein patterns of axenic cultures of *Symbiodinium microadriaticum*. *Proc. R. Soc. Lond.* **B207**:405–427.

Schoenberg, D.A. and Trench, R.K. (1980b). Genetic variation in *Symbiodinium* (= *Gymnodinium*) *microadriaticum* Freudenthal, and specificity in its symbiosis with marine invertebrates. II. Morphological variation in *Symbiodinium microadriaticum*. *Proc. R. Soc. Lond.* **B207**:429–444.

Schoenberg, D.A. and Trench, R.K. (1980c). Genetic variation in *Symbiodinium* (= *Gymnodinium*) *microadriaticum* Freudenthal, and specificity in its symbiosis with marine invertebrates. III. Specificity and infectivity of *Symbiodinium microadriaticum*. *Proc. R. Soc. Lond.* **B207**:445–460.

Smith, D.F. and Wiebe, W.J. (1977). Rates of carbon fixation, organic carbon release and translocation in a reef building foraminifera, *Marginopora vertebralis*. *Aust. J. Mar. Freshwater Res.* **28**:311–319.

Spero, H.J. (1987). Symbiosis in the planktonic foraminifer *Orbulina universa*, and the isolation of its symbiotic dinoflagellate, *Gymnodinium beii* sp. nov. *J. Phycol.* **23**:307–317.

Spero, H.J. and Parker, S.L. (1985). Photosynthesis in the symbiotic planktonic foraminifer *Orbulina universa* and its potential contribution to oceanic primary productivity. *J. Foram. Res.* **15**:273–281.

Spindler, M. and Hemleben, Ch. (1980). Symbionts in planktonic foraminifera (protozoa). In *Endocytobiology I* (eds W. Schwemmler and H.E.A. Schenk), pp. 133–140. Walter de Gruyter, Berlin.

Spindler, M., Hemleben, Ch., Salomons, J.B. and Smit, L.P. (1984). Feeding behavior of some planktonic foraminifers in laboratory cultures. *J. Foram. Res.* **14**:237–249.

Trench, R.K. (1979). The cell biology of plant–animal symbioses. *Ann. Rev. Plant Physiol.* **30**:485–532.

Trench, R.K. (1980). Uptake, retention and function of chloroplasts in animal cells. In *Endocytobiology I* (eds W. Schwemmler and H.E.A. Schenk), pp. 703–727. Walter de Gruyter, Berlin.

Trench, R.K. (1987). Dinoflagellates in non-parasitic symbioses. In *The Biology of Dinoflagellates* (ed. F.J.R. Taylor), pp. 531–570. Blackwell, Oxford.

Trench, R.K. (1988). Specificity in dinomastigote-marine invertebrate symbioses: An evaluation of hypotheses of mechanisms in producing specificity. In *Cell to Cell Signals in Plant, Animal and Microbial Symbioses* (eds S. Scannerini, D.C. Smith, P. Bonfante-Fasolo and V. Gianinazzi-Pearson), Vol. 17, pp. 325–346. NATO ASI Series H, Springer-Verlag, Heidelberg.

Trench, R.K. and Blank, R.J. (1987). *Symbiodinium microadriaticum* Freudenthal, *S. goreauii* sp. nov., *S. kawagutii* sp. nov. and *S. pilosum* sp. nov.: Gymnodiniod dinoflagellates symbionts of marine invertebrates. *J. Phycol.* **23**:469–481.

Zmiri, A., Kahan, D., Hochstein, S. and Reiss, Z. (1974). Phototaxis and thermotaxis in some species of *Amphistegina* (Foraminifera). *J. Protozool.* **21**:133–138.

7
Ecology and distribution of benthic foraminifera

JOHN W. MURRAY

1.0	Introduction	222
2.0	Sampling methods	223
3.0	Distribution	223
	3.1 Relationship with the substrate	224
	3.2 Density and biomass	225
	3.3 Patchiness	225
	3.4 Seasonality	227
	3.5 Dominance and diversity	227
	3.6 Population dynamics and production	229
4.0	Ecology	232
	4.1 Abiotic factors	232
	4.2 Biotic factors	232
	4.3 Characteristics of major environments	233
	4.4 Dead assemblages	243
5.0	Aspects of the test	243
	5.1 Function	243
	5.2 Ecophenotypes	244
	5.3 Morphogroups	244
	5.4 Growth and ontogeny	245
	5.5 Stable isotopes	246
6.0	Some aspects of life processes	246
	6.1 Feeding strategies	246
	6.2 Length of life	248
7.0	Application to palaeoecological studies	248
8.0	Summary, conclusions and future research	249
	Acknowledgements	250
	References	250

1.0 INTRODUCTION

The term benthos is used to describe those organisms which live on the sea floor. Most benthic foraminifera are of meiofaunal size (<2 mm), but some grow to 1 cm or more in diameter and these are the largest single-celled organisms known to date. Like planktonic foraminifera, the benthic ones are sensitive to water characteristics, but their distribution patterns are also controlled by substrate type.

The geological record of benthic foraminifera shows that they range from the Cambrian to the present and that their early evolution and diversification was slow. The first forms were single-chambered and had a test wall built of detrital material held together by an organic cement – an agglutinated wall structure (sometimes termed arenaceous). The earliest multi-chambered test may be of late Ordovician age but such tests were certainly present by the Middle Devonian (Brasier, 1982a). Calcareous walls first became common in the Devonian but modern porcellaneous and hyaline (glassy) types became common from the Mesozoic.

The majority of benthic taxa have long stratigraphic ranges and because of their differing ecological requirements they are facies-dependent. Nevertheless, they are useful in biostratigraphic studies, especially within single depositional basins, and they are excellent guides to palaeoecology. There are probably around 5000 extant species.

Early studies of benthic foraminifera were necessarily taxonomic and this phase continues today. Serious ecological studies commenced in the early 1950s, and the introduction of the rose Bengal staining technique to differentiate tests with protoplasm from those lacking it (Walton, 1952) was a landmark, for it permitted a distinction between "living" and dead individuals. Most of the ecological studies have been carried out by geologists who have sought to provide a database on modern forms with which fossil forms may be compared and interpreted. Experimental work based on cultures is of limited extent. The geographical distribution of benthic foraminifera is still poorly known, but we are slowly learning something of the dynamics of the foraminifera–environment system, including the role of post-mortem effects.

Reviews of the ecology on benthic foraminifera include Murray (1973, 1991), Boltovskoy and Wright (1976) and Buzas and Sen Gupta (1982). The only comprehensive review of the relationship between living and dead assemblages is that of Murray (1976). In this chapter, the principal taxonomic unit used is the genus (because of the large number of species) and the classification followed is that of Loeblich and Tappan (1988).

The literature on modern benthic foraminifera is enormous. In this brief review, it has been necessary to keep references to a minimum.

2.0 SAMPLING METHODS

Ideally, a sediment sampler which is sealed on the sea floor should be used, e.g. piston corer, multi-corer (Holme and McIntyre, 1984) or special grabs (Murray and Murray, 1987). However, many samplers are not sealed and a loss of surface sediment takes place during collection, e.g. most corers, grabs and dredges. In the intertidal zone and in shallow waters, samples can be taken by hand with care. Samples are normally 1 cm thick and should have an area of at least 30 cm to be representative. They are usually preserved in ethanol or buffered formalin.

In the laboratory, the samples are washed on a sieve (240 mesh, 63 μm), stained with rose Bengal (1 gm in 1 litre distilled water) for 30 min, washed again on a 240 mesh sieve and dried. Alternatively, the Sudan Black B staining method may be used, but this is slower and no more reliable than rose Bengal (see Walker *et al.*, 1974). Sometimes it may be necessary to float off the foraminifera in a heavy liquid such as trichloroethylene or carbon tetrachloride, but this should only be done in a fume chamber.

If rose Bengal staining has been carried out, tests with protoplasm, assumed to be alive at the time of collection, will be deep red, whereas empty tests, dead at the time of collection, will be unstained or have a light superficial pink coloration.

To obtain living individuals, place the freshly collected sediment in a dish with ambient sea water (see also Chapter 10). Place glass slides so that they are sloping into the sediment. Benthic foraminifera are commonly negatively geotactic and will climb up the sides of the dish and also up the sloping glass slides. The latter can be placed under a microscope to examine the soft parts of the foraminifera.

3.0 DISTRIBUTION

There is only limited understanding of the biogeographic distribution of benthic foraminifera and it is beyond the scope of this chapter to discuss the details. Instead, comments are made on a number of separate but interrelated aspects.

3.1 Relationship with the substrate

Perhaps the most common substrate is sediment, but benthic foraminifera also live on firm surfaces, such as shells or rocks, and on plants, such as seaweed or seagrass, in which case they are said to be phytal or epiphytic. Not all plants are suitable substrates and *Codium* and *Fucus* rarely support foraminifera (Matera and Lee, 1972). Those living on the substrate surface are epifaunal, whereas those that live within sediment are infaunal. Many species are free-living, i.e. able to move about using their pseudopodia, but others are attached, either by their pseudopodia (clinging) or by cementation (sessile). Examples are given in Table 1, from which it can be seen that *Elphidium*, an opportunistic shallow-water genus, has more than one mode of life. The majority of forms are free-living, either infaunal or epifaunal. Only in very high-energy environments is it necessary for infaunal taxa to be sessile and in the case of *Lepidotrochammina*, which is of scale-like form, it lives attached to large sand grains and shell debris.

The relationship with the substrate, the orientation of the test, and test form are closely linked with the physical energy of the environment and with feeding strategy. Brasier (1982a) generated theoretical models to demonstrate the evolution and range of morphology and its relation-

Table 1 Examples of the relationship of benthic foraminifera with the substrate

	Epifaunal	Infaunal
Free	*Quinqueloculina*	*Ammonia*
	Triloculina	*Cassidulina*
	Elphidium	*Elphidium*
		Uvigerina
Clinging	*Rosalina*	*Trochammina*
	Hanzawaia	*Elphidium*
	Pararotalia	
	Elphidium	
Sessile	*Cibicides*	*Lepidodeuterammina*
	Rosalina	
	Nubecularia	

ships to ecology. Free-living forms span a wider range of architecture than those constrained by a sessile mode of life.

3.2 Density and biomass

Density or standing crop is the number of individuals living on a unit area of sea-floor. Values are usually related to $10\,cm^2$ or $10\,cm^3$ of surface sediment. Density values range from 0 to several thousand per $10\,cm^2$. The variation is controlled by the size of the tests and the fertility of the area. For obvious reasons, high densities tend to be of small individuals. In general, densities of <10 characterize areas of low fertility, e.g. the oligotrophic deep-sea, and values of >1000 are typical of areas of high fertility, e.g. Mississippi Delta.

Where seasonal changes in shallow-water areas have been studied, a marked cyclicity has been observed which can be related to variations in food supply, rate of reproduction, survival of juveniles, etc. An example is given in Fig. 1.

Biomass is a measure of the quantity of organic material and it can be expressed in a variety of ways, including weight (wet or dry) or volume. Because of the small size of foraminifera, it is not easy to weigh the soft parts, but it is possible to estimate the volume by approximating the shape to an oblate or prolate sphaeroid or a cone (Murray, 1973). Average values are $<1\,mm^3$ per $10\,cm^2$ of sea-floor. Biomass and density are usually linked for a given environment (see Fig. 1).

3.3 Patchiness

Ecologists recognize three patterns of distribution: random, uniform and clumped. Benthic foraminifera are typically clumped because of the role of the micro-environments and the needs or consequences of reproduction. Most data are from intertidal or very shallow waters, and they confirm that even the common species show patchy distribution patterns on a scale of <1 to several metres. For example, Buzas (1968) related the clumped distributions in Delaware Bay to asexual reproduction. The limited data from deeper water suggest that the common species are generally distributed or have very large clumps (in excess of $2000\,m^2$), but the rare species show small-scale patchiness.

Epiphytic forms are invariably clumped in certain parts of each plant, often related to the exposure of their symbionts to light, and

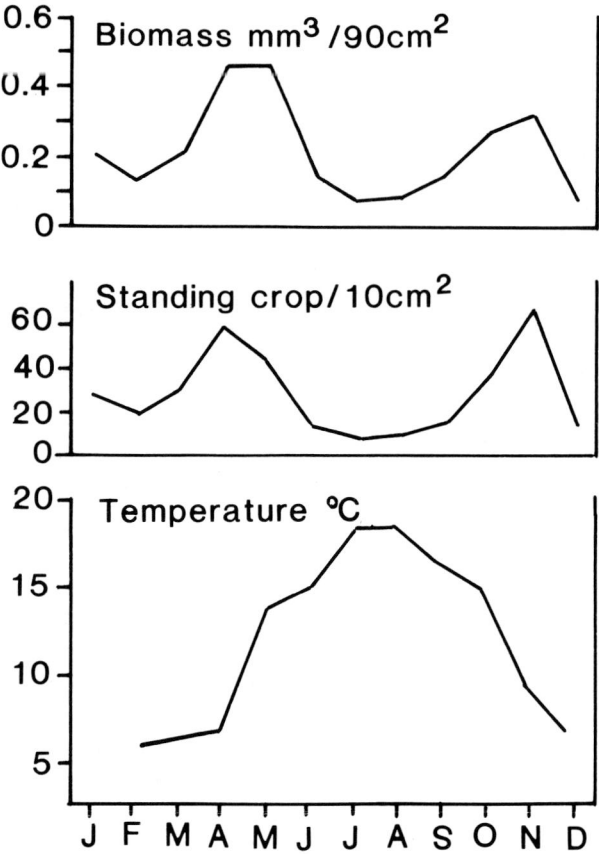

Fig. 1. Seasonal changes in biomass and density of the foraminiferal assemblages in relation to the temperature cycle in the Exe estuary, England (based on Murray, 1983).

the plants themselves may also be clumped. This leads to three-dimensional patchiness above the sea-floor, which could be an advantage to reduce competition for food.

Infaunal taxa extend several centimetres below the sediment–water interface. In one deep-sea example, 50–74% of the living assemblage was in the top 1 cm of sediment, although living forms were encountered down to the sampled depth of 5 cm (Gooday, 1986); some species prefer different infaunal depths (Corliss, 1985). In shallow waters, often only one form is dominant, e.g. *Ammobaculites* in Chesapeake Bay (Buzas, 1974) or *Fursenkoina* in the North Sea (Collison, 1980), and more individuals live at depths greater than 1 cm than in the top 1 cm (which is that part normally sampled in ecological studies).

3.4 Seasonality

Seasonal variation in density and biomass have already been discussed. In addition, in those parts of the sea subject to seasonal changes including seasonal blooms of diatoms, there may be a temporal sequence of generic replacement. For instance, a microtidal estuary that was more brackish in the winter and spring and least brackish in the autumn, was dominated by *Haynesina* in the winter, spring and summer, and by *Elphidium* in the autumn (Christchurch Harbour: Murray, 1968). Seasonality is not confined to shallow water, for in the north-east Atlantic the seasonal input of phytodetritus to the abyssal plain promotes the development of an opportunistic assemblage of *Epistominella* and *Alabaminella* (= *Eponides* of Gooday, 1988), which is scarcely present in the underlying sediment.

3.5 Dominance and diversity

In living assemblages, one species is normally more abundant than any other and is said to be dominant. Diversity refers to the number of different taxa in an assemblage. There is much debate among ecologists as how best to measure diversity in a statistically and biologically meaningful way. It is important to use a method which is not dependent on sample size, for intuitively we should expect more species in large samples. Two methods are commonly used for foraminiferal data, the α index of Fisher, Corbet and Williams, and the information function (both described in Murray, 1973). The α index can be calculated or read from a graph (Fig. 2). The information function $H(S)$ is the sum of the proportion (p = percent divided by 100) multiplied by the napierian log of the proportion of each species in the assemblage:

$$H(S) = - \sum_{i=1}^{s} p_i \ln p_i$$

The maximum value of $H(S)$ is attained when all the species in an assemblage have equal abundances.

These two measures are complimentary. The α index eliminates the effects of sample size while the information function gives an indication of the heterogeneity within an assemblage.

Two examples to show the relationship between dominance diversity and heterogeneity are given in Fig. 3. The low-diversity assemblage

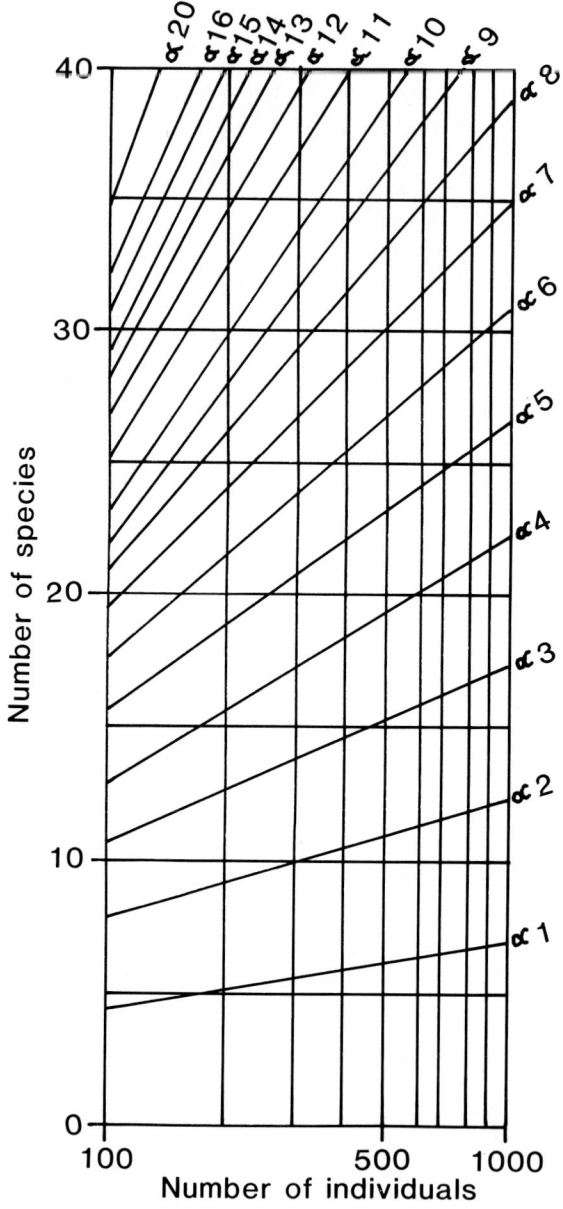

Fig. 2. Graph to show the relationship between the number of species and number of individuals in an assemblage and lines of equal α diversity index (from Murray, 1973).

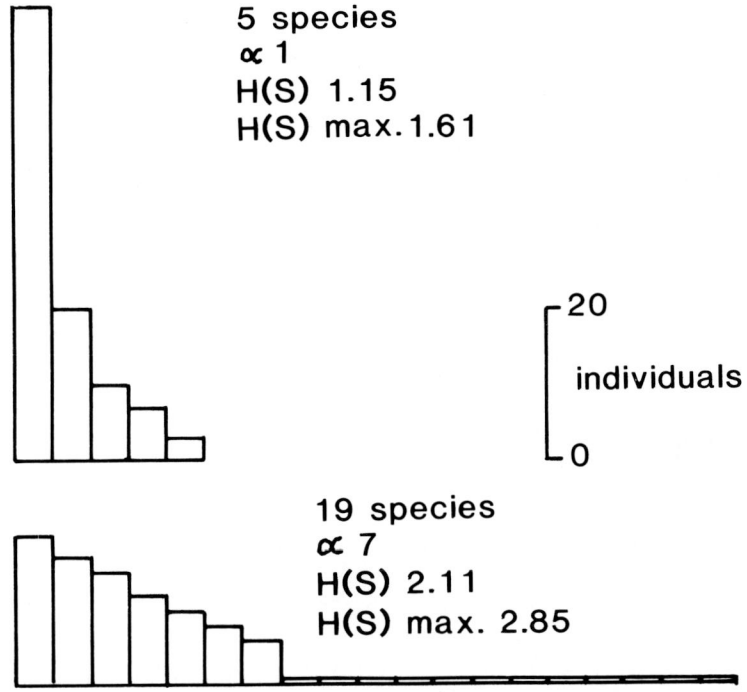

Fig. 3. Comparison of dominance and diversity in two theoretical assemblages each of 100 individuals.

has high dominance, low α and low $H(S)$. The higher-diversity assemblage has low dominance, a higher α value and higher $H(S)$. Living assemblages of benthic foraminifera show a clear pattern of α values, with the unstable, variable, marginal marine biotopes having low α values and the stable, less variable, normal marine environments having higher α values (Fig. 4).

3.6 Population dynamics and production

Each species of an assemblage has its own population of individuals. Population dynamics attempts to explain the causes of changes through time in the number of individuals of a species.

There are four parameters to be considered: birth (B), death (D), immigration and emigration, although, in practice, immigration and emigration can be discounted as foraminiferal movement is generally

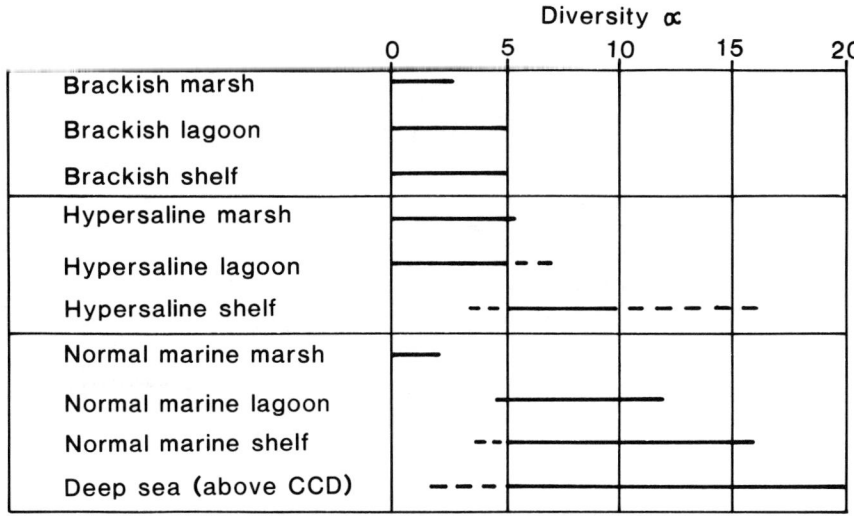

Fig. 4. Range of α diversity values for living benthic assemblages (modified from data in Murray, 1973).

slow and random. A simple population dynamics equation is:

$$N_{t+1} = N_t + B - D$$

where N_t = the number of individuals at time t, and N_{t+1} = the number of individuals one time-period later at time $t + 1$.

Of the three possible results, a population increase arises when births exceed deaths, a steady state arises when births balance deaths, and a decrease takes place when deaths exceed births. In seasonally influenced environments, there is likely to be fluctuations between increases and decreases in population both in the short- and long-term.

A population cannot increase in size indefinitely, because, as it increases, there is progressively more competition between individuals. Stability is achieved when each parent is replaced by one offspring, and this population size is termed the carrying capacity (K). Population growth is often simplified to the sigmoid logistic curve (Fig. 5).

Each phase of population growth is followed by a rapid decline (see Fig. 6) that is density-dependent due to intraspecific competition for food (diatoms), which is also density-dependent because of the limited availability of nutrients. The example given, *Nonion depressulus*, is a slightly euryhaline species living near the mouth of the Exe estuary. It is an opportunistic *r*-strategist, able to increase its numbers rapidly

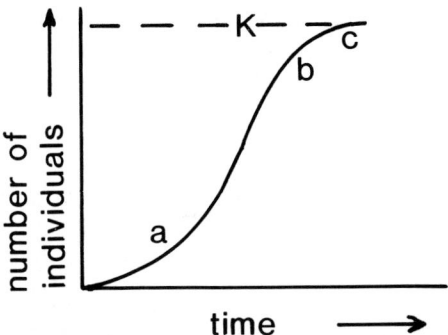

Fig. 5. A logistic curve of population growth. At a, population density is low and intraspecific competition is of minor importance. At b, population density is high and intraspecific competition causes density-dependent regulation of the net reproductive rate. At c, the population has reached the carrying capacity (K).

in response to a short-term improvement in the environment. The peak values of K vary from 45 to 65 per 10 cm^2.

Production is the number of tests contributed to a unit area of sea-floor during a chosen period of time. In all known examples, annual production greatly exceeds density observed at any one time. Most production values are between 100 and 1600 tests per 10 cm^2 per year (see Murray, 1983, for a summary of the data). Because many benthic

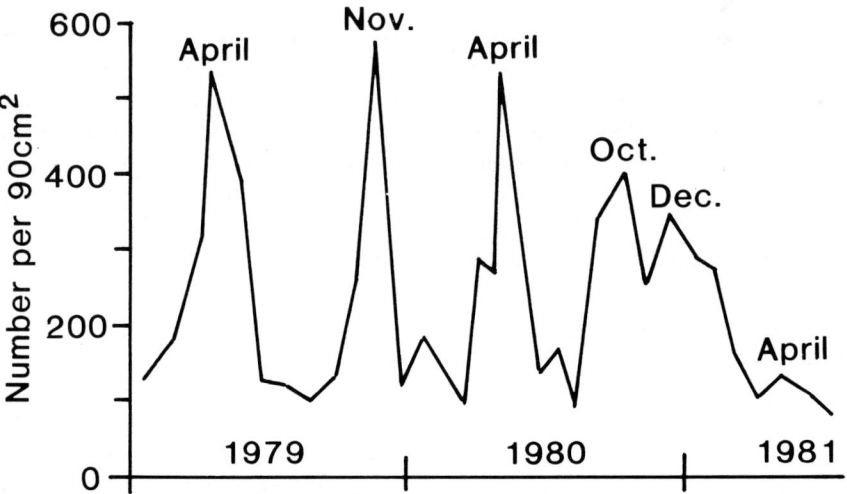

Fig. 6. Time-series data on *Nonion depressulus* (data from Murray, 1983).

foraminifera are calcareous, they contribute carbonate (mainly calcite) to sediments at a rate of 40–2800 gm/m^2/year (Hallock et al., 1986).

4.0 ECOLOGY

4.1 Abiotic factors

Physical and chemical factors which may influence the distribution of benthic foraminifera include temperature, salinity, substrate type, turbidity, light, nutrients, oxygen and tidal energy. Some factors (e.g. temperature and salinity) limit distributions, for beyond a certain threshold a given species will be unable to reproduce or may die. Others may influence abundance, e.g. the availability of food.

In many cases, these factors are interrelated and they collectively control distributions, e.g. water masses with their distinctive temperature, salinity, oxygen and carbonate saturation/undersaturation signatures. In other cases, one factor is clearly more important than any other, e.g. salinity between a shallow shelf and an adjacent estuary.

It has proved very difficult to determine the critical limiting conditions for individual species and indeed it may not be realistic to expect to do so. Except for a catastrophic event, there will inevitably be a lag between an environmental change and any response from an organism. Even if environmental parameters were measured continuously over a long period of time, it might still be difficult to define limiting conditions. This is because most species are living well within their tolerance limits but not always under optimum conditions. However, there are now sufficient data to determine which taxa are stenohaline, temperature or tropical, favour muddy low-energy environments, etc.

4.2 Biotic factors

Little is known of interspecific competition, except that it must involve space and food. These are probably interrelated in that the amount of space required per individual will decrease with increasing food supply. The epifaunal genus *Rosalina* is passive when food is abundant, but during times of shortage it actively grazes (Sliter, 1965). *Allogronia* is adversely affected by crowding, which causes reduced feeding and limits its reproductive capacity, whereas *Rosalina* and *Spiroloculina* are less affected (Muller, 1975). However, it is probable that most feeding strategies are non-competitive.

4.3 Characteristics of major environments

4.3.1 Marshes and mangals

Intertidal and supratidal vegetated areas are present in temperate latitudes where they are dominated by *Spartina, Salicornia*, etc., and in tropical–subtropical regions where they are dominated by bushes and trees collectively known as mangroves. Such areas have rather extreme environmental conditions, particularly with respect to temperature and salinity, both of which may have large diurnal and seasonal variations.

Marsh faunas are cosmopolitan. High marshes, reached only by the highest tides, have low-diversity assemblages ($\alpha < 2$) of *Trochammina* and *Jadammina* (agglutinated). Both are free-living herbivores or detritivores. *Jadammina* is epifaunal, whereas *Trochammina* is epifaunal to infaunal. High marshes are commonly brackish, but during dry summer months the pools may become hypersaline. Mid- to low marshes are brackish and have a hyaline fauna of *Elphidium* (infaunal or epifaunal, herbivore), *Haynesina* (infaunal, ? herbivore) and sometimes *Ammonia* (infaunal, ? herbivore), and agglutinated *Miliammina* (infaunal, detritivore). Densities are highly variable and range from 0 to several hundred per 10 cm^2. Marsh sediments are commonly muddy and have a high content of organic detritus; consequently, they are invariably anoxic just beneath the surface and there is partial or total post-mortem dissolution of calcareous tests in some areas (e.g. Cape Cod: Parker and Athearn, 1959) but not in others (e.g. South Texas: Phleger, 1966, Murray, 1976).

Mangrove swamps or mangals have received less attention. In Florida, *Ammonia* and *Elphidium* are the dominant forms, but in Brazil, *Arenoparrella* (agglutinated, epifaunal, herbivore) dominates with subsidiary *Miliammina* and other agglutinated genera (Zaninetti, 1979).

4.3.2 Lagoons and estuaries

These are enclosed areas showing the influence of both land and sea. The former may be the source of freshwater or it may just exert an influence on the temperature range. Estuaries are brackish and may be vertically mixed with a horizontal salinity gradient in meso- to macrotidal areas (tides >2 m) or be horizontally stratified with a landward penetrating tongue of sea water ("salt wedge") at the bottom, over which the freshwater flows seawards in micro- to mesotidal examples (tides <4 m). Lagoons may be brackish, normal marine, hypersaline or range from one extreme to the other during the course of a year. Tropical normal marine and hypersaline lagoons commonly have meadows of seagrass, which provide a substrate for epiphytic foraminifera.

Brackish lagoons and estuaries are commonly characterized by *Elphidium* (infaunal–epifaunal, herbivore), *Ammonia* and *Haynesina* (infaunal, ? herbivores) and sometimes by *Miliammina* (infaunal, detritivore), e.g. Christchurch Harbour (Murray, 1968) and San Antonio Bay (Parker et al., 1953). However, in the Chesapeake Bay area in turbid waters with organic-rich sediments, a totally different fauna is present dominated by *Ammobaculites* (infaunal, ? detritivore) (Ellison, 1972). Brackish lagoons have low diversity (Fig. 4).

Lagoons having normal marine conditions are rare, but the few known examples are dominated by *Quinqueloculina* (epifaunal, herbivore) with *Ammonia* (infaunal, ? herbivore) (see Phleger and Ewing, 1962). The diversity is similar to that of other normal marine environments (Fig. 4).

Hypersaline lagoons are confined to subtropical areas with a hot dry climate, e.g. the Arabian Gulf. The salinities are in excess of 40‰ and may reach 70‰ in the innermost parts. *Quinqueloculina* and *Triloculina* (epifaunal, herbivores) are commonly abundant. They may be accompanied by *Elphidium* (epifaunal, herbivore), *Ammonia* (infaunal, ? herbivore) and *Peneroplis* (epifaunal–epiphytic, herbivore with symbionts). Densities are very variable and may be very low on the sediment but high where seagrass is present. Diversity values are generally low (Murray, 1973).

Some lagoons are brackish at one season and hypersaline at another, and this creates very stressful conditions for the fauna. In the Texas lagoons (Phleger and Lankford, 1957), *Ammonia* (infaunal, ? herbivore) is dominant with subsidiary *Elphidium*. In Florida Bay, a region of carbonate sediment deposition, there are patches of turtle grass. The dominant form is *Quinqueloculina* (many of which are probably epiphytic) and *Sorites* is common on the broad blades of the grass (Lynts, 1962).

Using a ternary plot of wall structure, it can be seen that the different types of lagoon have distinctive fields which only partially overlap (Fig. 7). In general, the dead assemblages of lagoons and microtidal

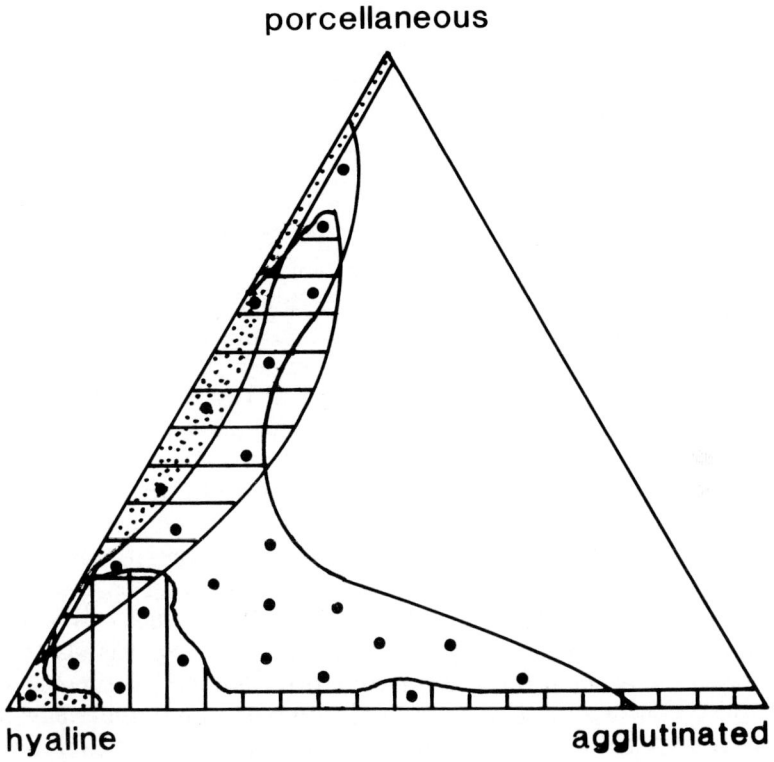

Fig. 7. Ternary plot of wall structure of lagoonal-living benthic assemblages. Small dots, hypersaline; horizontal lines, normal marine; vertical lines, brackish; large dots, brackish to hypersaline.

estuaries closely resemble the living except in those areas subject to powerful tidal currents. Meso- and macrotidal estuaries are swept by currents which transport in suspension dead tests < 200 μm in size from the shelf into the estuary where they are deposited on muddy intertidal flats (Wang and Murray, 1983). In fossil assemblages, these transported tests are easily recognized because they are small and size-sorted.

4.3.3 *Shelf seas*

Physiographically, the shelf extends from the shore to the shelf break, which may lie at any depth between ∼100 m and ∼600 m. However, in ecological terms, there is a tendency to regard the inner shelf as extending from the shore to 100 m and the outer shelf from 100 to 200 m.

It is difficult to generalize about the shelf environment as there are clearly major differences between polar and tropical examples. Some shelves have vertically mixed waters with no temperature gradient with increasing depth, but most have thermohaline stratification at least during the summer months. In this case, the surface waters become warm, often down to a depth of 40–50 m, and beneath this there is an abrupt change (at the thermocline) to much cooler and sometimes more saline waters. The significance of this is that those areas of shelf beneath a thermocline experience a very small annual range in temperature compared with adjacent shallow regions.

Although most shelves have normal marine salinities, there are exceptions, and these will be discussed first.

Some epicontinental shelves are brackish. The Baltic Sea is almost completely landlocked and it receives run-off from a great many rivers. The waters are stratified and even the most saline parts near the connection with the North Sea have salinities of only 11–22‰. The major taxa are *Elphidium*, with different species in waters of different salinity, and *Ammotium* (agglutinated, infaunal, detritivore), which occurs at the transition from the surface water layer to the bottom water, at a depth of 17–23 m in Kiel Bay (Lutze, 1965). *Ammotium* reproduces when temperatures fall below 3°C, during periods of high food supply and when the waters are more than 85% saturated with oxygen (Wefer, 1976). This is an opportunistic form that may be a recent arrival in the Baltic, for previous workers found it to be absent or very rare. Although the Baltic reaches depths of several hundred metres, it is in effect a brackish lagoon. It has very low diversity ($\alpha < 2$) and low and patchy densities of living individuals. Its waters are too cold for *Ammonia*.

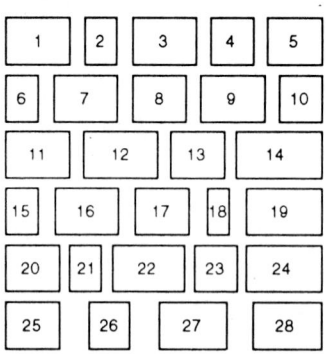

Plate 1. Benthic foraminifera. Marginal marine: 1,2,3, *Jadammina*; 4,5, *Miliammina*; 6, *Ammobaculites*; 7,8,9, *Trochammina*; 10, *Spirolina*; 11, *Elphidium*; 12,13,14, *Ammonia*; 15,16, *Haynesina*; 17,18, *Peneroplis*; 19, *Sorites*; 20,21, *Quinqueloculina*. Deep-sea: 22,23,24, *Nuttallides*; 25,26,27, *Fontbotia*; 28, *Epistominella*.

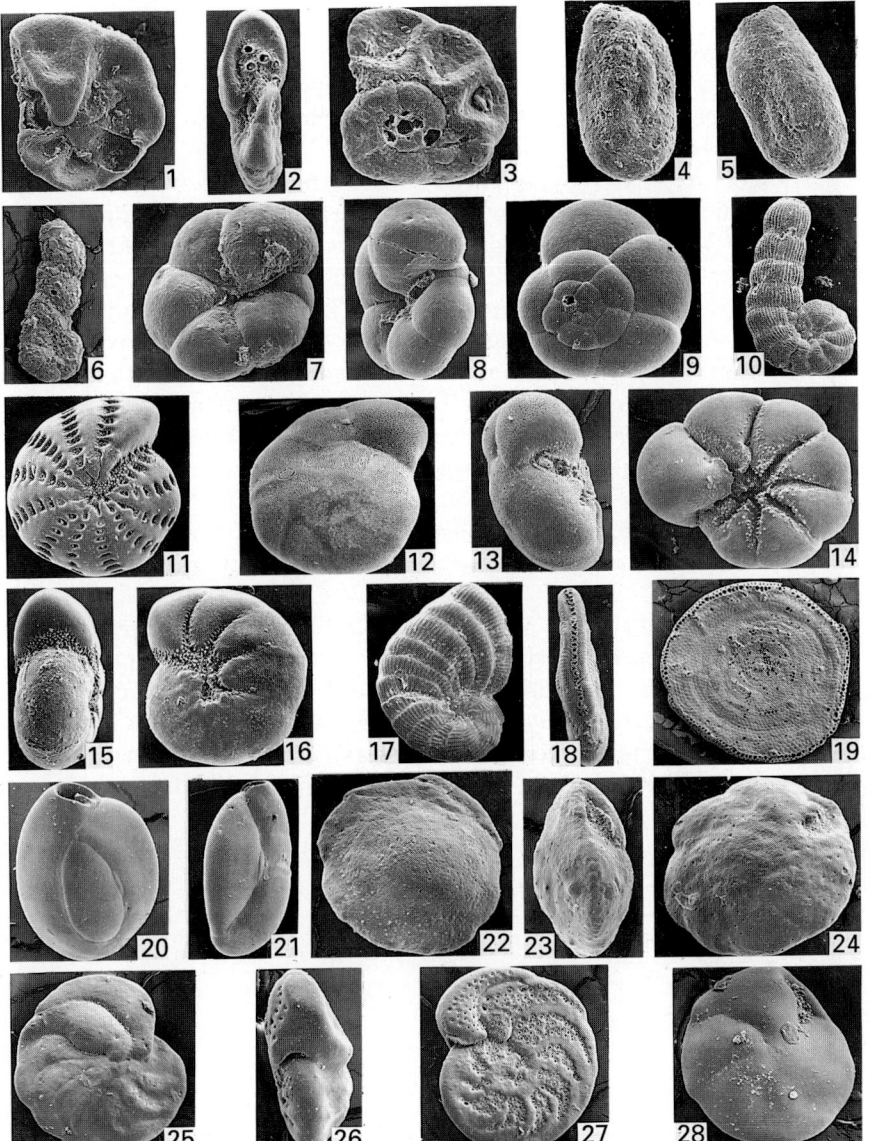

The Yellow Sea off China is semi-enclosed with salinities of 30–32‰ and an annual temperature range of 4–27°C. The dominant genus throughout is *Ammonia*, although different species characterize different sub-regions. *Elphidium* is also present. Diversity is modest [$H(S)$ 2.5–2.9 on average] and the assemblages consist mainly of hyaline forms with only a few porcellaneous and agglutinated taxa (Wang et al., 1985). Because this sea is only slightly brackish, it resembles similar nearshore areas of open shelves.

The most extensive hypersaline shelf is that of the Arabian Gulf, a land-locked area under an arid climatic regime. The salinity ranges from 38 to 40‰ over much of the Gulf, but increases to 42‰ along the Arabian shore. Low to high densities (15–1054 per 10 cm^2) have been reported from the central part. The most important genera living on muddy sediments are *Bulimina* (infaunal, detritivore) and *Nonionoides* (probably infaunal, detritivore) but numerous other genera are present and the diversity is high ($\alpha = 3$–16 but mostly $\alpha = 5$–10; data from Lutze, 1974). In sediments with coarser bioclastic material, the dead assemblages are composed of *Quinqueloculina* and *Textularia* (agglutinated with a calcareous cement). It seems likely that here these genera are epifaunal on hard substrates, e.g. bivalve shells, and that the living forms have not been adequately sampled by the techniques used.

Normal marine shelves are known from cold high latitudes to warm low latitudes. Some of the typical genera and their substrate preferences are listed in Table 2. Substrate type and the interrelated physical energy of the environment (tides, waves, currents) play an important role in determining the composition of assemblages. This can be seen from a comparative study of the Celtic Sea and English Channel which

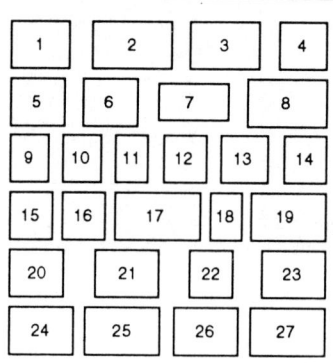

Plate 2. Benthic foraminifera. Shelf and slope: 1, *Saccammina*; 2, *Adercotryma*; 3,4, *Cribrostomoides*; 5, *Eggerella*; 6,7,8, *Textularia*; 9, *Reophax*; 10, *Bolivina*; 11, *Brizalina*; 12, *Bulimina*; 13, *Globobulimina*; 14, *Trifarina*; 15,16, *Fursenkoina*; 17,18,19, *Planorbulina*; 20, *Cancris*; 21, *Uvigerina*; 22,23, *Melonis*; 24,25, *Cassidulina*; 26,27, *Gavelinopsis*.

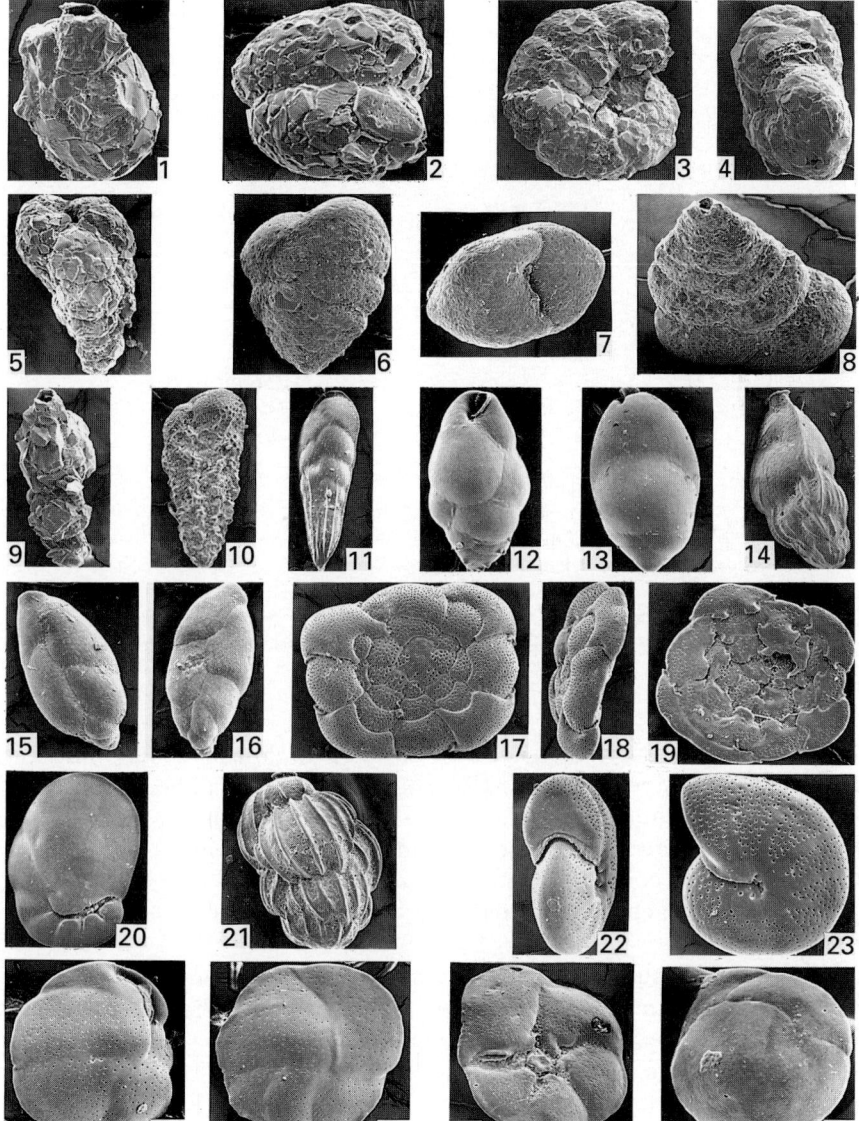

Table 2 Typical shelf genera with mode of life and feeding strategy

Wall structure	Low-energy muds, silts, fine sand	Medium- to high-energy sands, gravels
Agglutinated	Adercotryma (E,d)	Cribrostomoides (EC)
	Bigenerina	Gaudryina (EC)
	Cribrostomoides (EC)	Placopsilina (EA)
	Eggerella (I,d)	Textularia (EC)
	Reophax (I,d)	
	Saccammina (I,d)	
Porcellaneous	Cornuspira	Nubecularia (EA)
Hyaline	Bolivina (I,d)	Acervulina (EA)
	Brizalina (I,d)	Cibicides (EA)
	Buccella	Discorbis (EA)
	Bulimina (E,I)	Gavelinopsis (EA)
	Buliminella (I)	Patellina (EC)
	Cancris	Planorbulina (EA)
	Cassidulina (I,d)	Rosalina (EA)
	Fursenkoina (I)	Trifarina (I)
	Globobulimina (I)	
	Islandiella (I)	
	Nonionella (I)	
	Nonionoides (I)	
	Trifarina (I)	
	Uvigerina (I)	

Abbreviations: E, epifaunal free; EC, epifaunal clinging; EA, epifaunal attached; I, infaunal; d, detritivore.

have similar salinity, temperature and depth characteristics but low and high energy respectively. Sturrock and Murray (1981) found that the fine muddy sands of the Celtic Sea had an assemblage of free-living, infaunal or epifaunal taxa with a wide range of morphology, whereas the high-energy coarse sands with shells and pebbles from the English Channel had attached, epifaunal taxa, which commonly have a flattened form.

Fine-grained substrates sometimes have a high content of organic carbon ($>1\%$) and the principal genera are then *Bolivina, Brizalina, Bulimina, Globobulimina* or *Uvigerina*. In such sediments, the interstitial waters may have low oxygen concentrations and these genera are adapted to tolerate such conditions.

4.3.4 Bathyal and abyssal

The bathyal zone embraces the continental slope and rise, a depth range of 180–4000 m. The abyssal zone (>4000 m) includes the extensive abyssal plains. The sediments of these areas comprise a terrigenous component (clay, silt, sand) and a biogenic component which in most areas is calcareous (nannofossils = clay grade, planktonic foraminifera = sand grade) but, in others, notably in the Southern Ocean, is siliceous (diatoms, radiolaria, etc.).

Oceanic water masses are derived from two main sources: the Southern Ocean, which gives rise to Antarctic Bottom Water, and the Norwegian–Greenland Sea, which is the source of North Atlantic Deep Water. All other deep-sea bottom waters are derived from mixtures of these two or modifications of one or the other. They are characterized by their temperature, salinity and density differences, but each water mass has distinctive values for dissolved silica, saturation or undersaturation with respect to calcium carbonate, etc.

Where planktonic foraminifera are present in oceanic sediments, they generally outnumber benthic forms by 99:1. Further, the density of living benthic foraminifera is frequently low, especially beneath the oceanic central water masses. From a practical point of view, it is difficult to study living assemblages if living forms are rare. Instead, total assemblages (living plus dead) are examined.

The distribution of deep-sea benthic foraminifera has commonly been linked to the distribution of water masses, even though the differences between the latter are subtle. All the species which show such correlations are epifaunal (see Table 3). Although *Nuttallides* commonly has high dominance (50%), *Epistominella* and *Fontbotia* commonly have a maximum abundance of only ~25% and are accompanied by many other taxa as diversity is high.

It is by no means clear which attributes of a given water mass favour one or more genera. Bremer and Lohman (1982) consider that *Nuttallides* is related to the carbonate undersaturation of the water rather than to Antarctic Bottom Water *per se*. *Fontbotia* is extremely abundant in the Norwegian Sea, the source of North Atlantic Deep Water (Belanger and Streeter, 1980), but it is also the dominant genus in the Gulf of Mexico where it best correlates with the Mississippi Fan, an area of

Table 3 Correlation of bottom-water masses and benthic genera in the North Atlantic Ocean (based on Schnitker, 1980; Weston and Murray, 1983)

Genus	Water mass
Epistominella	North East Atlantic Deep Water
Fontbotia	North Atlantic Deep Water
Nuttallides	Antarctic Bottom Water

rapid sediment accumulation rich in organic detritus (Poag, 1984). As noted in Section 3.4, *Epistominella* responds to the seasonal input of phytodetritus. Thus, water mass alone is not the only ecological control in the deep sea.

Infaunal taxa such as *Melonis* and *Uvigerina* show less clear-cut relationships with water mass as might be expected. For these genera, the nature of the interstitial water is more important than that of the overlying bottom water.

Bathyal assemblages are normally fairly diverse (Fig. 4), although it must be remembered that the data include dead forms. Deep-sea sedimentation rates are slow (1 cm may represent 500 years' accumulation). The high diversity could be a product of numerous successive overlapping species patches rather than innate high living diversity. Until more detailed studies are undertaken, this will remain unresolved.

Deep-sea taxa are very long-ranging and evolution has been slow. It is probably that reproduction is almost exclusively asexual (the low density and high diversity mitigate against sexual reproduction) and this would not favour rapid evolution.

At great depth, dissolution of calcium carbonate takes place (CCD = calcite compensation depth). Below the CCD, exclusively agglutinated assemblages are found. The CCD is shallower in the Pacific Ocean than in the Atlantic and, whereas most deep-sea assemblages in the latter are calcareous, the Pacific is dominated by agglutinated assemblages.

4.3.5 Larger foraminifera

Although the majority of benthic foraminifera are small (<2 mm in length or diameter), some are large (>3 mm^3 in volume and 1 cm or more in diameter). These are collectively known informally as larger foraminifera. A few deep-water large taxa are known, but larger foraminifera proper are shallow-water dwellers, tropical in distribution and contain algal endosymbionts. Ross (1974) and Lee et al. (1979) have suggested that the evolution of larger foraminifera was in response to the needs of the algal symbionts. However, not all symbiont-bearing foraminifera are large.

Hallock (1985) used population growth models to address the problem "why are larger foraminifera large?" She concluded that in stable environments with limited food resources, delayed maturation and growth to larger size was advantageous. Where sunlight is available, algal symbiosis is also highly advantageous. Thus, possession of a large proloculus (>500 μm) greatly increases the chances of infant survival and the number of offspring is directly related to the size of the parent. Symbionts can only be passed directly from one generation to the next by asexual reproduction. In the warm, shallow waters favoured by larger foraminifera, metablic rates are high, and therefore considerable energy resources are needed just for survival. Algal symbiosis gives an energetic advantage, especially in nutrient-poor environments, for it recycles nutrients and reduces the chances of starvation (see also Chapters 3, 4 and 6). Furthermore, the heavily calcified tests deter predation. With symbionts, larger foraminifera can survive for several years, slowly accumulating the food reserves necessary for reproduction.

Larger foraminifera are highly adapted to low-latitude, shallow carbonate environments where terrestrial and seasonal influences are negligible. However, their specialized nature leads to extinction when environments change. Thus large size and adaptation to algal symbiosis have co-evolved many times during the Palaeozoic, Mesozoic and Cenozoic.

4.4 Dead assemblages

The dead assemblages preserved in bottom sediments show varying degrees of difference from the living assemblages from which they were drawn. The causes include production differences between species, seasonal succession of genera and species, loss or gain through current transport, and loss of calcareous tests through dissolution. However, notwithstanding this, in most cases, sufficient information is preserved to make a palaeoecological reconstruction possible (see Murray, 1976, for a review of this problem).

5.0 ASPECTS OF THE TEST

5.1 Function

It is clear that there must be functional advantages in possessing a test, but it is less clear as to what those advantages may be (see also Chapter 3). Marszalek *et al.* (1969) suggested five possibilities: protection against predation, shelter against unfavourable physical or chemical conditions, a receptacle for excreted matter, an aid to reproduction, and a buoyancy control.

There is evidence to support the concept of a shelter against unfavourable conditions, e.g. against certain wavelengths of light in shallow-water symbiont-bearing forms (Haynes, 1965) and against short-term osmotic changes. Otherwise, none of the other possibilities is supported by much evidence. In some cases, the morphology assists the feeding process, e.g. by raising the organism above the sediment surface in order to capture food in suspension.

Brasier (1982b) examined the architectural modifications which control the minimum line of communication between the aperture and the back of the proloculus (first chamber). He suggested that evolutionary rate, survival and lines of communication may be related, although in what way remains unclear. There have been repeated changes from longer to shorter minimum lines of communication in shallow-water tropical carbonate faunas but other forms have remained unchanged or have followed an opposite trend.

It has been suggested that the evolutionary sequence, from naked to tectinous to agglutinated to calcareous, represents "the accumulation of successively more sophisticated functions of the test through time" and that secretion of calcite was taken up "perhaps because it expended less energy than agglutination" (Brasier, 1986). Nevertheless, all wall types are still extant, so if there was any selective evolutionary advantage, it was not total.

5.2 Ecophenotypes

There is mounting evidence, especially from shallow-water environments, that populations of some species are adapted to the environmental conditions of their local habitat (ecophenotypes) and that some ecophenotypes are morphologically distinct. This has caused a great deal of taxonomic confusion where several different names have been used for the same species, e.g. as in *Elphidium*. Poag (1978) introduced the concept of paired ecophenotypes in marginal marine environments. In the calcareous forms, one ecophenotype has a small test, thin walls and few chambers (and lives under optimum conditions), whereas the other has a larger test, thick walls and more chambers (and lives under less favourable conditions).

5.3 Morphogroups

Forms showing similar growth plans and shapes may be grouped into morphogroups showing similar modes of life, feeding strategies, etc. This approach has been followed for agglutinated taxa by Jones and Charnock (1985). For example, their morphogroup A comprises tubular or branching, single- or multi-chambered tests having an erect, sessile, epifaunal or semi-infaunal life position and a suspension-feeding strategy, e.g. astrorhizids. Different environments are characterized by different morphogroups

Severin (1983) used the morphogroup approach to re-evaluate data on living assemblages from Texas. He divided the forms into six groups: straight-cylindrical, planoconvex, elongate-flattened, biconvex-keeled, tapered, and rounded-planispiral. The three lagoons had an abundance of planoconvex forms and, in two cases, also of straight-cylindrical forms. Only one of the shelf areas had common planoconvex forms and the rest were characterized by elongate-flattened, tapered or rounded-planispiral types. The advantage of such an approach is that it avoids problems of taxonomy, it is easy to separate the six groups, and it may even be possible to use an image analyser to carry out the task. It may provide a quick approach to palaeoecology, but it would clearly give a less refined interpretation than more detailed analyses.

5.4 Growth and ontogeny

Typically, growth is fast initially but then it slows to maturity (Fig. 8). Normally, reproduction would then ensue but unfavourable conditions may cause it to be deferred, in which case the mature individual may continue to grow slowly. The rate of growth is also influenced by the availability of food, temperature, salinity (Murray, 1963) and intensity

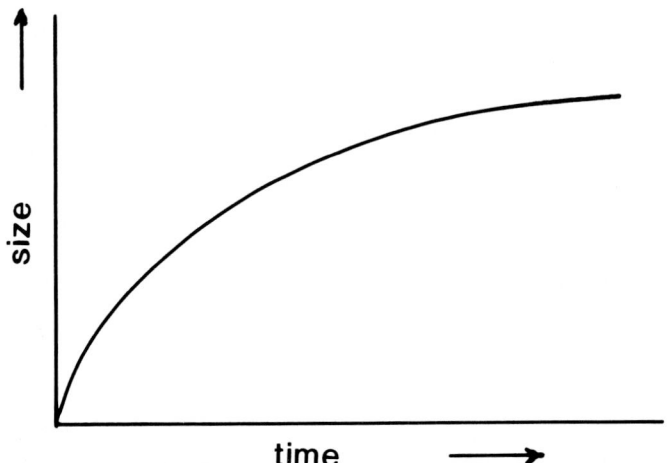

Fig. 8. A typical growth curve.

and duration of illumination (for forms with symbionts, see Röttger, 1972; Röttger et al., 1980; Hallock, 1981; see Chapters 3 and 6).

For many benthic foraminifera, the juveniles and adults are alike, although the latter, if calcareous, will have thicker walls and may be "ornamented". However, some genera show a change in growth plan during ontogeny so that juveniles and adults are dissimilar, e.g. *Spirolina*, which is planispiral when juvenile but develops a uniserial adult stage.

Some, mainly larger, foraminifera show dimorphism related to the alternation of generations (see Chapter 9). The adult gamont or A generation (product of asexual reproduction) has a large proloculus (megalospheric) but a small test, whereas the adult agamont or schizont or B generation (product of sexual reproduction) has a small proloculus (microspheric) and a large test. Asexual reproduction is more frequent than sexual reproduction, and therefore A forms greatly outnumber B forms.

5.5 Stable isotopes

The biological basis for this topic is discussed in detail in Chapter 4 and references therein. It suffices to note here that many symbiont-bearing larger foraminifera are depleted in ^{18}O and enriched or depleted in ^{13}C and vital effects play a major role in these variations. Smaller foraminifera from the shelf, slope and deep sea are enriched in ^{18}O and some show enrichment or near-equilibrium in ^{13}C. However, infaunal taxa are invariably depleted in ^{13}C and this may provide a method of determining the mode of life of extinct taxa and has been particularly applied to planktonic foraminifera (Hemleben et al., 1989; Charles, 1991; Ravelo, 1991; see also Chapter 7).

6.0 SOME ASPECTS OF LIFE PROCESSES

A detailed discussion of the various facets of cellular dynamics are given in Chapters 1-5. Certain points are highlighted here.

6.1 Feeding strategies

Benthic foraminifera live in all marine and marine-influenced environments, and part of the secret of their success is their development of many different types of feeding strategies. In addition, they must have evolved resource partitioning to minimize competition.

Herbivorous feeding is of necessity confined to those parts of the sea-floor within the photic zone. It is especially common in estuaries and lagoons. Feeding on detritus and on the bacteria associated with it can take place at any depth. In organic-rich sediments, detritus-feeding may be active not only at the surface but also infaunally, providing the interstitial pore-waters are not dysaerobic or anaerobic. Burrows made by the macrofauna take oxygen down into levels that are otherwise anoxic and provide local havens for infaunal taxa. In oligotrophic areas, the organic content of the sediment will be very low, and therefore infaunal detritivores will be rare or absent. Under quiet conditions, where the organic detritus settles slowly to the sea-floor, it may be advantageous to have an erect mode of life with pseudopodia spread out in the water to collect falling particles. This will also apply in areas subject to gentle currents, for foraminifera are only able to trap detritus brought to them as they have no means of filtering water.

Uptake of dissolved organic material may be a widespread phenomenon, although it has only been documented experimentally in a few cases (Schwab and Hofer, 1979). There are a few examples of carnivory and parasitism, but these are not important modes of feeding.

In shallow, tropical oligotrophic waters, many benthic foraminifera achieve nutrient conservation and recycling through their contained endosymbionts (Leutenegger, 1984). Because of the light requirements of their symbionts, the foraminifera have developed architectural plans best suited to achieve this. Many of the larger foraminifera are compressed and discoid for this reason, but where the tests are biconvex or inflated, the symbionts are displayed in the most appropriate parts of the test which may serve as a greenhouse (Haynes, 1965; see Chapter 3). Some symbiont-bearing foraminifera actively feed as well, and these forms are usually very successful for following this sensible strategy, e.g. *Archaias* (Lee and Bock, 1976).

Non-competitive feeding in living assemblages can be achieved by following different strategies or by feeding at different levels. For instance, in shallow water, the two dominant species may be a herbivore and a detritivore. Alternatively, detritivores may feed epifaunally or infaunally and within the latter at different depths beneath the surface. However, many forms are probably opportunistic omnivores, and therefore chance plays a part in the success or otherwise of individuals where food is in short supply.

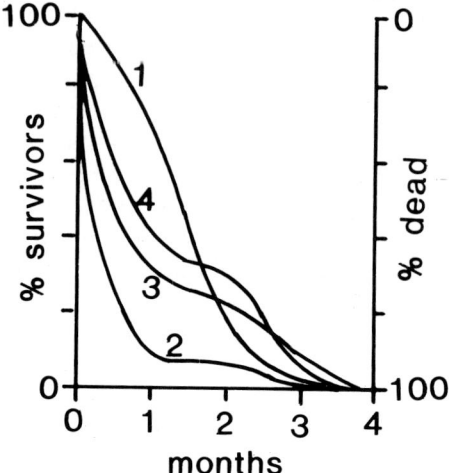

Fig. 9. Mortality/survivorship of *Nonion depressulus* at different seasons (from Murray, 1983). 1, April–May 1979; 2, November 1979–March 1980; 3, May–September 1980; 4, October 1980–February 1981.

6.2 Length of life

Foraminifera produce large numbers of offspring at each phase of reproduction, but there is a very high juvenile mortality so that few individuals survive to maturity (Fig. 9).

Because reproduction effectively terminates the life of an individual, the maximum length of life is from birth to reproduction. Under very favourable conditions, the length of life is a few weeks but it is commonly several months, sometimes a year and possibly up to 5 years in larger taxa. Annual reproduction characterizes those species which are living close to their limits of tolerance where conditions for reproduction are suitable only at one season. This is true of *Elphidium crispum*, a southern species which extends as far north as Britain and where reproduction is annual.

7.0 APPLICATION TO PALAEOECOLOGICAL STUDIES

Ecological data can be used to interpret the palaeoecology of fossil assemblages. This is easiest for the Cenozoic because many of the genera are still extant. However, non-taxonomic criteria like diversity, morphogroups, size distribution, etc., are useful palaeoecological criteria for assemblages of almost any geological age. Two contrasting examples are discussed below.

The middle Eocene Calcaire Grossier of the Paris Basin has assemblages dominated by porcellaneous forms (including *Orbitolites* and *Fasciolites*), and peneroplids, with hyaline *Discorbis, Cibicides, Nonion, Cancris* and polymorphinids. The α diversity ranges from 5 to 13, suggesting normal or slightly hypersaline shelf conditions. The modern analogue of *Orbitolites* is *Marginopora* and this, together with peneroplids, lives in shallow waters (0–35 m) in close association with seagrass and seaweeds in normal marine to hypersaline waters (36–50‰) and at temperatures of 15–30°C. Summer temperatures of at least 22°C are necessary for reproduction to take place. *Discorbis*, a shallow-water (0–18 m) dweller, requires temperatures of 18–26°C and a salinity of 35–56‰, and lives in association with seaweed, seagrass, and calcareous algae. Murray and Wright (1974) interpreted these assemblages as representing shallow (0–18 m), slightly hypersaline (36–40‰), warm (>22°C in the summer) waters with a flora of seagrass/seaweed.

The late Miocene of the Pacific Ocean has assemblages dominated by *Nuttallides* with an upper depth limit rising from 3500 to 2500 m between the mid- and late Miocene and the late Miocene and early Pliocene. Woodruff (1985) interpreted the assemblages as indicating the presence of Antarctic Bottom Water (AABW) (and hence the existence of an Antarctic ice cap) and also as showing increased production of AABW through time. Independent evidence from stable isotope studies suggests that the Antarctic ice cap may have been at its most extensive in the late Miocene.

8.0 SUMMARY, CONCLUSIONS AND FUTURE RESEARCH

Benthic foraminifera are extremely varied in morphology and very diverse. They have a long geological record and at present they colonize all marine and marginal marine environments, from the most extreme intertidal marsh to the deepest trenches of the ocean. The diversity is low in extreme, variable and ecologically stressed environments and is highest in stable, normal marine ecologically unstressed environments. Calcareous forms are present from intertidal to abyssal depths above the CCD. Agglutinated taxa are also present in these environments, but also at depths below the CCD and in other settings where carbonate dissolution takes place. Many genera show distribution patterns related to specific environments controlled by salinity, temperature, substrate, etc. Careful analysis of fossil assemblages enables fairly detailed palaeoecological reconstructions to be made for Cenozoic examples, but less so for Mesozoic and Palaeozoic ones.

Much work remains to be done. There are large areas of the world about which very little is known of the living assemblages, even from intertidal and lagoonal areas which are accessible and inexpensive to sample. Virtually nothing is known of seasonality in any environment except the marginal marine. The population dynamics and reproductive cycle is known for 20 or so species and all of these are very shallow-water examples. There is a need for experimental work on physiology, response to varying (or unvarying) environmental conditions, controls on the uptake of stable isotopes, method of construction of agglutinated tests, etc. All aspects of benthic foraminiferal ecology are still at an early stage of understanding and only when much more is understood of the modern forms will the interpretation of palaeoecology and evolutionary history make substantial progress.

ACKNOWLEDGEMENTS

I am most grateful to Mrs A Williams for typing the manuscript and Mr J. Jones for the photographs.

REFERENCES

Belanger, P.E. and Streeter, S.S. (1980). Distribution and ecology of benthic foraminifera in the Norwegian-Greeland Sea. *Mar. Micropaleontol.* **5**:401–428.

Boltovskoy, E. and Wright, R. (1976). *Recent Foraminifera.* W. Junk, The Hague.

Brasier, M.D. (1982a). Architecture and evolution of the foraminiferid test–a theoretical approach. In *Aspects of Micropalaeontology* (eds F.T. Banner and A.R. Lord), pp. 1–41. Allen and Unwin, London.

Brasier, M.D. (1982b). Foraminiferid architectural history: A review using the MinLOC and PI methods. *J. Micropalaeontol.* **1**:95–105.

Brasier, M.D. (1986). Why do lower plants and animals biomineralise? *Paleobiology* **12**:241–250.

Bremer, M.L. and Lohman, G.P. (1982). Evidence for primary control on the distribution of certain Atlantic Ocean benthonic foraminifera by degree of carbonate saturation. *Deep-Sea Res.* **29**:987–998.

Buzas, M.A. (1968). On the spatial distribution of foraminifera. *Contr. Cushman Fdn Foram. Res.* **19**:1–11.

Buzas, M.A. (1974). Vertical distribution of *Ammobaculites* in the Rhode River, Maryland. *J. Foram. Res.* **4**:144–147.

Buzas, M.A. and Sen Gupta, B.K. (1982). Foraminifera: Notes for a short course. *Studies in Geology* **6**:1–219. University of Tennessee, Department of Geological Sciences.

Charles, C.D. (1991). Late Quaternary Ocean Chemistry and Climate Change from an Antarctic Deep Sea Sediment Perspective. Ph.D. Thesis, Columbia University, New York.

Collison, P. (1980). Vertical distribution of foraminifera off the coast of Northumberland, England. *J. Foram. Res.* **10**:75–78.

Corliss, B.H. (1985). Microhabits of benthic foraminifera within deep-sea sediments. *Nature* **314**:435–438.

Ellison, R.L. (1972). *Ammobaculites*, foraminiferal proprietor of Chesapeake Bay estuaries. *Mem. Geol. Soc. Amer.* **133**:247–262.

Gooday, A.J. (1986). Meiofaunal foraminiferans from the bathyal Porcupine Seabright (northeast Atlantic): Size structure, standing crop, taxonomic composition, species diversity and vertical distribution in the sediment. *Deep-Sea Res.* **33**:1345–1373.

Gooday, A.J. (1988). A response by benthic foraminifera to the deposition of phytodetritus in the deep sea. *Nature* **332**:70–73.

Hallock, P. (1981). Light dependence in *Amphistegina*. *J. Foram. Res.* **11**:40–46.

Hallock, P. (1985). Why are larger foraminifera large? *Paleobiology* **11**:195–208.

Hallock, P., Cottley, T.L., Forward, L.B. and Halas, J. (1986). Population biology and sediment production of *Archaias angulatus* (Foraminiferida) in Largo Sound, Florida. *J. Foram. Res.* **16**:1–8.

Haynes, J. (1965). Symbiosis, wall structure and habitat in foraminifera. *Contr. Cushman Fdn Foram. Res.* **16**:40–43.

Hemleben, Ch., Spindler, M. and Anderson, O.R. (1989). *Modern Planktonic Foraminifera*. Springer-Verlag, New York.

Holme, N.A. and McIntyre, A.D. (1984). *Methods for the Study of Marine Benthos*. IBP Handbook No. 16, 2nd edn. Blackwell Scientific, Oxford.

Jones, R.W. and Charnock, M.A. (1985). "Morphogroups" of agglutinating foraminifera: Their life positions and feeding habits and potential applicability in (paleo)ecological studies. *Rev. de Paléobiol.* **4**:311–320.

Lee, J.J. and Bock, W.D. (1976). The importance of feeding in two species of soritid foraminifera with algal symbionts. *Bull. Mar. Sci.* **26**:530–537.

Lee, J.J., McEnery, M.E. and Kahn, E.G. (1979). Symbiosis and the evolution of larger foraminifera. *Micropaleontol.* **25**:118–140.

Leutenegger, S. (1984). Symbiosis in benthic foraminifera: Specificity and host adaptations. *J. Foram. Res.* **14**:16–35.

Loeblich, A.R. and Tappan, H. (1988). *Foraminiferal Genera and Their Classification*. Van Nostrand Reinhold, New York.

Lutze, G.F. (1965). Zür Foraminiferen-Fauna der Ostsee. *Meyniana* **15**:75–142.

Lutze, G.F. (1974). Benthische Foraminiferen in Oberflächen-Sedimenten des Persischen Golfes, Teil 1: Arten. *"Meteor" Forsch-Ergebnisse* **C17**:1–66.

Lynts, G.W. (1962). Distribution of recent foraminifera in Upper Florida Bay and associated sounds. *Contr. Cushman Fdn Foram. Res.* **13**:127–144.

Marszalek, D.S., Wright, R.C. and Hay, W.W. (1969). Function of the test in foraminifera. *Trans. Gulf Coast Assoc. Geol. Socs* **19**:341–352.

Matera, N.J. and Lee, J.J. (1972). Environmental factors affecting the standing crop of foraminifera in sublittoral and psammolittoral communities of a Long Island salt-marsh. *Mar. Biol.* **14**:89–103.

Muller, W.A. (1975). Competition for food and other niche-related studies of three species of salt-marsh foraminifera. *Mar. Biol.* **31**:339-351.
Murray, J.W. (1963). Ecological experiments on foraminiferids. *J. Mar. Biol. Assoc. U.K.* **43**:621-642.
Murray, J.W. (1968). The living foraminiferida of Christchurch Harbour, England. *Micropaleontol.* **14**:83-96.
Murray, J.W. (1973). *Distribution and Ecology of Living Benthic Foraminiferids.* Heinemann, London.
Murray, J.W. (1976). Comparative studies of living and dead benthic foraminiferal distributions. In *Foraminifera* (eds R.H. Hedley and C.G. Adams), Vol. 2, pp. 45-109. Academic Press, London.
Murray, J.W. (1983). Population dynamics of benthic foraminifera: Results from the Exe estuary, England. *J. Foram. Res.* **13**:1-12.
Murray, J.W. (1991). Ecology and paleoecology of benthic foraminifera. Longman, Harlow.
Murray, J.W. and Wright, C.A. (1974). Palaeogene foraminiferida and palaeoecology, Hampshire and Paris Basins and the English Channel. *Spec. Pap. Palaeont.* **14**:1-129.
Murray, W.G. and Murray, J.W. (1987). A device for obtaining representative samples from the sediment-water interface. *Mar. Geol.* **76**:313-317.
Parker, F.L. and Athearn, W.D. (1959). Ecology of marsh foraminifera in Poponesset Bay. *J. Paleontol.* **3**:333-343.
Parker, F.L., Phleger, F.B. and Peirson, J.F. (1953). Ecology of foraminifera from San Antonio Bay and environs, southwest Texas. *Cushman Fdn Spec. Pap.* **2**:1-75.
Phleger, F.B. (1966). Patterns of living marsh foraminifera in South Texas coastal lagoons. *Boln Soc. Geol. Mex.* **28**:1-44.
Phleger, F.B. and Ewing, G.C. (1962). Sedimentology and oceanography of coastal lagoons in Baja California, Mexico. *Bull. Geol. Soc. Am.* **73**:145-182.
Phleger, F.B. and Lankford, R.R. (1957). Seasonal occurrences of living benthonic foraminifera in some Texas Bays. *Contr. Cushman Fdn Foram. Res.* **8**:93-105.
Poag, C.W. (1978). Paired foraminiferal ecophenotypes in Gulf Coast estuaries: Ecological and paleoecological implications. *Trans. Gulf Coast Assoc. Geol. Socs* **28**:395-421.
Poag, C..W. (1984). Distribution and ecology of deep-water benthic foraminifera in the Gulf of Mexico. *Palaeogeog., Palaeoclimatol., Palaeoecol.* **48**:25-37.
Ravelo, A.C. (1991). Reconstructing the Tropical Atlantic Seasonal Thermocline using Planktonic Foraminifera. Ph.D. Thesis, Columbia University, New York.
Ross, C.A. (1974). Evolutionary and ecological significance of large, calcareous foraminiferida (Protozoa), Great Barrier Reef. In *Proceedings of the 2nd International Coral Reef Symposium*, Vol. 1, 327-333. Great Barrier Reef Committee, Brisbane, Australia.
Röttger, R. (1972). Analyse von Wachstumskurven von *Heterostegina depressa* (Foraminifera Nummulitidae). *Mar. Biol.* **17**:228-242.
Röttger, R., Irwan, A., Schmaljohann, R. and Franzisket, L. (1980). Growth of the symbiont-bearing foraminifera *Amphistegina lessonii* d'Orbigny and *Heterostegina depressa* d'Orbigny (Protozoa). In *Endosymbiosis and Cell Biology* (eds W. Schwemmler and H.A. Schenk), Vol. 1, pp. 125-132. Walter de Gruyter, Berlin.

Schnitker, D. (1980). Quaternary deep-sea benthic foraminifera and bottom water masses. *Ann. Rev. Earth Plant. Sci.* **8**:343–370.

Schwab, D. and Hofer, H.W. (1979). Metabolism in the protozoan *Allogromia laticollaris* Arnold. *Z. mikosk. anat. Forsch., Leipzig* **93**:715–727.

Severin, K.P. (1983). Test morphology of benthic foraminifera as a discriminator of biofacies. *Mar. Micropaleontol.* **8**:65–76.

Sliter, W. (1965). Laboratory experiments on the life cycle and ecological controls of *Rosalina globularis* d'Orbigny. *J. Protozool.* **12**:210–215.

Sturrock, S. and Murray, J.W. (1981). Comparison of low energy and high energy marine middle shelf foraminiferal faunas, Celtic Sea and Western English Channel. In *Micropalaeontology of Shelf Seas* (eds J.W. Neale and M. Brasier), pp. 250–260. Ellis Horwood, Chichester.

Walker, D.A., Linton, A.E. and Schafer, C.T. (1974). Sudan Black B: A superior stain to rose Bengal for distinguishing living from non-living foraminifera. *J. Foram. Res.* **4**:205–215.

Walton, W.R. (1952). Techniques for recognition of living foraminifera. *Contr. Cushman Fdn Foram. Res.* **3**:56–60.

Wang, P. and Murray, J.W. (1983). The use of foraminifera as indicators of tidal effects in estuarine deposits. *Mar. Geol.* **51**:239–250.

Wang, P., Min, Q. and Bian, Y. (1985). Distribution of foraminifera and ostracoda in bottom sediments of the northwestern part of the South Huanghai (Yellow) Sea and its geological significance. In *Marine Micropaleontology of China* (ed. P. Wang), pp. 93–114. China Ocean Press, Beijing/Springer-Verlag, Berlin.

Wefer, G. (1976). Umwelt, Produktion und Sedimentation benthischer Foraminiferen in der Westlichen Ostsee. *Rept. Sondersforschungsbereich 95, Kiel* **14**:1–103.

Weston, J.F. and Murray, J.W. (1983). Benthic foraminifera as deep-sea water-mass indicators. In *Benthos '83: Second International Symposium of Foraminifera*, pp. 605–610, Pau and Bordeaux.

Woodruff, F. (1985). Changes in Miocene deep-sea benthic foraminiferal distribution in the Pacific ocean: Relationship to paleoceanography. *Mem. Geol. Soc. Amer.* **163**:131–175.

Zaninetti, L. (1979). L'étude des foraminifères des mangroves actuels: réflexion sur les objectives et sur l'état des connaissances. *Arch. Sci. Genève* **32**:151–161.

8
Ecology and distribution of planktonic foraminifera

JOHN W. MURRAY

1.0	Introduction	256
2.0	Sampling methods	258
3.0	Distribution	259
	3.1 Patchiness	262
	3.2 Seasonality	263
	3.3 Depth distribution	264
	3.4 Sediment trap studies	265
4.0	Aspects of the test	267
	4.1 Ontogeny	267
	4.2 Test growth	268
	4.3 Coiling ratios	269
	4.4 Porosity	270
	4.5 Stable isotopes	270
	4.6 Morphological variation	273
5.0	Some aspects of life processes	273
	5.1 Buoyancy	273
	5.2 Feeding	274
	5.3 Symbiosis and oligotrophic conditions	275
	5.4 Length of life	276
	5.5 Effects of disturbance	276
6.0	Ecology	277
	6.1 Abiotic factors	277
	6.2 Biotic factors	278
7.0	Conclusions	280
8.0	Future research	280
	Acknowledgements	281
	References	281
	Glossary of terms	284

1.0 INTRODUCTION

The term plankton is used to describe those organisms which live within the water column and drift passively in the water currents. They range in size from a few micrometres to several centimetres and planktonic foraminifera fall in the intermediate range of ~60 μm to >1 cm, the upper limit being attained in certain spinose species. Some typical examples are illustrated in Plate 1.

It has been written of plankton that "The organisms are often better hydrographers than Man as it has been proved that minor differences or minor similarities in water bodies influence dispersal and speciation or variation as effectively as the larger differences" (Pierrot-Bults and van der Spoel, 1979, p. 357). This seems true of planktonic foraminifera.

The geological record of planktonic foraminifera shows them to have undergone frequent phases of diversification, notably in the mid-Cretaceous, late Cretaceous, Palaeocene, mid-Eocene, early Miocene and mid-Miocene. Many species have short stratigraphic ranges and the succession of overlapping species ranges makes them extremely useful in biostratigraphy. It is perhaps surprising that there are only around 100 extant species compared with perhaps more than 5000 species of extant benthic foraminifera.

Early ecological studies aimed to determine both the geographical distribution of the species and the abundance of individuals. Because the oceans cover 70% of the Earth's surface and the plankton is dispersed throughout the upper few hundred metres of the water column, this has been a mammoth undertaking. Nevertheless, the broad picture of the geographical distribution of plankton is now known, but data on abundance are far from complete and are less reliable. In

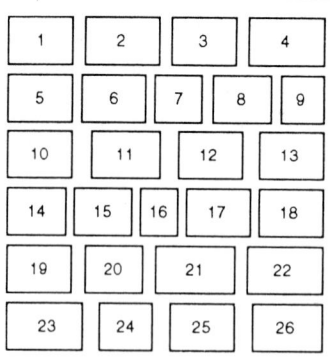

Plate 1. Planktonic foraminifera. 1,2, *Globigerina bulloides*; 3,4, *Neogloboquadrina pachyderma*; 5,6, *Neogloboquadrina dutertrei*; 7,8, *Globigerinoides sacculifer*; 9, cancellate wall of *G. sacculifer*; 10,11, *Globigerinoides ruber*; 12,13,14, *Globorotalia truncatulinoides*; 15,16,17, *Globorotalia menardii*; 18, *Sphaeroidinella dehiscens*; 19,20,21, *Globorotalia inflata*; 22, *Globorotalia hirsuta* with kummerform final chamber; 23, *Orbulina universa*; 24,25, *Globorotalia tumida*; 26, *Pulleniatina obliquiloculata*.

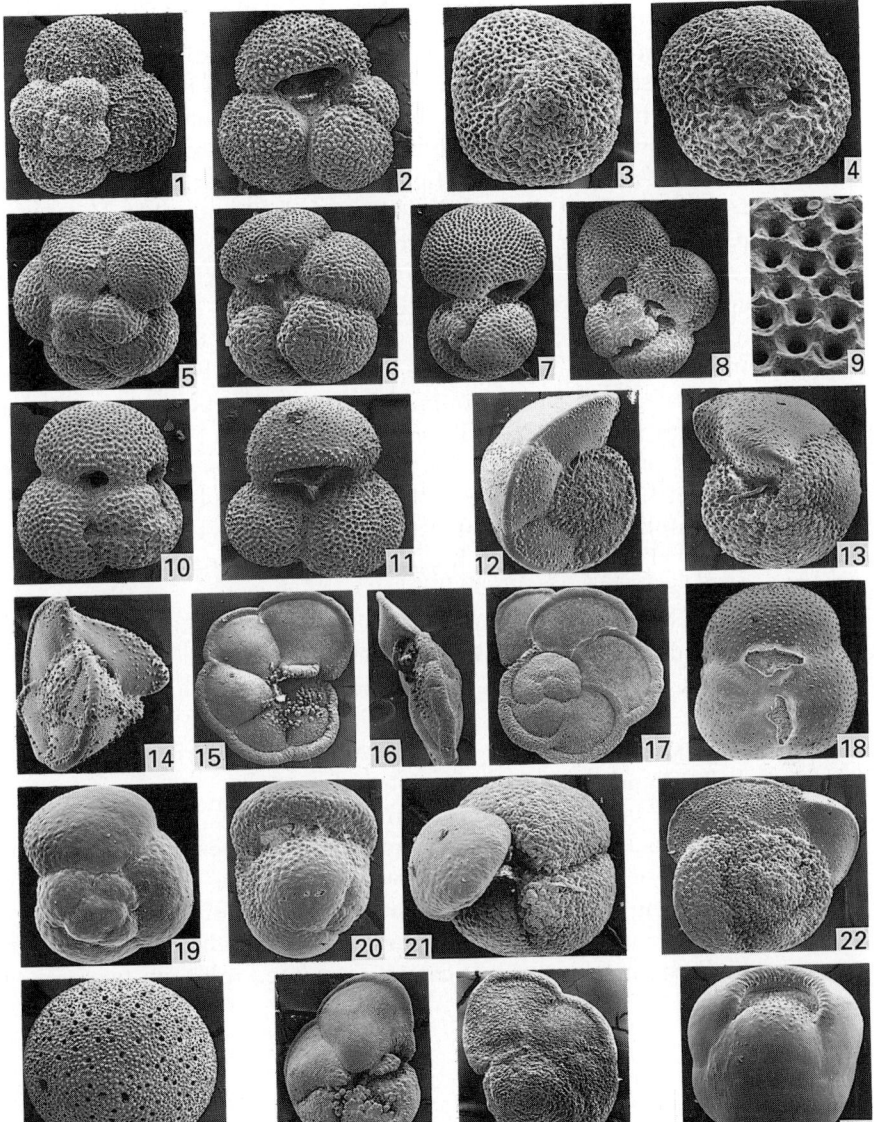

any case, abundance varies from season to season and perhaps over longer time periods and it is never likely to be possible to obtain a synoptic distribution for a single time period. Geographical distribution patterns have also been built up from a study of ocean floor sediments, although these are not the direct concern of this book.

In recent years, the main thrust of research has been towards trying to understand the dynamics of the ocean–plankton system. Since SCUBA divers have been able to catch individual planktonic foraminifera in small containers without causing them serious harm, it has become possible to culture them in the laboratory (see Chapter 10). However, at present, it is possible only to grow individuals from early stages to maturity and continuous cultures have not been achieved. This has opened up a whole new world allowing studies of reproduction, response to biotic factors (e.g. food, symbionts) and to abiotic factors (light, temperature, salinity, etc.). The process of chamber secretion has been observed and the isotopic signature of tests cultured under controlled conditions has been determined.

The integration of biological processes in living organisms and the supply of empty tests to the sea-floor has been advanced through time-series sampling of sediment traps to determine the seasonal and annual fluxes. This gives an interface with geological aspects of planktonic foraminifera.

Valuable reviews of the biology and ecology of planktonic foraminifera were published by Bé (1977, 1982), who made major contributions. Planktonic foraminifera have great potential in palaeoecological studies and this has so far been best realized through interdisciplinary climate reconstructions such as CLIMAP (McIntyre et al., 1976; McIntyre, 1981) and the Cenozoic isotopic record.

The taxonomy of planktonic foraminifera has not remained static. Bé (1967) provided a useful taxonomic key to species, but since then detailed studies of morphology and surface texture have been made using the scanning electron microscope. Banner (1982) prepared a key to genera and the most recent revision is that of Loeblich and Tappan (1988), which, with minor modification, has been adopted because of its simplicity.

2.0 SAMPLING METHODS

Traditionally, plankton samples are taken by slowly towing a fine mesh net through the water at a selected depth. There are great variations in the mesh sizes of the nets used. The nets become clogged after a

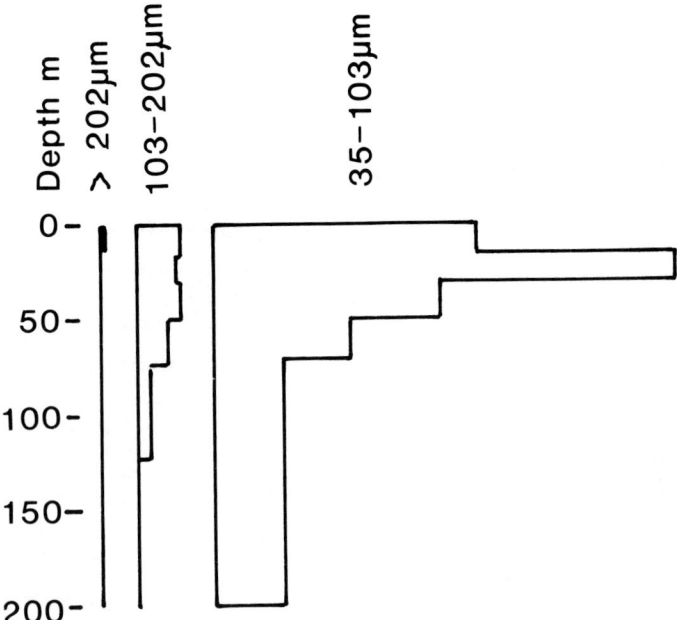

Fig. 1. Size distribution of planktonic foraminifera at an Indian Ocean site. Only 1.3% are > 202 μm (data from Bé and Hutson, 1977).

short time in areas of plankton abundance and much damage is done to fragile tests, although the species are still recognizable. In addition, some authors consider that the time of day or night influence the composition of a sample from near-surface waters. Despite these problems, the major patterns of both horizontal and vertical distribution had been mapped by the mid-1970s. Since that time, the use of SCUBA divers has enabled scientists to obtain undamaged spinose species from near-surface waters. These individuals are ideal for life and experimental studies in the laboratory. However, only large spinose forms are visible to divers. The majority of planktonic tests are very small (see Fig. 1) and if nets coarser than 100 μm or divers are used to sample plankton, these small forms are not captured.

3.0 DISTRIBUTION

The biogeographical provinces of living planktonic foraminifera are shown in Fig. 2. The nine geographical regions (from north to south) form five provinces because, with the exception of the tropical province,

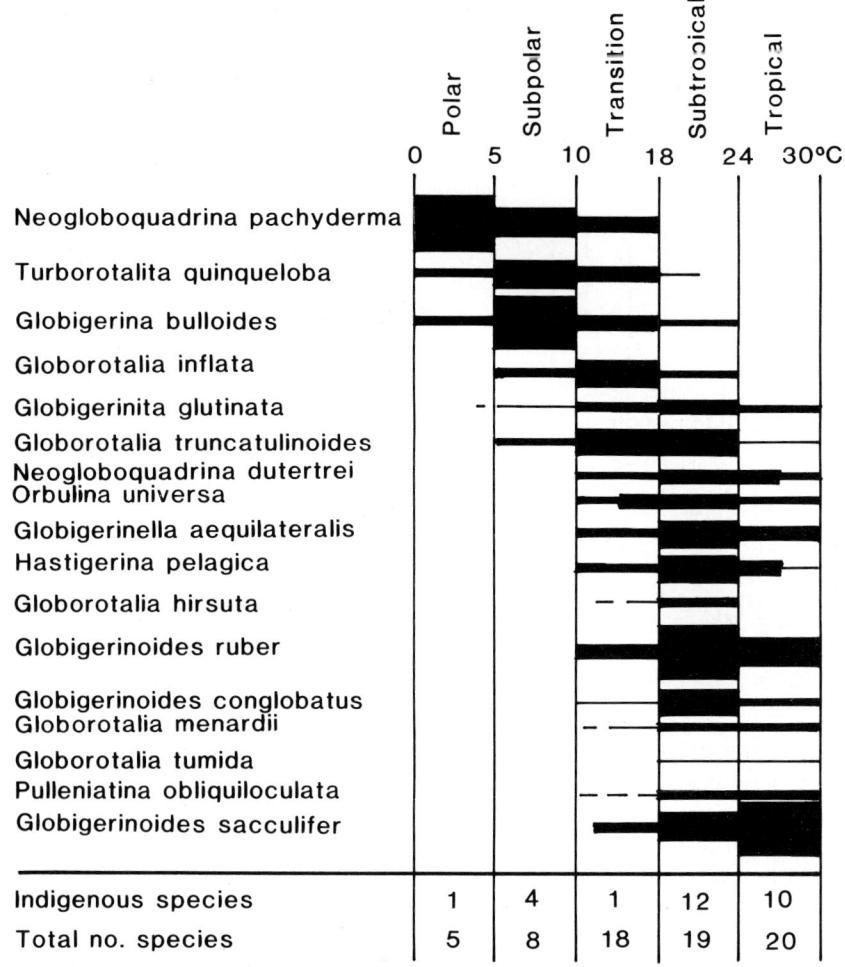

Fig. 2. The distribution and relative abundance of living planktonic foraminifera in the five faunal provinces (based on Bé, 1977).

each is represented in both the northern and southern hemispheres (Fig. 3). The cold regions are marked by low diversity and only a few indigenous species, whereas the warm regions have high diversity and many indigenous species (Bé, 1977). A few tropical species are absent from the Atlantic Ocean: *Bolliella adamsi, Neogloboquadrina hexagona* and *Globigerinoides conglobatus*. The coiling direction of *Neogloboquadrina pachyderma* is predominantly left in the polar provinces and right in the subpolar province. (Left coiling forms have the aperture on the left when the spire is uppermost.)

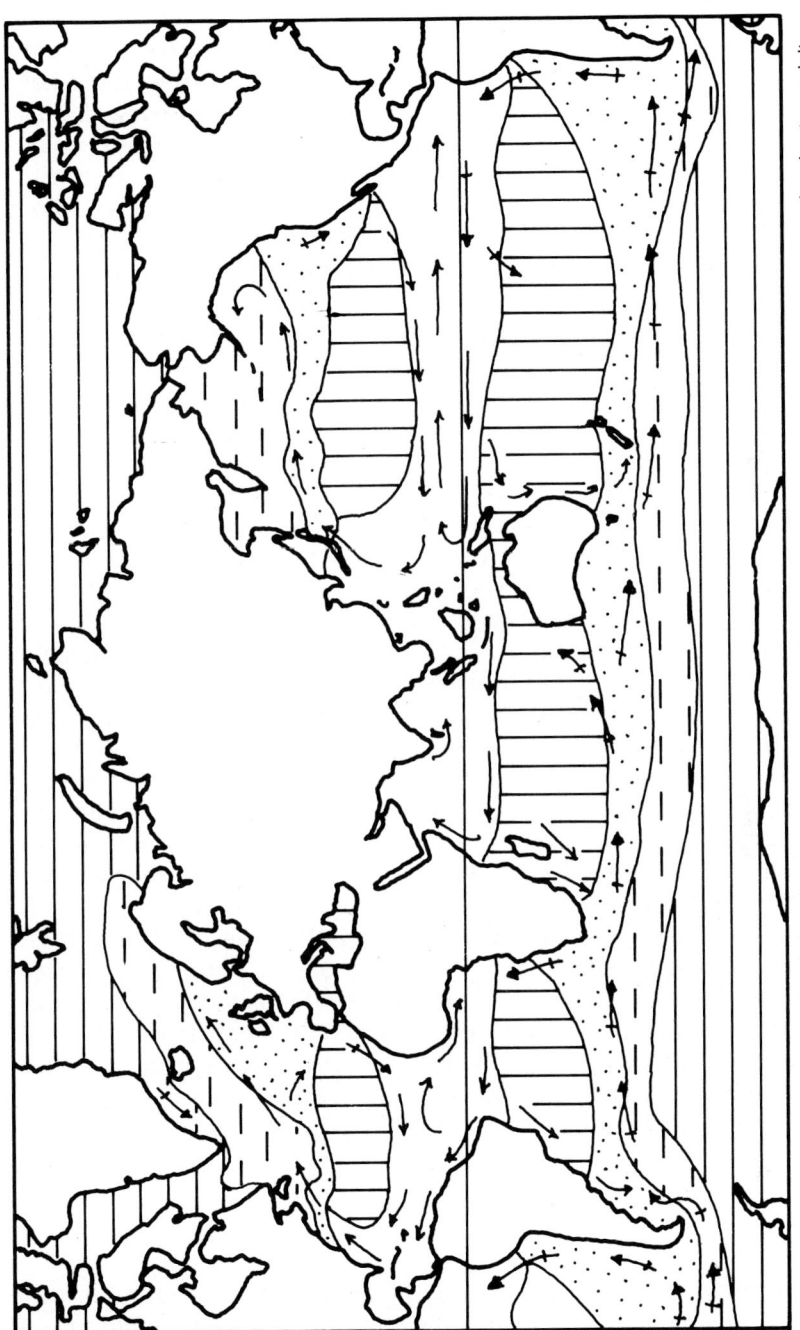

Fig. 3. Biogeographic faunal provinces of living planktonic foraminifera (based on Bé, 1977). Key to ornament: polar, horizontal lines; subpolar, horizontal dashes; transitional, stippled; subtropical, vertical lines; tropical, blank. Surface-water currents: arrows without tail crosses, warm; with tail crosses, cold.

It has been said that polar species show a disjunct distribution, i.e. two separate high-latitude occurrences separated by the tropical region from which they are absent. Bé (1977) considered that the most likely explanation of this was that continuous distributions during the last glacial event became separated as the surface waters warmed in the present interglacial. However, tropical distribution patterns during glacial times were as now: the polar and subpolar zones were enlarged (McIntyre et al., 1976). But perhaps the problem of disjunct distributions does not exist, for upwelling on the east side of oceans causes the penetration of cold-water species into low latitudes. Thus, Cifelli and Benier (1976) recorded polar and subpolar species as far south as 10°N in the eastern North Atlantic in a narrow upwelling zone along the coast of Africa, and McIntyre (1981) found these to be even more extensive during the last glacial maximum.

The modern biogeographical provinces show a clear correlation with the surface circulation pattern. In the North Atlantic, the tropical province is extended northwards by the Gulf Stream on the west side of the ocean and the North Atlantic Drift causes the transition zone to extend as far north as Norway (Fig. 3).

The abundance of planktonic foraminifera is controlled by the fertility of the ocean surface waters. Fertile areas are rich in nutrients, phytoplankton and zooplankton. They occur in major currents, boundary currents, divergence zones and upwelling zones. Very high densities are found in these areas ($>10\,000$ individuals per $1000\,m^3$: Bé, 1977). Low densities are typical of the oligotrophic central water masses (<1000 individuals per $1000\,m^3$) and continental shelves. In the latter case, turbidity and/or lowered salinities may be controlling factors. Abundant juvenile *Neogloboquadrina pachyderma* were reported from Antarctic ice (Lipps and Krebs, 1974). Although it was not certain whether they were alive or merely preserved in the ice, the authors considered it likely that they were feeding on the abundant diatoms. Further studies by Spindler and Dieckmann (1986) confirmed that only juveniles are present and that they are alive. It is considered that the juveniles are incorporated by frazil ice crystals growing in the water column. Within the ice, the foraminifera live in brine channels and feed on the diatoms until they are freed during a thaw.

3.1 Patchiness

The patchiness in the abundance of planktonic foraminifera off Barbados over a 5-month period is shown in Fig. 4. The possible causes

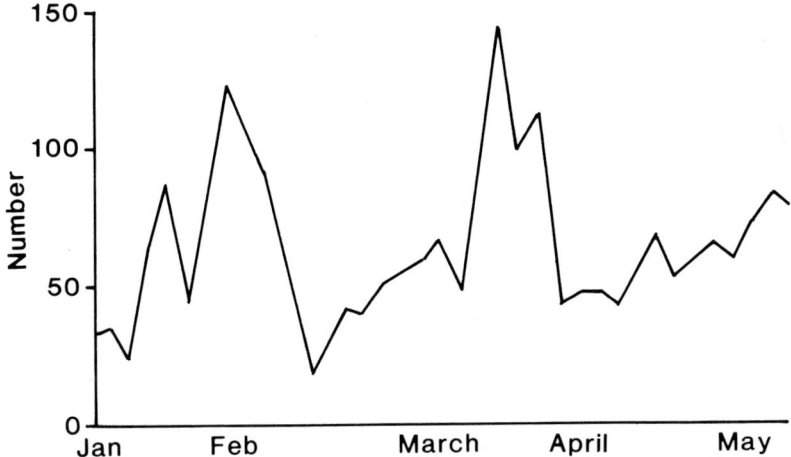

Fig. 4. Patchiness in the occurrence of spinose planktonic foraminifera collected by SCUBA diver off Barbados (after Hemleben and Spindler, 1983).

are (after Hemleben and Spindler, 1983):

1. Development of small-scale eddies.
2. Periodic intrusion of lower salinity ($<32‰$) surface water lenses (from the Amazon discharge) causing *Globigerinoides ruber* to replace *G. sacculifer*.
3. Changes in depth habitat related to gametogenesis (the reproductive cycle of symbiont-bearing spinose forms is controlled by the lunar cycle).
4. Changes in food supply or migration of foraminifera in pursuit of food.

To this list may be added the effects of predation, for Berger (1971a) observed that the abundance of foraminifera was inversely proportional to that of predators in a frontal region off Baja California.

3.2 Seasonality

Tropical and polar surface waters are little affected by seasonal temperature changes but in between there may be significant seasonal variation. Cifelli and McCloy (1983) considered that two different strategies were employed by planktonic foraminifera to overcome this. Temperate species tend to be capable of withstanding wide climatic variations so the same species are present throughout the year. Subtropical species are divided into two groups, each coping with a different set of conditions. In the Sargasso Sea, *Globorotalia truncatulinoides* is the winter species, whereas subtropical spinose species characterize the summer plankton.

3.3 Depth distribution

Most planktonic foraminifera live in the euphotic zone, because that is where there is the greatest abundance of zooplankton and phytoplankton food. Also, those with symbionts are limited to the uppermost waters because of the photosynthetic requirements of their algae. However, those without symbionts are not so restricted and they tend to live at greater depths. Nevertheless, individual species show wide ranges in vertical distribution varying with region and season. Bé (1977) stresses that depth stratification is only a statistical generalization (see Table 1). Some authors consider that there is a diurnal migration, rising during the day and falling at night, but others have found no evidence of this (see Boltovskoy, 1973, for a review).

In the North Atlantic, off Bermuda, a deep chlorophyll maximum layer (DCM) develops near the bottom of the euphotic zone, which

Table 1 Generalized depth habitats of planktonic foraminifera[a]

1. Shallow: predominantly in the upper 50 m
    Turborotalita quinqueloba sp
    Globigerinoides ruber sp sy
    Globigerinoides sacculifer sp sy
    Globigerinoides conglobatus sp sy

2. Intermediate: upper 100 m but predominantly 50–100 m
    Globigerina bulloides sp
    Hastigerina pelagica[b] sp sy
    Pulleniatina obliquiloculata
    Neogloboquadrina dutertrei
    Orbulina universa sp sy
    Globigerinita glutinata[b]

3. Deeper: adult stages predominantly > 100 m
    Globorotalia spp.
    Neogloboquadrina pachyderma (> 200 m)
    Sphaeroidinella dehiscens

[a] Based on Bé (1977, 1982). [b] 0–50 m, Hemleben and Spindler (1983).
Abbreviations: sp, spinose; sy, symbionts.

also coincides with a pycnocline (change in density). The DCM is a major food source for zooplankton, and planktonic foraminifera reach their maximum abundance in it. However, the level of the DCM varies from near the surface in winter to ~80 m in summer and the foraminifera move with it (Fairbanks and Wiebe, 1980).

The descent of adults during the life-cycle can take them into much cooler waters. Thus, *Globorotalia hirsuta* and *G. truncatulinoides*, subtropical species living in the Sargasso Sea off Bermuda, can tolerate a temperature range of 2–27°C (Hemleben *et al.*, 1985). In the Panama Basin, Bé *et al.* (1985) found that juveniles were 3–4 times as abundant as adults and that their depth distributions were similar. From this, they inferred that a substantial proportion of the gametes are released before the adult sinks below the euphotic zone.

3.4 Sediment trap studies

The rain of planktonic tests sinking towards the ocean floor can be sampled in a sediment trap, a device anchored above the sea-floor and left for a period of weeks. As pointed out by Berger (1971b), traps should collect a more reliable death assemblage than net tows or that preserved in the bottom sediment because they collect the true flux and they time-average small-scale patchiness. However, the record may be affected if there is current flow which causes turbulence resulting in the loss or gain of tests. In order to calibrate the timing of changes of assemblage composition in the trap with those in the surface waters, it is necessary to take into account the rate and duration of settling of tests.

An example from high latitudes is that of Reynolds and Thunell (1986) in the north-east Pacific near the northern edge of the North Pacific Drift. There, two morphotypes of *Neogloboquadrina pachyderma* were observed: those with a crystalline texture due to a calcite crust and predominantly sinistral (group A) and those with a reticulate texture and predominantly dextral (group B). The trap was moored at 3858 m and sampled on average every 15 days. Group A was dominant from March to May, whereas group B was dominant from June to February and most abundant from July to early January. The total flux of *N. pachyderma* tests has an annual maximum of 2365 specimens/ m^2/day. This forms 36% of the annual foraminiferal test flux.

Figure 5 shows the abundance of group A plotted against surface water temperature. However, at depths >10 m, the temperature is

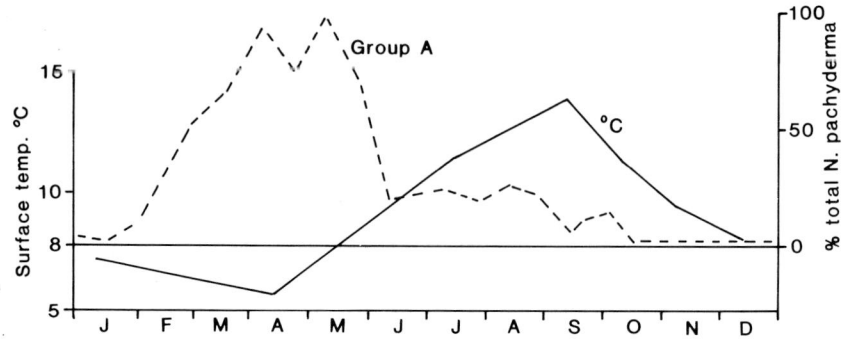

Fig. 5. A plot of group A *Neogloboquadrina pachyderma* against surface-water temperature over a period of 1 year (after Reynolds and Thunell, 1986).

<7.2°C throughout the year. The development of the calcite crust (group A) appears to be temperature-controlled (cold surface water). Group B is abundant during the period of thermal stratification (thermocline at 30 m).

A similar study in the Panama Basin (equatorial east Pacific) was beneath the Inter-Tropical Convergence Zone. Traps were placed at 890, 2590 and 3560 m and data gathered on the 12 most abundant species (Thunell and Reynolds, 1984). One group, comprising *Neogloboquadrina dutertrei, N. pachyderma, Globorotalia menardii, G. theyeri* and *Globigerinella aequilateralis,* had maximum fluxes during winter and early spring, and particularly during February–March. These forms live predominantly at or below the thermocline and during February–March upwelling causes the thermocline to rise to the surface. The second group had maximum fluxes during June–September with a secondary peak in the winter, e.g. *Globigerinoides ruber, G. sacculifer, Globigerinita glutinata* and *Globigerina bulloides.* The summer peaks correlate with a well-developed warm surface layer (temperature 27–28°C) and its associated phytoplankton bloom, while the winter peak is related to an increase in nutrients due to the upwelling. Only the group B morphotypes of *N. pachyderma* were found here and the presence of this essentially cold-water species in the equatorial region is due to the presence of cold upwelled water.

In the subtropical Sargasso Sea, seasonal changes in the composition of the foraminiferal flux is very marked (Deuser *et al.,* 1981). *Orbulina universa* shows a drop in winter, *Globigerina bulloides, Globorotalia hirsuta, G. truncatulinoides, G. inflata* and *Neogloboquadrina dutertrei* almost disappear during the summer and they peak during the winter and spring, *Pulleniatina obliquiloculata* was present only in winter, and *Globigerinoides ruber* and *Globigerinella aequilateralis* showed little change throughout the year.

Bé et al. (1985) concluded that the foraminiferal flux in the Panama Basin is higher than that of the tropical Atlantic by a factor of 2–5, and the tropical Atlantic is twice that of the central Pacific (Thunell and Honjo, 1981).

4.0 ASPECTS OF THE TEST

4.1 Ontogeny

Laboratory studies of the life-cycle of only a few planktonic species have so far been undertaken, and therefore ontogenetic changes have had to be studied indirectly. Most species show five phases of growth: prolocular, juvenile, neanic, adult and terminal. Brummer et al. (1986) recognized three species groups:

Group 1. *Globigerinella aequilateralis, Globigerinoides conglobatus, G. sacculifer* and *Orbulina universa*. The juveniles have a few large pores concentrated along the sutures, a smooth wall with sparse delicate spines, and an extra-umbilical aperture. At a size of $\sim 80\,\mu m$, there is an abrupt change to the neanic stage: densely distributed spines cover the chambers with pores in between giving rise to the cancellate (honeycomb) wall texture. During this stage, the number of chambers per whorl decreases and the primary aperture moves towards its final position. At $\sim 200\,\mu m$, the test becomes adult with the addition of secondary apertures.

Group 2. *Globorotalia menardii* and *Neogloboquadrina dutertrei*. The ontogenetic sequence is similar to that of group 1 except that the wall is pustulate instead of spinose, and it is not cancellate. A keel develops in *G. menardii* during the neanic stage.

Group 3. *Candeina nitida, Globigerinita glutinata* and *G. juvenilis*. These retain most of their morphological features throughout ontogeny.

The classification of planktonic foraminifera is based on adult features, but the growth phases of *Globigerinoides ruber* span three adult genera. The juvenile is like an adult globorotaliid, the neanic phase is like *Globigerina* and the adult is *Globigerinoides*. These morphological changes are matched by behavioural differences. The juveniles feed mainly on algae and small protozoans but only the neanic phase and adults have sufficiently strong spines and rhizopods to capture and feed on copepods, their main source of food (Brummer et al., 1987).

4.2 Test growth

The rate of growth of the test is controlled by the availability of food. The rate of addition of a new chamber decreases with age. Thus, an individual 325 μm in diameter took 1.6 days to form its first chamber in culture but 2.3–6.2 days for later chambers (Bé et al., 1981). The process of chamber formation in *Globorotalia hirsuta* and *G. truncatulinoides* takes 5–6 h (Bé et al., 1979). The amount of calcium carbonate deposited per new chamber is 12.8 μg (initial diameter 457 μm and chamber growth 166 μm) to 20.8 μg (initial diameter 519 μm and chamber growth 211 μm) in *Globigerinoides sacculifer* (Anderson and Faber, 1984). The latter species secretes its chambers at night (Hemleben et al., 1987).

In the Sargasso Sea, *Globorotalia truncatulinoides* and *G. hirsuta* tolerate a temperature range of 2–27°C and high abundances are found over a range of 2–23°C (Hemleben et al., 1985). Normal test walls are deposited at temperatures above 10°C but below this a secondary thickening ("calcite crust") is laid down (Bé et al., 1979). The calcite crust is not associated with gametogenesis (Hemleben et al., 1985). Surface-water adult *G. truncatulinoides* have a diameter of ∼400 μm but those from a sediment trap were ∼650 μm in diameter, which suggests that at least two chambers are added during the descent into deeper water (Hemleben et al., 1985).

Sphaeroidinella dehiscens is characterized by an extremely smooth final test wall layer which is deposited over a spinose *Globigerinoides sacculifer* precursor, and Bé and Hemleben (1970) consider it to be an aberrant development of the latter.

The logarithmic growth of planktonic tests leads to an increase in size without a change in shape. Tests of this kind have been termed normalform. However, if growth slows down as maturity is reached, the final chamber will be the same size or smaller than the previous one and growth is no longer logarithmic. Such tests are termed kummerform. They are thought to arise in response to environmental stress (see review of Kennett, 1976).

In *Globigerinoides sacculifer*, modifications to the test take place before gametogenesis. First, the spines are resorbed and the pseudopodia which are supported by them (rhizopodia) become wrapped over the test surface where they secrete secondary calcite. This causes the pores to become constricted, the inter-pore ridges to change from sharp to broad and rounded, and the development of "spine holes" about 0.5 μm in diameter due to the continued calcitic growth of spine bases. In addition, a sac-like final chamber is added. Gametogenic calcification adds $\sim 28\%$ weight (Bé, 1980). It takes place below the euphotic zone as the individual is sinking due to loss of buoyancy. Bé (1982) estimated that an adult 500 μm in diameter and 24 μg in weight living in the euphotic zone would sink to 800 m and continue to calcify in 24 h. The gametogenic calcite is enriched in ^{18}O from the cold deeper waters (Duplessy et al., 1981b). Individuals which die without undergoing gametogenesis have thin walls and remnants of spines. The frequency of occurrence of gametogenic tests in deep-sea sediments could be used as an index of fertility in the absence of any post-mortem dissolution (Bé et al., 1981).

4.3 Coiling ratios

Trochospiral tests may be dextral (right-coiling) or sinistral (left-coiling). Commonly, both morphotypes are present in a single species and have no apparent ecological significance. However, in certain species, the patterns of distribution of dextral and sinistral forms differ (see Kennett, 1976, for a review). The best documented case is that of *Neogloboquadrina pachyderma*. In subpolar regions it is predominantly dextral, whereas in polar regions it is predominantly sinistral. Although this may seem to be a temperature-related effect, this is not absolutely clear-cut. The picture may be obscured because in many cases it is the distribution of tests in the bottom sediment (rather than the plankton) which is correlated with surface water temperatures. Consequently, there is no agreement between authors who have studied different parts of the ocean as to the critical temperature at which the population changes from predominantly sinistral to predominantly dextral. Furthermore, the presence of some dextral individuals in a predominantly sinistral assemblage, and *vice versa*, requires an explanation. Although many hypotheses have been put forward, the only one which satisfactorily explains the patterns is that coiling is genetically controlled and that it is maintained by restricted genetic exchange between individuals in separate water masses (Brummer and Kroom, 1988).

The application of coiling ratios to palaeoecology or biostratigraphy should be exercised with caution. The two morphotypes sometimes differ in size, thickness of wall and therefore in settling velocity (Reynolds and Thunell, 1986) and it has been shown that, in sediments, the coiling ratios vary according to the size-fraction studied (more dextral in the coarser fraction in the Indian Ocean: Vella, 1974).

4.4 Porosity

There are conspicuous differences in both the number and size of pores in the test wall of different species, and individual species show differences related to ontogeny and secondary wall thickening. In the first systematic study of porosity, Bé (1968) examined 22 species and measured pore diameter on the inner wall surface. Mean porosity values were calculated as a percentage of open pore area per unit area of test. There is a latitudinal correlation with equatorial forms having high porosities and high-latitude forms having low porosities, e.g. *Neogloboquadrina pachyderma* 2%, *Globigerinoides sacculifer* 18%. Of greater interest is that there is latitudinal variation in porosity within a single species. *Orbulina universa* in the Indian Ocean has $>10\%$ porosity in low latitudes and $<5\%$ in high latitudes (Bé et al., 1973; Fig. 6). These differences have been attributed to buoyancy and the density of water, temperature, light, etc., but no convincing explanation has yet been put forward which also accounts for the variations in porosity during ontogeny (as described in Hemleben et al., 1985).

4.5 Stable isotopes

An understanding of the isotopic signature of plankton is essential for unravelling the isotopic record held in fossil forms in oceanic sediment successions. The stable isotopes of oxygen give data on water temperature and, for fossil forms, on the volume of water locked up in ice, while those of carbon reveal details of the carbon cycle and fertility of the waters (e.g. Hemleben et al., 1989).

Three different approaches have been followed: cultured material, living plankton from the ocean and surface sediment samples. These give different results which are only partially understood.

If the isotopic composition of a foraminiferal test is the same as that of the water in which it lived, it is said to be in isotopic equilibrium. If it is not, fractionation has occurred and it is said to be in isotopic

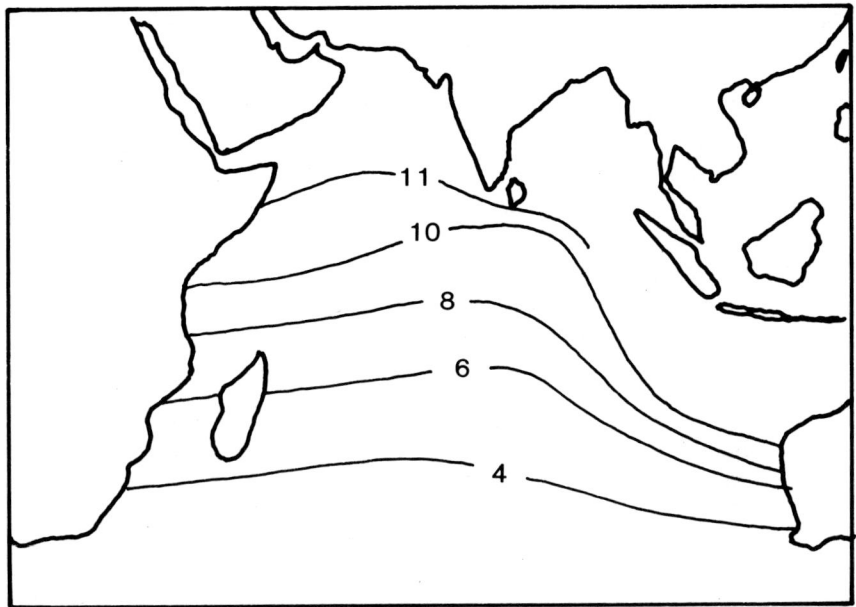

Fig. 6. Variation in test porosity of *Orbulina universa* in the upper 300 m of the Indian Ocean (after Bé et al., 1973).

disequilibrium. However, the oxygen isotopic composition of sea water varies with both salinity and temperature. As temperature is a very variable feature in all parts of the oceans except the polar regions, this introduces complexities. A foraminifer which adds to its test throughout its life but changes its depth habitat, from warm surface waters when young to deeper cool waters prior to gametogenesis, will contain a mixture of different isotopic compositions. As analyses average the values of several individuals, the resultant value may not correspond with that of any of the ambient waters.

The results from culturing *Orbulina universa* and *Globigerinoides sacculifer* show equilibrium oxygen isotopic values, and thus this may be the best way of establishing the relationship between temperature and test $\delta^{18}O$ (Bouvier-Soumagnac and Duplessy, 1985; Erez and Luz, 1983).

Analyses of plankton tow specimens commonly show disequilibrium values which are isotopically light (by -0.35 to $-2.24‰$) (Hecht, 1976; Duplessy et al., 1981a; Kahn and Williams, 1981). Seasonal studies have shown that there is a direct relationship with the surface water temperature, with depleted values during the summer months and enriched values during the winter months (Williams et al., 1979, 1981). However, the validity of these observations has been questioned by

Erez and Luz (1983), who believe that in removing unwanted soft parts from plankton by improper combustion techniques, the composition of the test is made to appear isotopically light.

The $\delta^{18}O$ values of *Neogloboquadrina pachyderma* collected in sediment traps in the Drake Passage are $+3.25‰$ PDB for tests of all sizes and they represent equilibrium values. There is little difference in temperature with depth in this cold water area (Wefer et al., 1982). In the seasonally variable Sargasso Sea, sediment traps yielded some species whose tests were deposited in isotopic equilibrium with the surface waters (e.g. *Globigerinoides ruber*, average yearly; *Neogloboquadrina dutertrei*, December–May; *Pulleniatina obliquiloculata*, late winter), those showing disequilibrium (e.g. *Globigerina bulloides*), and those showing addition of calcite crusts in deeper cool waters (e.g. *Orbulina universa* and *Globorotalia truncatulinoides*) (Deuser et al., 1981). Calcification below the photic zone represents about 45% of total test weight (Erez and Honjo, 1981).

The results have profound implications for the use of oxygen isotope values in palaeotemperature values. An optimistic view is that it may be possible to determine annual temperature ranges in some regions. Deuser et al. (1981) consider that *Globigerinoides ruber* provides a good record of the average yearly surface temperature, while *Pulleniatina obliquiloculata* provides a winter minimum. The total annual range would be twice the difference between the two temperatures.

The carbon-stable isotopes have been less studied. One of the main results to date is the recognition of a size effect, with young individuals depleted with respect to the $\delta^{13}C$ of total dissolved CO_2 (Duplessy et

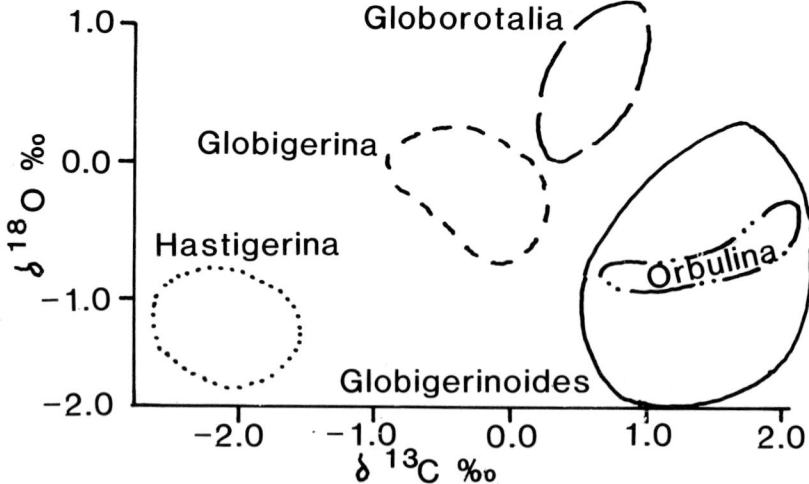

Fig. 7. Plot of $\delta^{18}O$ vs $\delta^{13}C$ for sediment trap samples (after Erez and Honjo, 1981).

al., 1981a; Williams et al., 1981; Bouvier-Soumagnac and Duplessy, 1985). This is a vital effect brought about by the incorporation of isotopically light carbon from the activity of symbionts. The effect dies out with increasing water depth as the activity of the symbionts is reduced (Kahn and Williams, 1981). Specimens from surface sediments are significantly enriched in $\delta^{13}C$ because they are generally larger than those from plankton tows (Williams et al., 1981).

Plots of $\delta^{18}O$ and $\delta^{13}C$ for genera show distinct fields which may indicate a genetic/biological control on the composition (Fig. 7; Erez and Honjo, 1981). Recently, additional evidence has been gained for the value of using $\delta^{13}C$ and $\delta^{18}O$ measurements as tracers of oceanographic parameters based on shell composition of living planktonic foraminifera from tropical (Ravelo, 1991) and polar (Charles, 1991) regions.

4.6 Morphological variation

As has already been noted, individual species show morphological changes during ontogeny, variations in wall porosity, texture and thickness attributed to environmental causes, some develop a stunted final chamber (kummerform) and others a modified final chamber during gametogenesis. In addition, there may be variations in the shape of the adult tests. Two or even three sub-populations may be recognized, each having slightly different ecological requirements (see Kennett, 1976, for a review). In many cases, these studies are based on sediment samples and in some instances the preferred habitat of life is inferred from their stable isotopic composition. Thus, Healy-Williams et al. (1985) recognized three morphotypes of *Globorotalia truncatulinoides* in the southern Indian Ocean, with different shapes (highly conical, subtropical; compressed elongate, subantarctic) and different isotopic signatures ($\delta^{13}C$ indicating different vital effects between subtropical and subantarctic forms).

5.0 SOME ASPECT OF LIFE PROCESSES

5.1 Buoyancy

Bé et al. (1985), from their own data together with those recalculated from Fok-Pun and Komar (1983), estimated dry weight densities to vary from $0.10 \, g/cm^3$ for *Globigerinoides ruber*, to $0.15 \, g/cm^3$ for *G. sacculifer* and *Globorotalia hirsuta* to $0.5 \, g/cm^3$ for *Orbulina universa*.

In water, these values would be 1.10, 1.15 and 1.5 g/cm^3, the latter figure being the average suggested by Berger and Piper (1972). As sea water has a density close to 1.02 g/cm^3, the excess density of the foraminiferal tests over the water displaced is ~ 0.1 to ~ 0.5 g/cm^3. Thus, to maintain a state of neutral buoyancy, where the weight of the organism exactly equals the weight of water displaced, it would be necessary to incorporate some low-density material. Anderson and Bé (1976) suggested that the vacuolated fibrillar system may serve a buoyancy function. Alternatively, there may be gas bubbles, as is perhaps the case in *Hastigerina pelagica* with its bubble capsule, or oil, which also serve as a food reserve. However, the density of oil is not low enough to offset the excess weight of robust tests, and thus for most adult planktonic foraminifera there should be a tendency to sink. Dead tests sink 500–2000 m/day (Bé *et al.*, 1985), and therefore living individuals must clearly have a mechanism for maintaining their chosen position in the water column, presumably without the expenditure of much energy. The presence of long spines and rhizopodia increases the diameter as much as ten-fold, and this in turn should reduce the tendency to sink. Non-spinose forms perhaps rely on the pseudopodia alone. Turbulence cannot be a major factor, as otherwise dead tests would also remain in suspension. In later life, prior to reproduction, some species secrete additional calcite which aids them to sink rapidly to the chosen depth for reproduction.

Brasier (1986) has speculated on the relationship between test morphology and growth plan and their effect on density and buoyancy.

5.2 Feeding

The widespread distribution of planktonic foraminifera in oceanic surface waters testifies to their success in feeding and their importance in the food web. They are opportunistic and prey on whatever food organisms they encounter (omnivorous: Anderson *et al.*, 1979; Bé, 1982; see also Chapter 10). The pseudopodia form a dense sticky net in which the food becomes ensnared. The food may be other zooplankton, phytoplankton or organic detritus. It may swim or drift into the pseudopodial net, but the latter does not function as a suspension feeding mechanism for it has no power to draw water through its mesh.

In some species, the pseudopods are supported by spines and this greatly increases the effective size of the organism for the diameter from spine tips is commonly ten times that of the test proper. Once a motile organism has become caught up in the pseudopodial net, it is slowly drawn towards the test aperture. The common food is copepods.

They are ruptured by small pseudopods and pieces of muscle and other tissue are taken into food vacuoles for digestion.

Calanoid copepods are more readily caught and digested faster than cyclopoid copepods. *Globigerinoides sacculifer* is more successful than *G. ruber* at catching them. Harpacticoid copepods are too strong for foraminifera to catch. On average, each foraminifer catches one copepod per day (Spindler *et al.*, 1984). Copepods are secondary producers and they predominate in the oligotrophic (food-poor) central oceanic water masses. Most of the foraminifera in these regions are spinose taxa (*Globigerinoides* spp., *Orbulina universa*, *Hastigerina pelagica*) and contain symbionts which may supply additional nutrients to their hosts (Gastrich and Bartha, 1988; see also Chapter 6).

Non-spinose species are also omnivorous and they are especially abundant in eutrophic (nutrient-rich) waters, where there is high phytoplankton productivity, for example, in upwelling regions and boundary currents (e.g. *Globorotalia hirsuta, G. inflata, G. menardii, G. truncatulinoides, Neogloboquadrina dutertrei, Pulleniatina obliquiloculata* and *Globigerinita glutinata*).

The life span of an individual is the duration from birth to death or gametogenesis (since in foraminifera reproduction leads to the vacation of the parent test). Reproduction normally takes place as soon as the individual reaches maturity. On theoretical grounds, a well-fed individual should mature sooner than a poorly fed one and experimental studies have confirmed that this is so. *Globigerinoides sacculifer* fed daily had a life span of 14–16 days compared with 20–28 days for those fed every third day (Bé *et al.*, 1981; Caron *et al.*, 1982). Individuals kept in continuous darkness reproduced after 4.2 days on average, compared with 9.1 and 15.6 days for those in low and high light intensities (see also Chapter 9).

The significance of these observations for ecology is that there is a rapid turnover during time of plentiful food, whereas during times of food shortage the symbionts probably help to sustain individuals but the onset of maturity and reproduction is delayed. In the experiments, unfed individuals survived 46–54 days. Thus, adverse periods can be weathered and opportunistic feeding leading to reproduction and an increase in numbers can take place rapidly in response to a bloom in food.

5.3 Symbiosis and oligotrophic conditions

Near-surface dwelling spinose species commonly have symbionts (see Table 2) and some live in oligotrophic areas such as the central water masses. In these areas, although the external environment is poor in nutrients, the internal environment of the foraminiferal cell releases waste products from the digestion of prey and these may serve as nutrients for the symbionts. These, in turn, make a considerable amount of carbon available to the host and this may assist calcification (Gastrich and Bartha, 1988). The presence of symbionts assists foraminifera to survive in areas of very low food supply.

The photosynthetic activity of the symbionts is a form of primary production, and Spero and Parker (1985) described the symbiotic relationships as "... small packets of intense photosynthetic activity in the euphotic zone". They considered that in most oligotrophic environments with 1–5 foraminifera ($>200\,\mu m$) per m^3, the contribution of symbiont productivity would approach 1% of the total primary productivity, but Gastrich and Bartha (1988) describe it as "negligible" (see also Chapter 6).

5.4 Length of life

Although planktonic foraminifera readily undergo gametogenesis in laboratory cultures, so far no-one has succeeded in culturing a species through an entire life-cycle. Therefore, knowledge of the frequency with which planktonic foraminifera reproduce is circumstantial (see also Chapter 5, 6, 9 and 10). From calculations of rates of sedimentation, Berger (1971b) inferred a life span of "no more than about 2 weeks". However, Spindler et al. (1979) and Hemleben et al. (1987) have observed that reproduction in *Hastigerina pelagica* and *Globigerinoides sacculifer* is linked to the lunar cycle. If these are representative of warm-water species, it would appear that the length of life and frequency of reproduction is around 28 days. In polar regions subject to ice cover, reproduction may be less frequent and length of life therefore longer.

Experimental studies on *Globigerinoides sacculifer* showed that it had the shortest life span in continuous darkness and the longest in high illumination (Bé et al., 1981). Individuals placed in continuous darkness underwent gametogenesis within 3–4 days (Caron et al., 1982). The symbionts appear to play a regulatory role in this, for they influence the host's response to the amount of light available (Bé et al., 1982).

5.5 Effects of disturbance

During collection in a plankton net, spinose species normally shed their spines, but they are able to regenerate them again in culture vessels. If *Globigerinoides ruber* is disturbed with a pipette, it sheds its spines but *G. sacculifer* does not. Non-spinose species retract their rhizopods when disturbed, but return to normal after less than 1 min (Hemleben and Spindler, 1983).

The spines are highly flexible and can soon be straightened after being bent or twisted. Each spine is optically a single crystal of calcite and it is enveloped by rhizopodia possessing large numbers of microtubules. These are thought to provide the force necessary to straighten the spines (Bé *et al.*, 1977).

6.0 ECOLOGY

6.1 Abiotic factors

Among the abiotic factors which may influence the distribution of planktonic foraminifera are temperature, salinity, water density, nutrients, light, turbidity and currents. Some are limiting, i.e. beyond a certain threshold, a given species will be unable to reproduce or may die. Others control abundance, e.g. availability of nutrients. Some are directly interrelated, e.g. density is controlled by temperature and salinity. Others may be indirectly related, e.g. most nutrient-rich waters are cold.

It is a complex matter to try to determine the controls on the distribution and abundance of individual species. All attempts to do so are based on the premise that individuals are in equilibrium with their environment. This may not be true, because if a species is responding to an environmental stimulus, there will inevitably be some delay. Also, very rarely are environmental parameters measured exactly where the plankton sample is collected and even if they were they would be instantaneous observations giving no clue to diurnal or longer-term variations. Nevertheless, attempts have been made to define the optimum conditions for individual taxa (Table 2) from which it can be seen that the examples selected have different optimum requirements.

In an attempt to reconstruct Pleistocene sea surface conditions, Kipp (1976) related the abundance of foraminifera in core top sediment samples to oceanographic data in 1° quadrangles of sea surface. He then used a transfer function to estimate palaeoceanographic conditions from fossil assemblages. The transfer function consists of a set of equations relating a set of oceanographic parameters to faunal data. Kipp used the following parameters: temperature and salinity at the surface and at 100 m for winter, spring, summer and autumn. Statistically

Table 2 Estimated optima and standardized ranges of selected Indian Ocean species[a]

Species	Sea surface		100 m
	Temperature (°C)	Salinity (‰)	$PO_4 - P$ (mg at /l)
Neogloboquadrina pachyderma	4.8 ± 4.7	34.0 ± 0.4	1.7 ± 0.5
Globigerina bulloides	13.4 ± 7.8	34.8 ± 0.7	1.0 ± 0.6
Globorotalia truncatulinoides	20.3 ± 2.8	35.6 ± 0.3	0.3 ± 0.3
Globigerinoides sacculifer	25.0 ± 3.2	34.9 ± 0.5	0.5 ± 0.3

[a] Data from Bé and Hutson (1977).

independent parameters were determined by calculating linear correlation coefficients between all pairs of variables. The four covariance groups – surface temperatures, temperatures at 100 m, surface salinities and salinities at 100 m – showed high within-groups correlation, but between-groups correlation was lower. A comparison of the measured and estimated parameters for a given faunal assemblage showed standard errors of 1.165–1.698°C for temperature and 0.308–0.389‰ for salinity.

This approach has been used in CLIMAP studies. During the last glacial maximum, there was an expansion of polar water and an equatorwards shift of the polar front to above latitude 40° in both hemispheres. However, large areas in the tropical and subtropical regions had warm sea surface temperatures much as now (McIntyre, 1981). The biogeographical pattern of the planktonic foraminifera (Fig. 8) shows an extension of polar faunas and narrowing of the subpolar and transitional zones, but limited change in the tropics.

6.2 Biotic factors

Biological factors include symbiotic relationships, food supply and diet, predation, interfaunal relationships and productivity (Bé and Hutson, 1977). Symbiosis and food supply have already been discussed. Neither is competitive as planktonic foraminifera appear to rely on the food coming to them by chance. There is no evidence that they actively pursue prey.

Predation by other organisms on planktonic foraminifera must take its toll, but there is no evidence that it is species-selective, although it may well be size-selective.

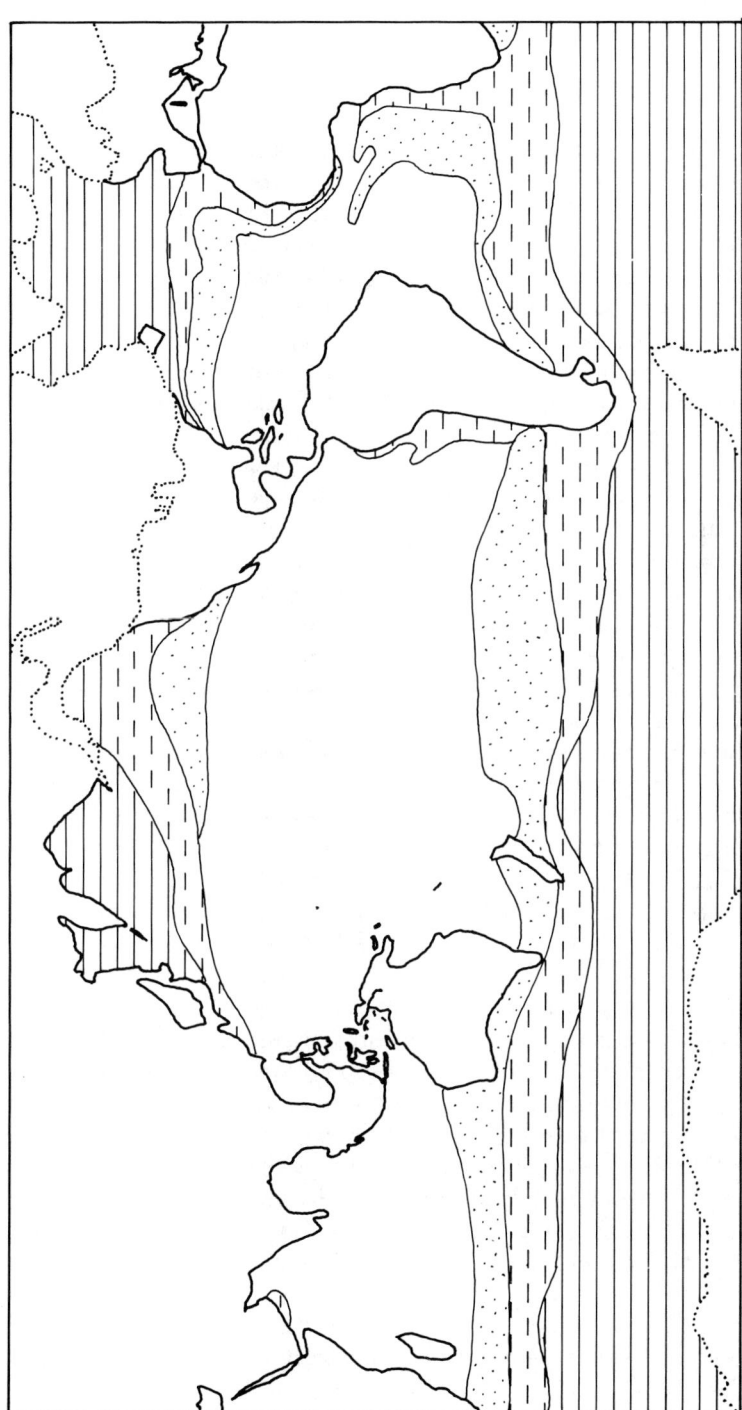

Fig. 8. Distribution of planktonic foraminifera at the last glacial maximum (based on McIntyre, 1981). Shading as in Fig. 3. Land area delimited for lowered sea-level. Areas of ice defined by dotted line.

7.0 CONCLUSIONS

It could be argued that planktonic foraminifera have developed uncompetitive strategies between species. This is achieved by each having a slightly different set of optimum conditions for reproduction and perhaps by subtle partitioning of space both vertically and horizontally. During the course of seasonal changes, species take it in turn to peak in abundance as their individual optimum conditions are experienced. This is a very simplistic view and in reality the situation is more complex for certain species are always rare.

For warm water planktonic foraminifera, reproduction appears always to be sexual and to be controlled by a lunar cycle. The juveniles live in the euphotic zone and initially are herbivorous. They also commonly contain symbionts. Chambers are added at intervals of several days and usually during the night. Once large enough, they become omnivorous and feed on copepods. Towards the end of their adult life, they start to sink and descend into colder water and into darkness. They may add additional chambers and a calcite crust, and then they reproduce. The life-cycle perhaps takes one lunar month. It is likely that there is very high infant mortality and this is allowed for by the prolific production of gametes.

The fact that planktonic foraminifera appear to reproduce principally, if not solely, through sexual reproduction, perhaps partly explains why they have undergone continuous and commonly rapid evolution through geological time.

8.0 FUTURE RESEARCH

A primary objective must be to culture one or more species through several generations. Then it may be possible to investigate the full range of physiological and morphological attributes under known environmental conditions. Further study of how much of the variation in carbon isotopic composition is due to vital effects (especially the presence or absence of symbionts and their metabolic state at the time the calcite was deposited) and how much can be attributed to environmental effects will improve the accuracy of the technique. It is also possible that trace element geochemistry of the test will reveal something of the control of environmental variables. Much awaits attention and the input of new approaches by imaginative young researchers.

ACKNOWLEDGEMENTS

I am grateful to Mrs A. Williams for typing the manuscript, Mr J. Jones for preparing the photographs, Drs O.R. Anderson and J. Lee for helpful comments and Professor F.T. Banner for advice on current taxonomic usage.

REFERENCES

Anderson, O.R. and Bé. A.W.H. (1976). The ultrastructure of a planktonic foraminifer, *Globigerinoides sacculifer* (Brady), and its symbiotic dinoflagellates. *J. Foram. Res.* **6**:1–21.

Anderson, O.R. and Faber, W.W. (1984). An estimation of calcium carbonate deposition rate in a planktonic foraminifer *Globigerinoides sacculifer* using ^{45}C as a tracer: A recommended procedure for improved accuracy. *J. Foram. Res.* **14**:303–308.

Anderson, O.R., Spindler, M., Bé, A.W.H. and Hemleben, Ch. (1979). Trophic activity of planktonic foraminifera. *J. Mar. Biol. Assoc. U.K.* **59**:791–799.

Banner, F.T. (1982). A classification and introduction to the Globigerinacea. In *Aspects of Micropalaeontology* (eds F.T. Banner and A.R. Lord), pp. 142–239. Allen and Unwin, London.

Bé, A.W.H. (1967). Foraminifera families: Globigerinidae and Globorotaliidae. Fiches d'identification due zooplancton. Conseil perm. pour l'explor. Mer, Zooplankton. Sheet 108.

Bé, A.W.H. (1968). Shell porosity of recent planktonic foraminifera as a climate index. *Science* **161**:881–884.

Bé, A.W.H. (1977). An ecological, zoogeographic and taxonomic review of recent planktonic foraminifera. In *Oceanic Micropalaeontology* (ed. A.T.S. Ramsay), pp. 1–100. Academic Press, London.

Bé, A.W.H. (1980). Gametogenic calcification in a spinose planktonic foraminifer *Globigerinoides sacculifer* (Brady). *Mar. Micropaleontol.* **5**:283–310.

Bé, A.W.H. (1982). Biology of planktonic foraminifera. In *Foraminifera: Notes for a Short Course* (ed. T.W. Broadhead), pp. 51–92. Studies in Geology Vol. 6. University of Tennessee, Knoxville.

Bé, A.W.H. and Hemleben, Ch. (1970). Calcification in a living planktonic foraminifer, *Globigerinoides sacculifer* (Brady). *N. Jb. Geol. Paläont. Abh.* **134**:221–234.

Bé, A.W.H. and Hutson, W.H. (1977). Ecology of planktonic foraminifera and biogeographic patterns of life and fossil assemblages in the Indian ocean. *Micropaleontol.* **23**:369–414.

Bé, A.W.H., Harrison, S.M. and Lott, L. (1973). *Orbulina universa* d'Orbigny in the Indian Ocean. *Micropaleontol.* **19**:150–192.

Bé, A.W.H., Hemleben, Ch., Anderson, O.R., Spindler, M., Hacunda, J. and Tuntivate-Choy, S. (1977). Laboratory and field observations of living planktonic foraminifer. *Micropaleontol.* **23**:155–179.

Bé, A.W.H., Hemleben, Ch., Anderson, O.R. and Spindler, M. (1979). Chamber formation in planktonic foraminifera. *Micropaleontol.* **25**:294–307.

Bé, A.W.H., Caron, D.A. and Anderson, O.R. (1981). Effects of feeding frequency on life processes of the planktonic foraminifer *Globigerinoides sacculifer* (Brady) in laboratory culture. *J. Mar. Biol. Assoc. U.K.* **61**:257–277

Bé, A.W.H., Spero, H.J. and Anderson, O.R. (1982). Effects of symbiont elimination and reinfection on the life processes of the planktonic foraminifer *Globigerinoides sacculifer*. *Mar. Biol.* **70**:73–86.

Bé, A.W.H., Bishop, J.K.B., Sverdlove, M. and Gardner, W.D. (1985). Standing stock, vertical distribution and flux of planktonic foraminifera in the Panama Basin. *Mar. Micropaleontol.* **9**:307–333.

Berger, W.H. (1971a). Planktonic foraminifera: Sediment production in an oceanic front. *J. Foram. Res.* **1**:95–118.

Berger, W.H. (1971b). Sedimentation of planktonic foraminifera. *Mar. Geol.* **11**:325–358.

Berger, W.H. and Piper, D.J.W. (1977). Planktonic foraminifera: Differential settling, dissolution and redeposition. *Limnol. Oceanogr.* **17**:275–287.

Boltovskoy, E. (1973). Daily vertical migration and absolute abundance of living planktonic foraminifera. *J. Foram. Res.* **3**:89–94.

Bouvier-Soumagnac, Y. and Duplessy, J.C. (1985). Carbon and oxygen isotopic composition of planktonic foraminifera from laboratory culture, plankton tows and recent sediment: Implications for the reconstruction of paleoclimatic conditions and of the global carbon cycle. *J. Foram. Res.* **15**:302–320.

Brasier, M.D. (1986). Form, function, and evolution in benthic and planktic foraminiferid test architecture. *Systematics Ass. Spec. Pub.* **30**:251–268.

Brummer, G.J.A. and Kroom, D. (1988). Genetically controlled planktonic foraminiferal coiling ratios as tracers of past ocean dynamics. In *Planktonic Foraminifera as Tracers of Ocean–Climate History: Ontology, Relationships and Preservation of Modern Species and Stable Isotopes, Phenotypes and Assemblage Distribution in Different Water Masses*, pp. 293–297. Free University Press, Amsterdam.

Brummer, G.J.A., Hemleben, Ch. and Spindler, M. (1986). Planktonic foraminiferal ontogeny and new perspectives for micropalaeontology. *Nature* **319**:50–52.

Brummer, G.J.A., Hemleben, Ch. and Spindler, M. (1987). Ontogeny of extant spinose planktonic foraminifera (Globigerinidae): A concept exemplified by *Globigerinoides sacculifer* (Brady) and *G. ruber* (d'Orbigny). *Mar. Micropaleontol.* **12**:357–381.

Caron, D.A., Bé, A.W.H. and Anderson, O.R. (1982). Effects of light intensity on life processes of the planktonic foraminifer *Globigerinoides sacculifer* in laboratory culture. *J. Mar. Biol. Assoc. U.K.* **62**:435–451.

Charles, C.D. (1991). Late Quaternary Ocean Chemistry and Climate Change from an Antarctic Deep Sea Sediment Perspective. Ph.D. Thesis, Columbia University, New York.

Cifelli, R. and Benier, C.S. (1976). Planktonic foraminifera from near the West African coast and a consideration of faunal parcelling in the North Atlantic. *J. Foram. Res.* **6**:258–273.

Cifelli, R. and McCloy, C. (1983). Planktonic foraminifera and euthecosomatus pteropods in the surface waters of the North Atlantic. *J. Foram. Res.* **13**:91–107.

Deuser, W.G., Ross, E.H., Hemleben, Ch. and Spindler, M. (1981). Seasonal changes in species composition, numbers, mass, size, and isotopic composition of planktonic foraminifera settling into the deep Sargasso Sea. *Palaeogeog., Palaeoclimatol., Palaeoecol.* **33**:103–127.

Duplessy, J.C., Bé, A.W.H. and Blanc, P.L. (1981a). Oxygen and carbon isotopic composition and biogeographic distribution of planktonic foraminifera in the Indian Ocean. *Palaeogeog., Palaeoclimatol., Palaeoecol.* **33**:9–46.

Duplessy, J.C., Blanc, P.L. and Bé, A.W.H. (1981b). Oxygen-18 enrichment of planktonic foraminifera due to gametogenic calcification below the euphotic zone. *Science* **213**:1247–1250.

Erez, J. and Honjo, S. (1981). Comparison of isotopic composition of planktonic foraminifera in plankton tows, sediment traps and sediment. *Palaeogeog., Palaeoclimatol., Palaeoecol.* **33**:129–156.

Erez, J. and Luz, B. (1983). Experimental paleotemperature equation for planktonic foraminifera. *Geochim. Cosmochim. Acta* **47**:1025–1931.

Fairbanks, R.G. and Wiebe, P.H. (1980). Foraminifera and chlorophyll maximum: Vertical distribution, seasonal succession and palaeoceanographic significance. *Science* **209**:1524–1526.

Fok-Pun, L. and Komar, P.D. (1983). Settling velocities of planktonic foraminifera: Density variations and shape effects. *J. Foram. Res.* **13**:60–68.

Gastrich, M.D. and Bartha, R. (1988). Primary productivity in the planktonic foraminifer *Globigerinoides ruber* (d'Orbigny). *J. Foram. Res.* **18**:137–142.

Healy-Williams, N., Ehrlich, R. and Williams, D.F. (1985). Morphometric and stable isotopic evidence for subpopulations of *Globorotalia truncatulinoides*. *J. Foram. Res.* **15**:242–253.

Hecht, A. (1976). The oxygen isotopic record of foraminifera in deep-sea sediment. In *Foraminifera* (eds R.H. Hedley and C.G. Adams), Vol. 2, pp. 1–43. Academic Press, London.

Hemleben, Ch. and Spindler, M. (1983). Recent advances in research on living planktonic foraminifera. *Utrecht Micropaleontol. Bull.* **30**:141–170.

Hemleben, Ch., Spindler, M., Breitinger, I. and Deuser, W.G. (1985). Field and laboratory studies on the ontogeny and ecology of some globorotaliid species from the Sargasso Sea off Bermuda. *J. Foram. Res.* **15**:254–272.

Hemleben, Ch., Spindler, M., Breitinger, I. and Ott. R. (1987). Morphological and physiological responses of *Globigerinoides sacculifer* (Brady) under varying laboratory conditions. *Mar. Micropaleontol.* **12**:305–324.

Hemleben, Ch., Spindler, M. and Anderson, O.R. (1989). *Modern Planktonic Foraminifera*. Springer-Verlag, New York.

Kahn, M. and Williams, D.F. (1981). Oxygen and carbon isotopic composition of living planktonic foraminifera from the north-east Pacific Ocean. *Palaeogeog., Palaeoclimatol., Palaeoecol.* **33**:47–69.

Kennett, J.P. (1976). Phenotypic variation in some Recent and Late Cenozoic planktonic foraminifera. In *Foraminifera* (eds R.H. Hedley and C.G. Adams), Vol. 2, pp. 11–170. Academic Press, London.

Kipp, N.G. (1976). New transfer function for estimating past sea-surface conditions from sea-bed distribution of planktonic foraminiferal assemblages in the North Atlantic. *Mem. Geol. Soc. Am.* **145**:3–41.

Lipps, J.H. and Krebs, W.N. (1974). Planktonic foraminifera associated with Antarctic sea ice. *J. Foram. Res.* **4**:80–85.

Loeblich, A.R. and Tappan, J. (1988). *Foraminiferal Genera and Their Classification*. Van Nostrand Reinhold, New York.

McIntyre, A. (co-ordinator) (1981). Seasonal reconstructions of the Earth's surface at the last glacial maximum. *Geol. Soc. Amer.* **MC-36**.

McIntyre, A., Lipp, N.G., Bé, A.W.H., Crowley, T., Kellogg, T., Gardner, J.V., Prell, W. and Ruddiman, W.F. (1976). Glacial North Atlantic 18,000 years ago: A CLIMAP reconstruction. *Mem. Geol. Soc. Am.* **145**:43-76.

Pierrot-Bults, R.C. and van der Spoel, S. (1979) General conclusions in *Zoogeography and Diversity in Plankton* (eds S. van der Spoel and A.C. Pierrot-Bults), pp. 356-358. Edward Arnold, London.

Ravelo, A.C. (1991). Reconstructing the Tropical Atlantic Seasonal Thermocline using Planktonic Foraminifera. Ph.D. Thesis, Columbia University, New York.

Reynolds, L.A. and Thunell, R.C. (1986). Seasonal production and morphologic variation of *Neogloboquadrina pachyderma* (Ehrenberg) in the northeast Pacific. *Micropaleontol.* **32**:1-18.

Spero, H.J. and Parker, S.L. (1985). Photosynthesis in the symbiotic planktonic foraminifer *Orbulina universa*, and its potential contribution to oceanic primary productivity. *J. Foram. Res.* **15**:273-281.

Spindler, M. and Dieckmann, G.S. (1986). Distribution and abundance of the planktonic foraminifer *Neogloboquadrina pachyderma* in sea ice of the Weddell Sea (Antarctic). *Polar Biol.* **5**:185-191.

Spindler, C., Hemleben, Ch., Bayer, U., Bé, A.W.H. and Anderson, R.O. (1979). Lunar periodicity of reproduction in the planktonic foraminifer *Hastigerina pelagica*. *Mar. Ecol. Prog. Ser.* **1**:61-64.

Spindler, M., Hemleben, Ch., Salomons, J.B. and Smit, L.P. (1984). Feeding behaviour of some planktonic foraminifers in laboratory cultures. *J. Foram. Res.* **14**:237-249.

Thunell, R.C. and Honjo, S. (1981). Planktonic foraminiferal flux to the deep ocean: Sediment trap results from the tropical Atlantic and the central Pacific. *Mar. Geol.* **40**:237-253.

Thunell, R.C. and Reynolds, LA. (1984). Sedimentation of planktonic foraminifera: Seasonal changes in species flux in the Panama Basin. *Micropaleontol.* **30**:243-262.

Vella, P. (1974). Coiling ratios of *Neogloboquadrina pachyderma* (Ehrenberg): Variations in different size fractions. *Bull. Geol. Soc. Am.* **85**:1421-1424.

Wefer, G., Suess, E., Balzer, W., Liebzeit, G., Müller, P.J., Ungerer, C.A. and Zenk, W. (1982). Fluxes of biogenic components from sediment trap deployment in circum-polar waters of the Drake Passage. *Nature* **299**:145-147.

Williams, D.F., Bé, A.W.H. and Fairbanks, R.G. (1979). Seasonal oxygen isotopic variations in living planktonic foraminifera off Bermuda. *Science* **206**:447-449.

Williams, D.F., Bé, A.W.H. and Fairbanks, R.G. (1981). Seasonal stable isotopic variations in living planktonic foraminifera from Bermuda plankton tows. *Palaeogeog., Palaeoclimatol., Palaeoecol.* **33**:77-102.

GLOSSARY OF TERMS

Cancellate Texture like a honeycomb.

Keel A marginal thickening of the test.

Kummerform Having a smaller final chamber than should be the case with logarithmic growth.

Proloculus The first chamber.

Pustulose Wall texture with short knobs (pustules) (as distinct from spines).

9
Life-cycles of foraminifera

JOHN J. LEE, WALTER W. FABER JR, O. ROGER ANDERSON and JAN PAWLOWSKI

1.0	Historical synopsis and classical life-cycle	285
2.0	Terminology and overall model of life-cycle options in foraminifera	289
	2.1 Terms associated with sexual reproduction	289
	2.2 Terms associated with asexual reproduction or the asexual phase	291
3.0	Life-cycle programmes in foraminifera	292
	3.1 Classical life-cycles: Gametogamy	294
	3.2 Paraclassical life-cycles: Gamontogamy	297
	3.3 Paraclassical life-cycles: Autogamy	302
	3.4 Larger foraminifera: Trimorphic, paratrimorphic life-cycles	305
	3.5 Apogamic life-cycles	310
4.0	Planktonic foraminifera	312
	4.1 Gamete production	313
	4.2 Gametogenesis in planktonic foraminifera	314
	4.3 Reproductive cycles	315
	4.4 Ontogeny	316
	4.5 Five-stage concept of planktonic foraminiferal ontogenesis	317
5.0	Areas for further research	319
	Plate section: Figures 12–45	321
	References	328

1.0 HISTORICAL SYNOPSIS AND CLASSICAL LIFE-CYCLE

Interest in the life-cycles of foraminifera stretches back to the mid-nineteenth century (Schultze, 1854), if not earlier. By the end of the century, dimorphism (form A-megalospheric *vs* form B-microspheric) was well recognized in many distinctly related species (reviewed and tabulated by Lister, 1895). Although others (e.g. D'Archaic and Haime, De Hantken and De La Harpe) had commented on the recognizable differences between the two forms, Lister (1895) credited Munier-Chalmas (1880) and Schlumberger (1875) (Munier-Chalmas and Schlumberger, 1883) for the first clear recognition that each species of foraminifera consists of two sets of individuals: (1) those with a large proloculus (*mégalosphère*) and which attain an overall small size, and (2) those with a small proloculus (*micrósphère*) which attain a larger overall size. Using picro-carmine, counterstained with methylene blue, Lister was able to characterize the nuclear cytology of both forms of *Elphidium crispum* and developed what we now recognize as the classical life-cycle (Fig. 1).

Although later cytological studies showed that the position of meiosis was incorrect (Le Calvez, 1938), Lister's (1895) and Schaudinn's (1895a,b) original conceptual framework was a good model life-cycle

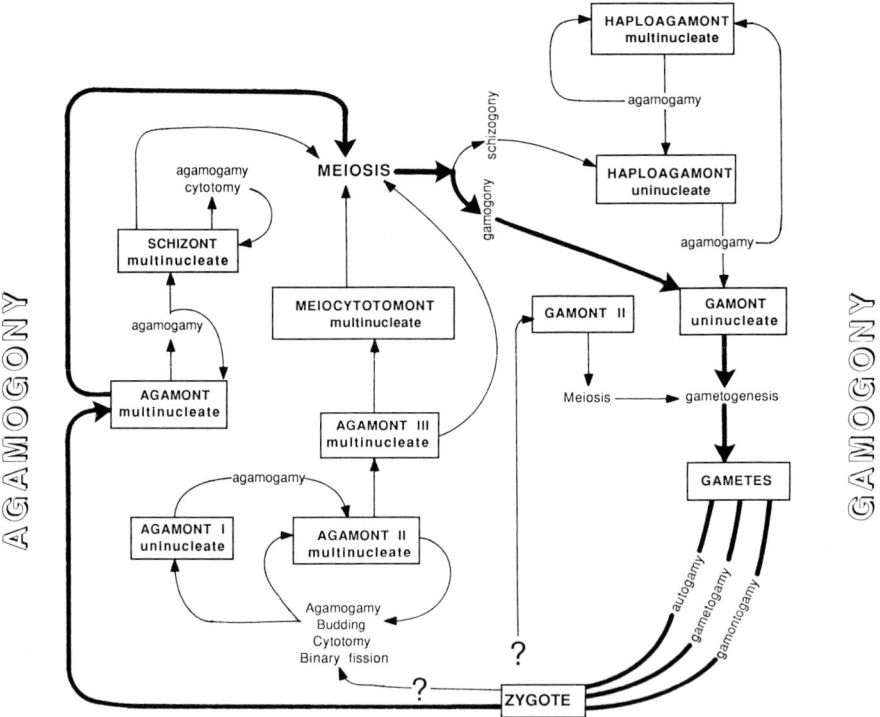

Fig. 1. A diagram of the life-cycle options found in foraminifera. The steps in the *classical* and *paraclassical* life-cycles are indicated by the bold arrows. In the classical and paraclassical cycles, there is a regular alternation between a haploid, uninucleate, megalospheric *gamont* and a diploid, multinucleate, microspheric *agamont*. In the classical life-cycle, sexual reproduction is gametogamic. In paraclassical cycles, sexual processes are either gamontogamic or autogamic.

for a foraminifer with regular alternation of sexually reproducing and asexually reproducing generations (metagenesis). The *gamont* is now recognized as a haploid, often megalospheric, smaller, sexually reproducing, uninucleate individual for most of its life span. The *agamont* is diploid, microspheric, larger, multinucleate and reproduces asexually after meiotic divisions of the nuclei.

Shortly after the classical life-cycle concept was proposed, it became obvious to contemporary workers that there were some species which did not seem to fit into the classic life-cycle. Rhumbler (1909) suggested that a third, biologically different form, may be part of the life-cycle of some species. In foraminifera with such a cycle, the third form comes between the agamont and the gamont. Microspheric agamonts produce megalospheric forms which are not gamonts but which are asexually reproducing *schizonts*. This life-cycle modification, later dubbed *trimorphic* (Fig. 2), was further developed by Hofker (1930a,b). On the basis of his biometric studies of living and dead specimens of *Streblus flavensis* and *Quinqueloculina circularis*, he suggested that the succession of generations was obligate (*Holotrimorphic*). Microspheric agamonts (form B) asexually produce megalospheric forms, *schizonts* (A2), which reproduce sexually to complete the cycle. The proloculus of form A1 is smaller than that of form A2 but larger than that of form B.

A later study of the life-cycle of *Planorbulina mediterranensis* by Le Calvez (1938), led to further modification of life-cycle concepts in foraminifera and paved the way for the much broader view of the life-cycles in the group as we partially understand them today. In his 1938 study, Le Calvez found that "megalospheric" agamonts of *P. mediterranensis* can undergo repetitive asexual generations before producing sexually reproducing gamonts. Because the succession of the three generations was not obligate, the concept of a *paratrimorphic* life-cycle (Fig. 2) was introduced, although he recognized that the three biologically distinct generations he found did not necessarily result in distinct morphotypes. Perhaps his greatest contribution to the evolving conceptual understanding of the life-cycles of foraminifera was his emphasis on describing the biological characteristics of each generation (e.g. ploidy, type of reproduction).

Studies in the late 1930s and 1950s (e.g. Myers, 1935, 1936; Grell, 1954, 1957b) led to further erosion of the dimorphic and trimorphic concepts as exclusive foci for organizing life-cycle interpretations (Hedley, 1964; Boltovskoy and Wright, 1976). Grell (1954, 1957a,b, 1958a, 1979), in his extensive studies of the tiny *Rotaliella heterocaryotica, R. roscoffensis, Metarotaliella* sp. and *Rubratella intermedia*, found no differences in prolocular sizes of the different generations. Myers' (1935,

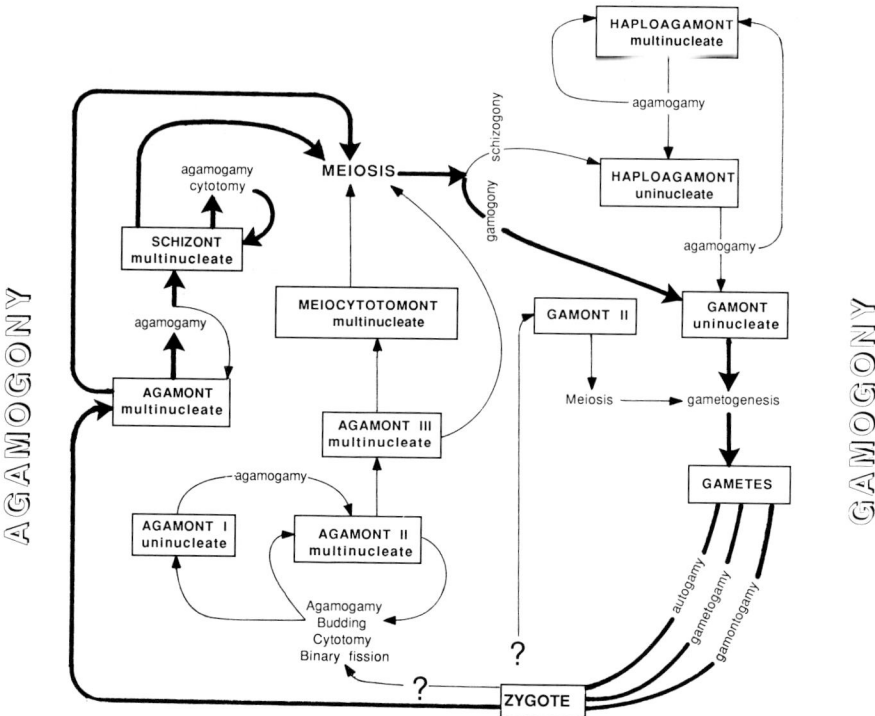

Fig. 2. A diagram of the life-cycle options found in foraminifera with *trimorphic* and *paratrimorphic* life-cycles indicated by the bold arrows. In the trimorphic life cycle, there is a regular sequence of generations beginning with a haploid, uninucleate, megalospheric *gamont*, followed by gametogenesis, syngamy and the development of a diploid, multinucleate, microspheric *agamont*, which after agamogony gives rise to a diploid, multinucleate, megalospheric *schizont*. The schizont undergoes meiosis and gamogony to give rise to the gamont to complete the life-cycle. In the paratrimorphic life-cycle, the schizont can also undergo agamony or cytotomy to produce another generation of schizonts or it can undergo meiosis and gamogony to produce a gamont.

1936) observations that the proloculus of the gamonts of *Spirillina vivipara* and *Patellina corrugata* were smaller than those of the agamonts, reminded workers in the field that Schaudinn (1895a) had found that some of the proloculi of megalospheric forms of *Elphidium crispum* are smaller than the largest proloculi of microspheric forms. Le Calvez (1938) suggested that the number of nuclei enclosed during the cytokinesis of agamonts had a direct relationship to the size of the proloculus of the offspring. This idea is supported by aspects of the life-cycle of the tiny new species, *Rotaliella* sp. (Elat isolate, Pawlowski and Lee, in prep.). In this species, the agamont has a variable number of generative nuclei (1–3). After meiosis, karyokinesis and cytokinesis, either 4, 8 or 12 gamonts are produced per parent. The size of the proloculus of a cohort of 12 gamonts is about half the size of those producing 4 gamonts. The proloculi of those in a cohort of 8 gamonts are intermediate.

While for historical and partially logical reasons we recognize classical dimorphic, or modified paratrimorphic, life-cycles as central paradigms for foraminifera, we also recognize additional variations in life-cycles which heretofore have not yet been woven into a conceptual framework which allows direct comparison of this aspect of the biology of the group. It is our aim in this review to show not only the comparatively rich diversity of foraminiferal life-cycles, but also look critically at the present gaps in our knowledge which should be targets for future studies.

2.0 TERMINOLOGY AND OVERALL MODEL OF LIFE-CYCLE OPTIONS IN FORAMINIFERA

An overall model of the life-cycle options of foraminifera is presented in Fig. 3. Wherever possible we have drawn on terminology that has traditionally been used in the study of foraminifera and which is in consonance with the terminology generally in wide usage in protozoology or protistology (e.g. Grell, 1967; Corliss and Lom, 1985; Margulis *et al.*, 1990). For the sake of clarity, the terms are defined briefly below. An extensive and very thoughtful discussion of possible life-cycle options in foraminifera is found in Arnold's (1955b) paper on *Allogromia laticollaris*.

2.1 Terms associated with sexual reproduction

Autogamy A kind of sexual reproduction in which zygotes are formed by the fusion of two haploid nuclei from the same parental cell (synonym: pedogamy).

Gamete A mature haploid sex cell that unites with another sex cell of the same (isogamete) or different (anisogamete) type to form a diploid zygote.

Gametogamy The fusion of haploid gametes, released by two or more parents into the medium, to form a diploid zygote (partial synonym: syngamy).

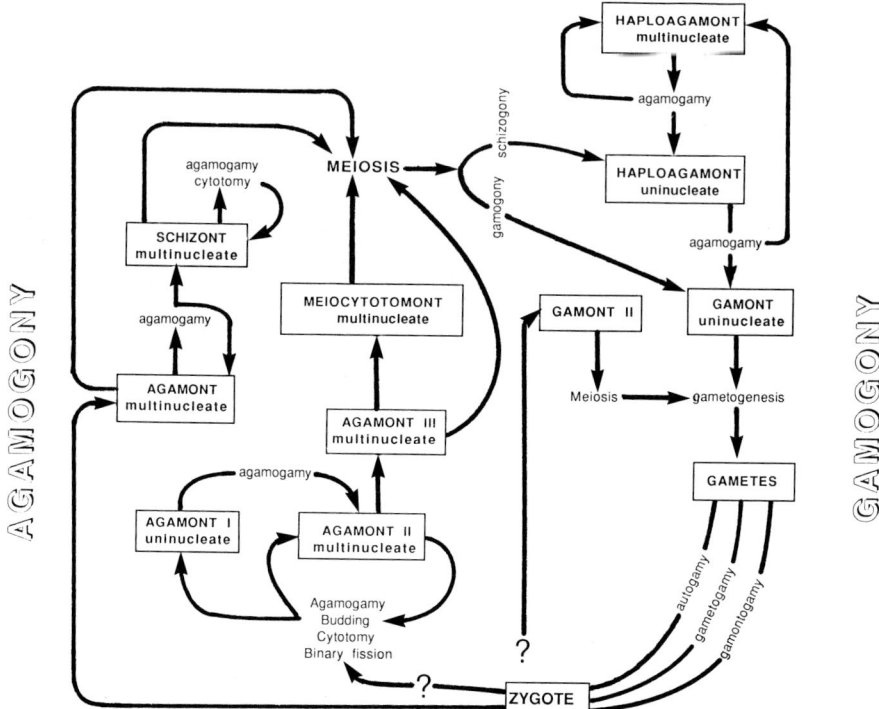

Fig. 3. A composite diagram with all the known life-cycle options which have been found in different species of foraminifera. There may be homology between *agamont II* and *agamont* and between *agamont III* and *schizont* but this is not yet verified or disproved.

Gamont That stage in the life-cycle which produces gametes. The gamont in foraminifera is haploid, uninucleate for most of its part in the life-cycle, and usually megalospheric and smaller than other life-cycle stages.

Gamont II A term for a postulated diploid gamont of planktonic foraminifera which forms after syngamy and which undergoes meiosis and gametogenesis to produce gametes. It is possible that this might be an undiscovered aspect in the life-cycles of other foraminifera.

Gamontogamy The union or pairing of two or more gamonts with the subsequent production of gametes and their fusion. Gamontogamy may involve the formation of a gamontocyst ("nuptial chamber") which encircles the joined gamonts. In one case, *Rubratella intermedia*, gamonts of unequal size join in gamontogamy (synonym: plastogamy).

Meiocytotomont A rarely formed cell in the life-cycle of *Allogromia laticollaris* (CSH strain, Fig. 40). This giant multinucleate cell undergoes meiosis and cytotomy to form multinucleate haploid gamonts (synonym: gametocytotomont) (McEnery and Lee, 1976).

2.2 Terms associated with asexual reproduction or the asexual phase

Agamogony   A series of asexual cell divisions producing cells that are neither gametes nor capable of forming gametes.
Agamont   An asexually reproducing organism at a stage in the life-cycle which does not produce gametes and does not have structures associated with sexual reproduction. In foraminifera, this multinucleated, diploid, microspheric stage is commonly understood to be formed by the growth of a zygote. In trimorphic life-cycles, the agamont is larger than the shizont.
Agamont I   An asexually reproducing organism in an apogamic life-cycle characterized by being uninucleate and diploid (e.g. Fig. 8).
Agamont II   An asexually reproducing organism in an apogamic life-cycle characterized by being multinucleate and diploid (e.g. Fig. 8).
Agamont III   Gamont mother cell in the *Allogromia laticollaris* life-cycle (McEnery and Lee, 1976; Fig. 8).
Binary fission   Cell division in which the filial products are equal, or nearly so, in size (e.g. Fig. 30).
Brood chamber   A permanent internal alteration in the chamber formation pattern of larger foraminifera resulting in larger outer chambers within which the internal formation (cytokinesis) of filial cells is accomplished (synonyms in other protists: crypt in chonotrichs; brood pouch, embryo sac and marsupium in suctorians).
Budding   Cell division in which the filial products are extremely unequal in size. The smaller of the filial cells is known as a bud. Both exogenous (external, evaginative) and endogenous (internal, invaginative) budding are found in allogromids (e.g. Figs 8, 29).
Cytokinesis   Cytoplasmic division exclusive of nuclear division. Not necessarily directly coupled with karyokinesis in foraminifera.
Cytotomy   Cell division of a multinucleated polyoral cell in which cell division (cytokinesis) is asynchronous and gives rise to unequal-sized filial cells (e.g. Figs 31, 32).
Gamogony   The process of the formation of gamonts by schizogony.
Generative nuclei   Small compact nuclei in heterokaryotic foraminifera which are the antecedents of the nuclei of the next generation (e.g. Fig. 26) (synonym: micronuclei in ciliates).

Haploagamont A haploid agamont capable of asexually producing more of its own kind by agamogony or by undergoing gamogony to produce gamonts. This term replaces schizont-4 used by Arnold (1955b) for a stage in the life-cycle of the strain of *A. laticollaris* he studied. There is a contradiction in Arnold's combination of terms (schizo-gamont) associated in the mononucleated stage in the life-cycle. We would call a cell which produces more of its own kind by agamogony a haploagamont, and a cell which produces gametes a gamont.

Heterokaryotic Foraminifera with nuclear dimorphism, and two or more morphologically and genetically differentiated nuclei in a common cytoplasm (e.g. Figs 25–27).

Karyokinesis The division of a nucleus to form two filial nuclei. In foraminifera, karyokinesis and cytokinesis are not necessarily directly coupled.

Plasmotomy The fission of a multinucleate protistan cell in which karyokinesis is temporally separated from cytokinesis. Subcategories are budding, cytotomy and binary fission. This term has been applied in foraminiferal life-cycles to the process which asexually produces filial offspring from a multinucleate parent, which itself remains viable and capable of regrowing and reoccupying the partially emptied parental test.

Schizogony Multiple cell disivion in which the cytokinesis of the parental cell is synchronous (or nearly so) and the filial cells are equal (or nearly so) in size. The term has a slightly different meaning among the "sporozoan" groups, where a distinction is made between merogony, gametogony and sporogony.

Schizont The stage or form in the life-cycle that undergoes schizogony (synonyms in other protistan groups: meront and tomont). In foraminifera with trimorphic life-cycles, the schizont is megalospheric, diploid, multinucleate and undergoes gamogony. In a paratrimorphic life-cycle, it may undergo repeated shizogony (agamogony) before undergoing meiosis and production of gamonts (gamogony).

Somatic nuclei The larger of the two kinds of nuclei in foraminifera. The site of mRNA synthesis (e.g. Figs 24, 27) (synonym: macronucleus in ciliates).

3.0 LIFE-CYCLE PROGRAMMES IN FORAMINIFERA

1. *Classical* (gametogamic). A life-cycle in which there is regular alternation between a haploid, uninucleate, gamont stage and a

diploid, multinucleate, agamont stage (Fig. 1). Free-swimming gametes are produced by the gamont and undergo syngamy in the open sea (example: *Elphidium crispum*).

2. *Paraclassical* (gamontogamic). A life-cycle in which there is regular alternation between a haploid, uninucleate, gamont stage and a diploid, multinucleate, agamont stage (Fig. 1). Two – and rarely three – gamonts move together so that their umbilical surfaces appose. A pseudocyst ("nuptual chamber") may be formed around the mating gamonts. Gametes are released within the pseudocyst where they undergo syngamy (example: *Glabratella sulcata*).

3. *Paraclassical* (autogamic). A life-cycle in which there is regular alternation between a haploid, uninucleate, gamont stage and a diploid, multinucleate, agamont stage (Fig. 1). The gamonts may be gamontogamic or not. The essential feature of this life-cycle is that gametes from the same parent fuse (example: *Rotaliella heterocaryotica*).

4. *Trimorphic*. A life-cycle in which there is regular alternation between three stages (Fig. 2). A megalospheric, haploid, uninucleate, gamont stage produces gametes which fuse to form a microspheric, diploid, multinucleate, agamont stage. The agamont asexually (agamogony) produces megalospheric, diploid, multinucleate schizonts, which, in turn, undergo meiosis and gamogony (schizogony) to complete the life-cycle when gamonts are formed (example: *Quinqueloculina circularis*).

5. *Paratrimorphic*. A life-cycle which is a variant of the trimporhic in which the succession of the two asexually reproducing stages (agamont and schizont) is not necessarily regularly followed (Fig. 2). Cytotomy or agamogony can produce successive asexual generations of schizonts. Cytotomy (plasmotomy) is also an option in some of the foraminifera with this life-cycle pattern (example: *Planorbulina mediterranensis*).

6. *Apogamic*. A life-cycle in which asexual reproduction predominates. Sexual reproduction is rare in such cycles or never occurs at all (Fig. 4). All forms of plasmotomy (budding, cytotomy and binary fission) have been observed in some organisms with this life-cycle option (e.g. Fig. 8). In *Allogromia laticollaris*, some of the asexual pathways are diet-dependent (example: some strains of *A. laticollaris*).

7. *Gamic*. A life-cycle which involves successive sexual generations (Fig. 5). Asexual stages are unknown or rare (example: *Hastigerina pelagica*).

8. *Metagenic*. A life-cycle which has regular alternation between haploid sexual and diploid asexual generations. A term which has been borrowed from plant biology and which has been loosely applied to

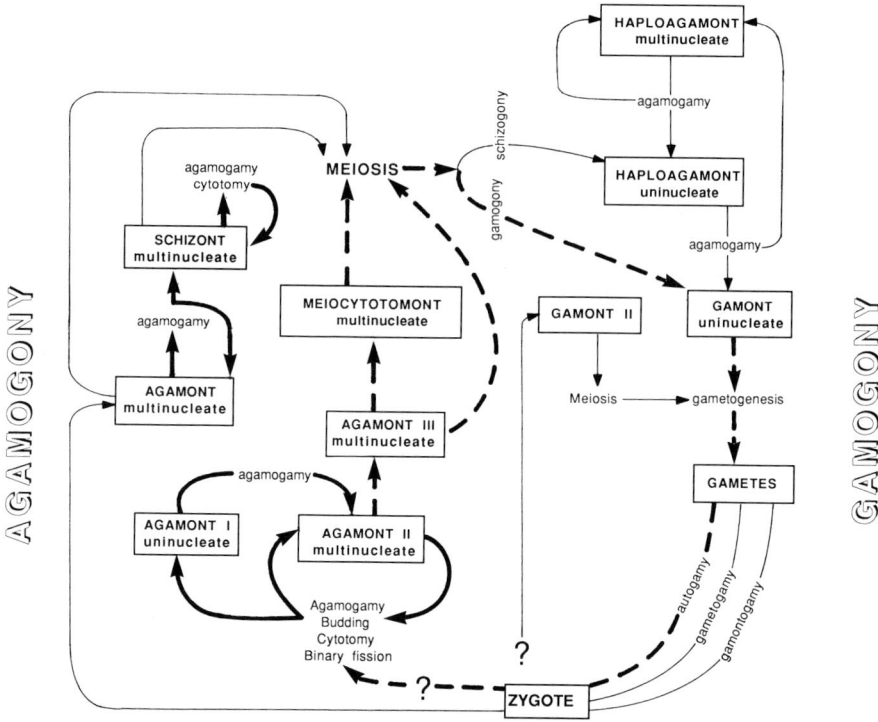

Fig. 4. A diagram of the life-cycle options found in foraminifera with several *apogamic* life-cycle options outlined by bold arrows. The life-cycle of isolates of *Allogromia laticollaris*, which are basically apogamic, are also shown in Fig. 7. Apogamic life-cycles are completed by many repeated generations of asexually reproducing individuals; sexual reproduction is rare or never occurs in these species or strains of foraminifera.

foraminifera. Strictly, it should be applied only to classical and paraclassical life-cycles, but usage among students of foraminiferan life-cycles has broadened the term to include trimorphic and paratrimorphic life-cycles.

3.1 Classical life-cycles: Gametogamy

The length of time needed to complete the classical life-cycle of *Elphidium crispum* has been shown to be environmentally sensitive. Monthly samples taken by Myers (1943a) in the vicinity of Plymouth, U.K., showed seasonal changes in growth and reproduction which correlated well with measurable ecological conditions in the sea. Additional samples were taken along the Mediterranean coast of

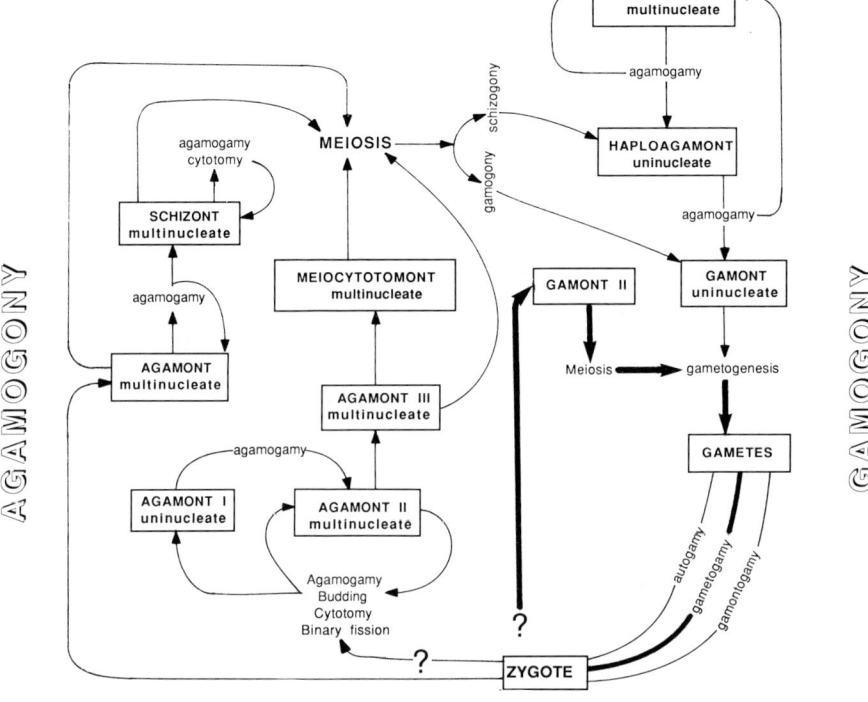

Fig. 5. A diagram of the life-cycle options found in foraminifera with a postulated gamic life-cycle outlined by bold arrows. The life-cycle illustrated has a *gamont* that produces gametes which undergo syngamy to form a *zygote*. The rest of the life-cycle is not known but postulated here. The zygote grows into a *gamont II*, which undergoes meiosis prior to gametogenesis. Alternatively, one could speculate that meiosis is post-zygotic and occurs before gamont II grows and matures.

France, the coast of southern California and the western portion of the Java Sea. Myers concluded that in temperate regions the life span of individual foraminifera is usually 1 year. The whole life-cycle (both sexual and asexual phases) is completed in 2 years in tide pools, but may take longer (3–4 years) in deeper waters. On the other hand, growth is faster in tropical waters, and Myers (1943a) concluded that the life-cycle is completed in 1 year. He also concluded that growth and reproduction of *Elphidium crispum* closely parallels the phytoplankton cycle in the sea. Other species of *Elphidium* – *E. insertum* and *E. translucens* – have seasonal peaks of reproduction in northeastern U.S. salt marshes (Matera and Lee, 1972) but their life-cycles remain to be worked out.

Lee *et al.* (1963) found a paraclassical life-cycle in the agnotobiotic cultured populations of *Rosalina leei* they worked with in the laboratory (Figs 6, 12 15, 17), but Hedley and Wakefield (1967) isolated only apogamic clones of the same species from the cultures they started from collections obtained at Plymouth, U.K. and Wellington, New Zealand. While the details of the life-cycles of many species of foraminifera

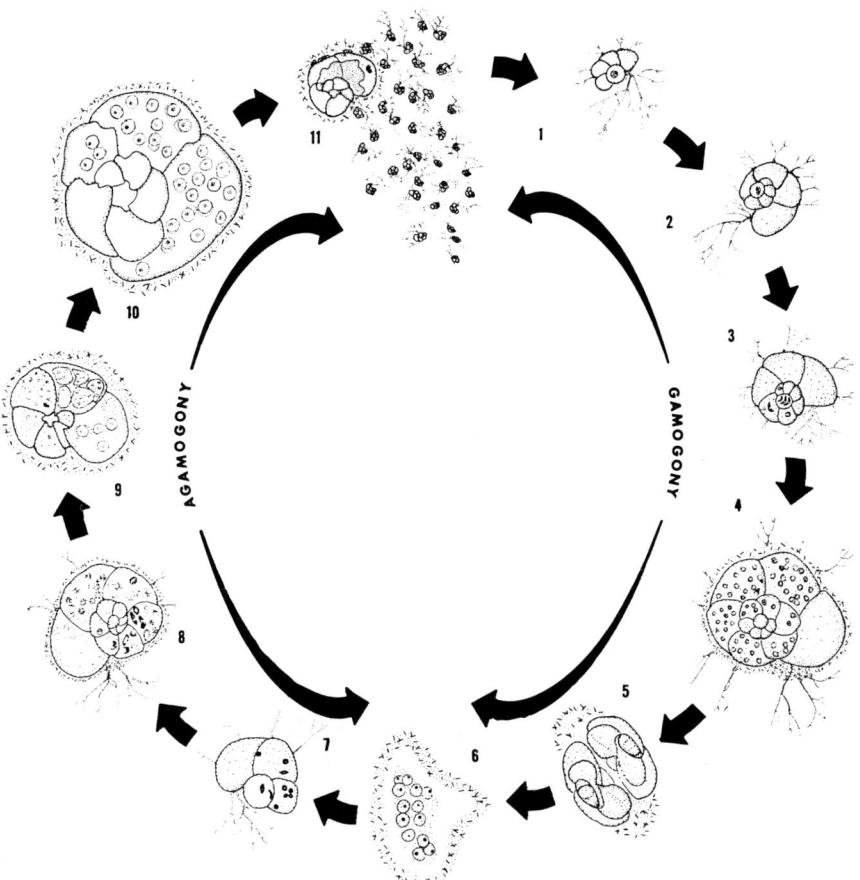

Fig. 6. A diagrammatic representation of the life-cycle of one strain of *Rosalina leei*. 1, Juvenile gamont; 2, mature gamont; 3, gamont with pregametogenic mitosis; 4, gamont with pregametic nuclei; 5, two gamonts in gamontogamy which have formed a "nuptial cyst" around their apposed ventral surfaces; 6, aggregated gametes or zygotes in the chamber between the two gamonts; 7, young agamont; 8, simultaneous meiotic divisions in a mature agamont; 9, wave of cytokinesis spreading from the youngest chamber during the process of gamogony; 10, proloculi of newly formed gamonts just prior to their release from the agamont. Reproduced with permission from Lee and McEnery (1970).

are only partially known, it is believed that many other species of foraminifera have classical or paraclassical metagenic life-cycles (Table 1). Among these are *Myxotheca arenilega* (Føyn, 1936), *Iridia lucida* (Le Calvez, 1938), *Planorbulina mediterranensis* (Føyn, 1936), *Discorbis vilardeboanus* (Føyn, 1937), *Discorbis bertheoti* (Myers, 1940), *Tretomphalus bulloides* (Myers, 1943a), *Triloculina rotunda* (Le Calvez, 1938) and *Quinqueloculina suborbicularis* (Le Calvez, 1939).

3.2 Paraclassical life-cycles: Gamontogamy

The sexual reproduction initiated by the union of gamonts which join together by their apertural sides to mutually exchange gametes is called gamontogamy. The aggregates of two or more tests were first reported in foraminifera by Moebius (1880). Schaudinn (1895b) introduced a name, plastogamy, for what he thought was a form of asexual reproduction of foraminifera. For a long time, the aggregates of gamonts were interpreted as a result of budding (Heron-Allen, 1915). The sexual aspects of this reproduction were discovered in *Patellina corrugata* by Myers (1935) and Le Calvez (1938), who called it "syzygy". Later, the gamontogamy was studied in *Spirillina vivipara* (Myers, 1936), *Glabratella patelliformis* (Myers, 1940; Le Calvez, 1952), *Glabratella mediterranensis* (Le Calvez, 1950), *Rubratella intermedia* (Grell, 1958a), *Glabratella sulcata* (Grell, 1958b), *Rosalina leei* (Lee et al., 1963), *Metarotaliella parva* (Weber, 1965) and *Glabratella ornatissima* (Lipps and Erskian, 1969; Erskian and Lipps, 1987).

The term "gamontogamy" was used by Grell (1973), who suggested that this form of sexual reproduction evolved either from gametogamy or autogamy. The presence of free-swimming, triflagellated gametes in some gamontogamous species of the genus *Glabratella* would indicate their close relationship with some gamontogamous species, especially of the genus *Discorbis*, in which all *Glabratella* were once classified (Loeblich and Tappan, 1964). The gamontogamous species with ameboid gametes (*S. vivipara, P. corrugata, Metarotaliella* spp., *Rubratella intermedia*) possibly evolved from autogamous species, some of them being capable occasionally of forming gamontogamous pairs (Grell, 1954). Gamontogamy appears to be an adaptation to life in turbulent shallow waters, where the chances of gametes surviving and copulating are limited (Lipps and Erskian, 1969).

The process of the formation of gamontogamous aggregates of *G. mediterranensis* was described in detail by Le Calvez (1950) and in *G. ornatissima* by Lipps and Erskian (1969). When the gamonts reach

Table 1 Life-cycle of foraminifera

	Methods[a]		Life-cycle			Type of sexual reproduction	Gametes[b]	Nuclear dualism	References
	NS	TEM	Metagenic	Apogamic	Gamic				
Allogromida									
LAGYNIDAE									
Myxotheca arenilega	X		X			Gametogamy	Biflag.		Føyn (1936), Grell (1958c), Schwab (1969), Angell (1971), Hedley et al. (1968)
Boderia turneri	X		X			Gametogamy	Biflag.		
ALLOGROMIIDAE									
Allogromia laticollaris	X	X	X	X		Autogamy (gametogamy)	Amoeb.		Arnold (1955b), Lee and McEnery (1970), McEnery and Lee (1976), Schwab (1976)
Heterotheca lobata	X		X			Gametogamy	Biflag.		Grell (1988)
Nemogullmia longevariabilis	X		X			Gametogamy	Biflag.		Nyholm (1956)
Textulariida									
HEMISPHAERAMMINIDAE									
Iridia lucida	X	X	X			Gametogamy	Biflag.		Le Calvez (1938), Cesana (1972, 1975, 1978)
RHABDAMMINIDAE									
Haliphysema tumanowiczii	X			X					Hedley (1958)
SACCAMMINIDAE									
Saccammina alba	X		X	X		Gametogamy	Biflag.		Hedley (1962), Goldstein (1988), Goldstein and Barker (1990)
Psammophaga simplora	X		?			Gametogamy	Amoeb.		Arnold (1982, 1984)
TROCHAMMINIDAE									
Trochammina cf. *T. quadriloba*	X		X						Salami (1976)
Spirillinida									
SPIRILLINIDAE									
Spirillina vivipara	X		X			Gamontogamy	Amoeb.		Myers (1936)

Taxon		Reproduction	Gamete	Reference
PATELLINIDAE				
Patellina corrugata	X	Gamontogamy	Amoeb.	Myers (1935), Le Calvez (1938), Grell (1959)
Miliolida				
NUBECULARIIDAE				
Calcituba polymorpha	X			Arnold (1967)
SPIROLOCULINIDAE				
Spiroloculina hyalina	X			Arnold (1964)
PENEROPLIDAE				
Peneroplis pertusus	X	Gametogamy	Uniflag.	Winter (1907)
Lagenida				
ELLIPSOLAGENIDAE				
Oolina marginata	X			Le Calvez (1947)
Globigerinida				
GLOBIGERINIDAE				
Globigerinoides sacculifer	X ?	Gametogamy	Biflag.	Bé and Anderson (1976), Bé et al. (1983)
HASTIGERINIDAE				
Hastigerina pelagica	X X	Gametogamy	Biflag.	Spindler et al. (1978, 1979), Hemleben et al. (1979)
Rotaliida				
BOLIVINIDAE				
Bolivina doniezi	X	Autogamy (?)	?	Sliter (1970)
DISCORBIDAE				
Discorbis vilardeboanus	X	Gametogamy	Biflag.	Føyn (1936), Le Calvez (1950)
ROSALINIDAE				
Neoconorbina orbicularis	X X	Gamontogamy	Amoeb.	Le Calvez (1938)
Rosalina leei	X X	Gametogamy	Biflag.	Lee et al. (1963)
Rosalina globularis	X	Gametogamy	Biflag.	Sliter (1965)
Tretomphalus bulloides	X			Myers (1943b)

Table 1 continued

	Methods[a]		Life-cycle			Type of sexual reproduction	Gametes[b]	Nuclear dualism	References
	NS	TEM	Metagenic	Apogamic	Gamic				
ROTALIELLIDAE									
Rotaliella heterocaryotica	X		X			Autogamy	Amoeb.	X	Grell (1954)
Rotaliella roscoffensis	X		X			Autogamy	Amoeb.	X	Grell (1957a)
Metarotaliella parva	X		X			Gamontogamy	Amoeb.	X	Weber (1965)
Metarotaliella simplex	X		X			Gamontogamy	Amoeb.	X	Grell (1979)
GLABRATELLIDAE									
Glabratella patelliformis	X		X			Gamontogamy	Triflag.		Myers (1940)
Glabratella mediterranensis	X		X			Gamontogamy	Triflag.		Le Calvez (1950)
Glabratella sulcata	X		X			Gamontogamy	Triflag.	X	Grell (1958b)
Glabratella ornatissima			X			Gamontogamy	?		Lipps and Erskian (1969), Erskian and Lipps (1987)
CIBICIDAE									
Cibicides lobatulus	X		X			?	?	X	Voronova (1976, 1978) Nyholm (1961)
PLANORBULINIDAE									
Planorbulina mediterranensis	X		X	X		Gametogamy	Biflag.		Le Calvez (1938)
ASTERIGERINATIDAE									
Rubratella intermedia	X		X			Gamontogamy	Amoeb.		Grell (1958a)
ELPHIDIIDAE									
Elphidium crispum	X		X			Gametogamy	Biflag.		Lister (1895), Jepps (1942)
NUMMULITIDAE									
Heterostegina depressa	X	X	X	X		Gametogamy	Biflag.		Röttger et al. (1986, 1989, 1990a)

[a]NS, nuclear staining; TEM, transmission electron microscopy.
[b]Uniflag, uniflagellated; Biflag, biflagellated; Triflag, triflagellated; Amoeb., amoeboid.

maturity, they creep towards each other and lift their tests to position their apertural sides face to face. They then firmly cling together and join their tests by an organic membrane which contains a nonsulphated acid mucopolysaccharide (Lipps and Erskian, 1969). In *R. leei*, the gamonts build a "pseudocyst" composed of diatoms and organic debris around their tests (Lee et al., 1963; Fig. 6). In *G. patelliformis*, the fused pairs of gamonts can move for a short time before they attach firmly to the substratum (Myers, 1940). Later, the ventral walls of pairing gamonts, as well as the inner walls separating the chambers, dissolve, providing a large space within which the gametes can copulate. The gametes are ameboid or flagellated. The number of ameboid gametes varies from 6 in *M. simplex* (Grell, 1979) to 16 in *S. vivipara* (Myers, 1935). The number of triflagellated gametes can exceed 25 per gamont in *G. sulcata* (Grell, 1958b). The embryonic agamonts probably secrete an enzyme which degrades the membrane to liberate themselves from the united gamonts (Lipps and Erskian, 1969).

Laboratory experiments on the gamontogamic reproduction of *D. mediterranensis* and *P. corrugata* showed that in both species the gamonts are sexually differentiated, i.e. only the gamonts of opposite types ("+" and "−") can mate (Le Calvez, 1950; Grell, 1957b, 1960). In *P. corrugata*, which often form aggregates of more than two gamonts (from 2 to 14), the gametes are also sexually differentiated. Grell (1960) demonstrated a presence of left-over gametes in aggregates composed of three gamonts, and concluded that only the gametes produced by gamonts of different mating types can fuse.

The sexual differentiation of gametes was also suggested in the case of *R. intermedia* (Grell, 1958a), and *M. parva* (Weber, 1965). In *R. intermedia*, the mating of gamonts of different size is common. Because the size of gametes and their nuclei is proportional to the size of the gamont, it was easy to observe that only gametes of different sizes fused, suggesting that they came from different gamonts. In *M. parva*, the differentiation of gametes was deduced from the even number of left-over gametes (Weber, 1965). It was demonstrated that in this species the gamonts are not sexually differentiated, but each gamont forms gametes of type "+" and "−". The gametes of type "+" fuse only with the gametes of type "−" of another partner. The sexual differentiation of gametes probably occurred during the last synchronous mitotic division (Grell, 1973).

Some experiments using *M. parva* showed that three stages can be distinguished in the development of gamonts, each with a different capability to induce and stimulate pairing (Weber, 1965). Juvenile gamonts were capable of stimulating pairing with mature and older gamonts, but they could not induce pairing between themselves. Mature gamonts had both inducing and stimulating capabilities. Older gamonts could pair but were not capable of stimulating pairing, and therefore they were unable to form aggregates with the gamonts of the same age.

Mating between juvenile forms seems possible in other foraminifera. Le Calvez (1950) observed that the gamonts of *G. mediterranensis* started to form aggregates as early as the 4–6 chamber stage. He had to separate them to keep them growing. He suggested that this precocious sexuality was due to a high concentration of chemo-attractive substances secreted by the gamonts in culture dishes.

The gamontic stage is also very short in *Rotaliella* sp. (Elat strain, Pawlowski and Lee, in prep.; Figs 16, 18–22). In this species, the gamonts mate only a few hours after they leave the reproductive cyst and build their first chamber. They probably lose the capability to grow, for if we separate them they die. This seems to be exceptional, however, for in *P. corrugata*, the gamonts which do not participate in mating can remain alive for several months (Grell, 1957b).

3.3 Paraclassical life-cycles: Autogamy

Autogamy is a process of self-fertilization, in the course of which the gametes from the same gamont fuse between themselves. This is a case of an obligatory monoecy, because in most other species of foraminifera, and most other protists, there are self-sterility barriers which prevent the self-fertilization of gametes produced by the same gamont (Grell, 1973). Autogamy is believed to be an evolutionary advanced rather than a primitive mode of reproduction in foraminifera (Grell, 1979). There are examples of autogamy in both monothalamous (*A. laticollaris*: Arnold, 1955b; McEnery and Lee, 1976) and polythalamous (*Rotaliella* spp.: Grell, 1954, 1957a) foraminifera.

The life-cycles of both *Rotaliella heterocarytocia* and *R. roscoffensis*, two tiny shallow-water species, consist of the regular alternation of asexually and sexually reproducing generations (Grell, 1954). Generation time is relatively short (5–6 days) and both generations are morphologically identical. Three to four asynchronous mitotic divisions were observed in a mature gamont. The last division was synchronous, so that even numbers of gametic nuclei were formed. The amoeboid isogametes fuse in pairs within the parental test. The differences in nuclear structure of gametes suggested they were sexually differentiated (Grell, 1973).

A. laticollaris has an essentially apogamic life-cycle (Figs 7, 8). Autogamic sexual reproduction is rare. Arnold (1955b) described the formation of ameoboid gametes in haploid individuals by an unusual process of nuclear degeneration (*Zerfall*) in which large quantities of chromosome-free nuclear remnants are released into cytoplasm. The gamonts were also produced occasionally in two strains of *A. laticollaris* examined by Lee and McEnery (1970). They were uninucleated for most of their lives. Several hundreds to thousands of gamontic nuclei and small uninucleated gametes were found in mature gamonts. The fusion of gametes was rarely observed, but the authors suggested that autogamy was more probable than gamontogamy.

Autogamic reproduction can be properly identified only with detailed cytological study. It is difficult to distinguish between apogamy and the heterophasic alternation of schizogony and autogamy in two recently studied species, *Bolivina doniezi* (Sliter, 1970) and *Trochammina* cf. *T.*

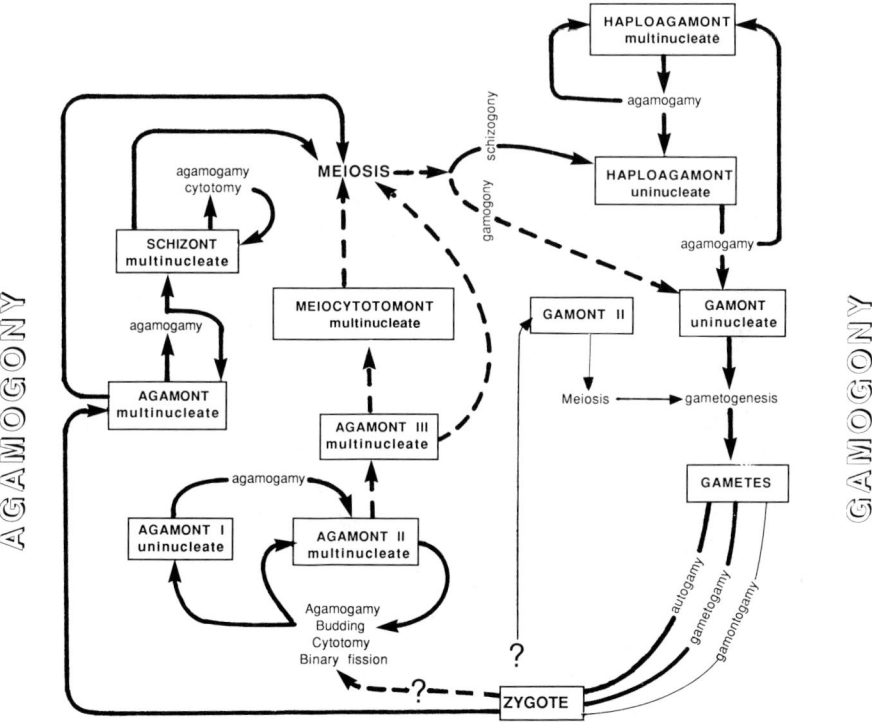

Fig. 7. A diagram of the life-cycle options found in foraminifera with the portions of the options found in various strains of *Allogromia laticollaris* indicated by bold arrows (see the text for differences between the strains).

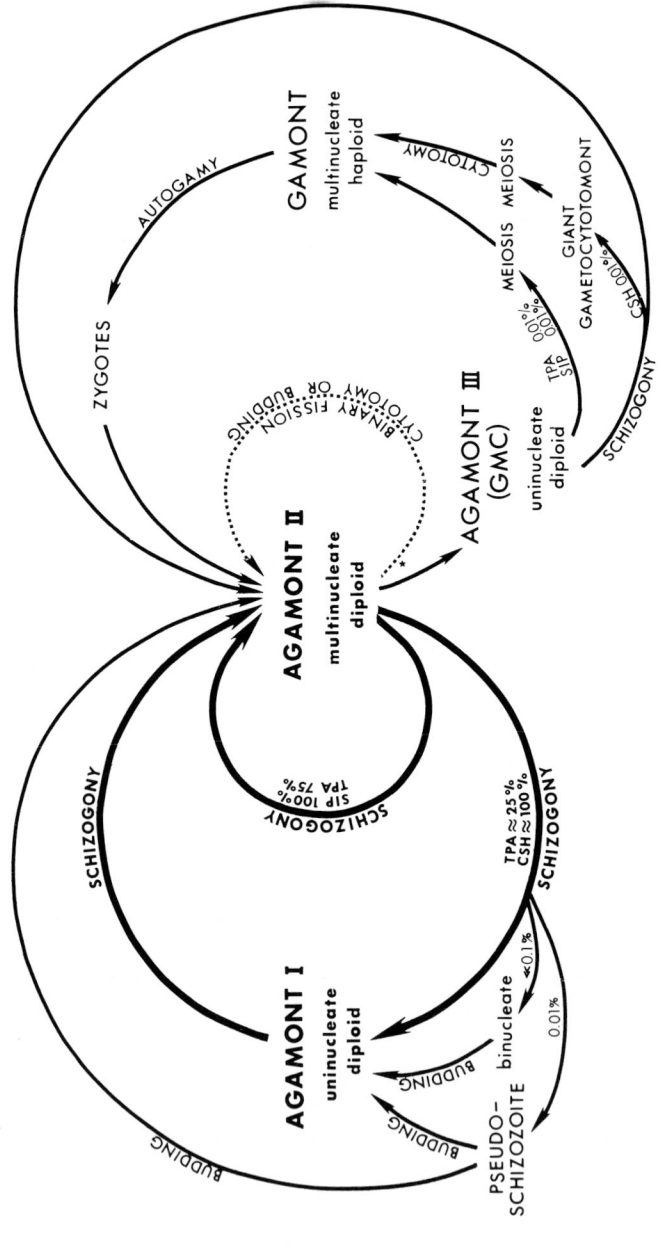

Fig. 8. A diagram of the life-cycle options of three strains (SIP, TPA, CSH) of *Allogromia laticollaris*. Reproduced with permission from McEnery and Lee (1976).

quadriloba (Salami, 1976). Because the cytological details of only a few species are known throughout their entire life-cycles, the real occurrence of autogamy in foraminifera is difficult to evaluate.

3.4 Larger foraminifera: Trimorphic, paratrimorphic life-cycles

A trimorphic life-cycle distinguishes between three generations: agamonts which are microspheric, diploid and multinucleate; schizonts which are megalospheric, diploid and multinucleate; and gamonts which are megalospheric, haploid and uninucleate (Fig. 2). The agamonts attain larger overall final test sizes than either of the megalospheric specimens, although the gamonts tend to have larger prolocular sizes and larger overall test sizes than schizonts. These designations were inferred from the existence of large microspheric and small megalospheric tests found in the fossil record (Rhumbler, 1909). Paratrimorphism is the succession of the three generations in which alternation is not obligatory (Le Calvez, 1938). Although little is actually known of the life-cycles of larger foraminifera, this concept has been embraced for the group despite the diversity of genera (and families) involved and the lack of empirical evidence.

The occurrence of microspheric specimens (agamonts) is rare for the larger foraminifera. Sexual reproduction has been observed in only two species, *Peneroplis pertusus* (Winter, 1907) and *Heterostegina depressa* (Röttger et al., 1984, 1986, 1989). Winter (1907) described uniflagellated 1-μm isogametes for *P. pertusus*, whereas Le Calvez (1938) reported biflagellated gametes. *H. depressa* has biflagellated 2.85-μm isogametes (Röttger et al., 1989). After formation, several gametes fused within the gamont test, suggesting autogamy, whereas others swam away from the test and fused (Röttger et al., 1989). These studies show sexual reproduction is possible, but further research is required to understand this more fully.

Schizogony dominates asexual reproduction. Schaudinn (1895b) considered three possible modes of schizogony: (1) completely internal to the maternal test, with embryonic tests being released; (2) internal cytokinesis, external embryonic shell formation; (3) completely external. Larger foraminifera tend to display either completely internal or completely external schizogony. The number of embryos formed may be a species characteristic but may possibly be related to the size of the mature adult (Muller, 1974; Röttger, 1974; Hallock et al., 1986).

Peneroplis exhibits internal schizogony (Winter, 1907) where the embryo develops a proloculus and a flexostyle prior to emergence from the maternal test (Figs 23, 24). Winter (1907) described nuclear separation and growth prior to cytokinesis. Lister (1895) gave the size of the prolucli of *Peneroplis pertusus* as 27–34 µm. We observed 26 mature *P. planatus* reproduce. The majority of the maternal tests were megalospheric (i.e. schizonts), although a few were microspheric (agamonts) (Faber and Lee, 1991). We observed no difference in the prolocular sizes of the broods which averaged 62.5 µm. These measurements correspond with Winter's (1907) illustrations of *P. pertusus* (based on current taxonomy, and his descriptions and illustrations, we feel Winter actually worked with *P. planatus* and not *P. pertusus*).

The soritids grow brood chambers: the final 1–5 chambers of *Sorites orbiculus* and *Amphisorus hemprichii* (Kloos and MacGillavry, 1978; Kloos, 1981, 1984); the final 6–10 chambers of *Cyclorbiculina compressa* (Lutze and Wefer, 1980); and the final 4–9 chambers of *Marginopora vertebralis* (Ross, 1972). The cytoplasm migrates into these chambers and undergoes multiple fission (Kloos, 1981). For *S. orbiculus*, 18–207 embryonic individuals are released from the maternal test, each embryo possessing a proloculus and the flexostyle which are formed separately (Kloos and MacGillavry, 1978), whereas *M. vertebralis* releases 60–150 embryos consisting of a proloculus, flexostyle and a deutoconch (Ross, 1972). Both Lacroix (1941) and Hofker (1976) found bimodal size ranges for the megalospheres of *S. orbiculus*. Kloos and MacGillavry (1978) found no size difference between the parental and brood proloculi, suggesting their material consisted solely of schizonts.

The calcarinids form a brood chamber covering the spiral side of the maternal tests (Sakai and Nishihira, 1981; Krüger, 1990). The maternal cytoplasm enters the brood chambers through the canal systems (Röttger *et al.*, 1990b). The prolocular sizes of megalospheric specimens of *Calcarina gaudichaudii* collected in the field ranged from 62.5 to 100 µm, whereas the juveniles produced from one agamont were all 125 µm long (Röttger *et al.*, 1990b).

Heterostegina depressa, Amphistegina lobifera and *A. lessonii* have completely external schizogony, their cytoplasm flows out of the maternal test and undergoes multiple fission (Röttger, 1974; Hallock *et al.*, 1986). Between 200 and 3000 embryos are formed. The embryos form a 2- to 3-chambered stage prior to initial calcification of the test (Röttger, 1974). Recently, Röttger *et al.* (1990a) demonstrated empirically that the agamont could form a schizont instead of solely forming gamonts, based on the observed size difference between the two megalospheric proloculi. The complete details of any of the life-cycles of the species of larger foraminifera have yet to be worked out, causing many problems and gaps in our knowledge of their biology. Although we speculate that they have trimorphic life-cycles, proof is lacking. Several of the characters used to distinguish between the three generations may be invalid.

The clear separation between prolocular size is obvious between microspheric and megalospheric specimens. Other morphological features also differ, such as the differences in the subdivision of the chambers into chamberlets by secondary septa in *H. depressa* (Röttger, 1974). However, the separation of the schizont and gamont based on size differences of the megalospheres is less certain (Table 2). Megalosphere size may be influenced by environmental parameters. Fermont (1977) showed an increase in prolocular diameter with depth to 80 m in *H. depressa*. Ross (1972) found a large variability in prolocular sizes between the parental tests of *M. vertebralis* and a small difference between the brood prolocular sizes. This confuses the bimodal prolocular size variations seen between schizonts and gamonts, which may not apply to this species. Similarly, overall test size is not a good indication of reproductive behaviour. Some species (e.g. *A. hemprichii*) show a clear size difference between microspheric and megalospheric specimens (Ross, 1972). In general, test sizes in larger foraminifera are dependent on growth rates (Hallock, 1985), which, in turn, are affected by a wide range of environmental parameters. In *P. planatus*, there is a large overlap between the overall test sizes of microspheric and megalospheric specimens, with no bimodality seen in the megalospheric specimens. Reproductively mature specimens must be considered in isolation, because younger specimens obscure the differences. However, adult tests which do not undergo reproduction and simply die may also create variability. In the asexual reproduction of *P. planatus*, the release of the embryos occurs along the edges of the specimens, where the test becomes partially dissolved, thin, brittle and broken. The destruction of the maternal test may add another bias between the schizonts and gamonts in the sediments. The correlation of prolocular size with other biometric measurements or physiological observations may provide a better reference for the distinction between schizonts and gamonts, especially for species which lose their embryonic apparatus (like *M. vertebralis*) or have hidden embryonic apparatus (like *P. planatus*). Although *H. depressa* schizonts and gamonts overlap in megalospheric prolocular size, number of opercular chambers and overall test sizes, there is a clear distinction between their growth rates (Röttger *et al.*, 1986; see Figs 3, 4). However, Röttger *et al.* (1989, 1990a) explained

Table 2 Prolocular size and number of asexually reproduced offspring for a few species of larger foraminifera

Species	Prolocular size (μm)			No. of offspring	References
	Agamont	Schizont	Gamont		
Calcarina gaudichaudii	8.5–12.5	62.5–100	125	1000–3000	Kruger (1990), Röttger et al. (1990a)
Heterostegina depressa	27	84	135	1000–3000	Röttger (1974), Röttger et al. (1986)
Sorites orbiculus	16	52–64	70–86	—	Hofker (1976)
Marginopora vertebralis	—	41.5–115.8	—	18–207	Kloos and MacGillavry (1978), Kloos (1984)
	—	120–250	—	60–150	Ross (1972)
Cyclorbiculina compressa	—	120–150	30–50	500–1000	Lutze and Wefer (1980)
Archaias angulatus	—	96–144	—	300–1000	Hallock et al. (1986)

why it is important to observe the reproductive behaviour of megalospheric specimen in order to label them schizonts or gamonts.

The nuclear characteristics used to separate the generations are also suspect. Ploidy has never been examined for any larger foraminifera. The assignment of haploid/diploid is inferred from other known foraminiferal life-cycles. One assumes the gametes are haploid, and the agamont, which is formed from the fusion of two gametes, would be diploid. Meiosis is assumed to occur before the formation of the gamont by multiple fission, in either the agamont or the schizont. This results in a haploid gamont. Kloos and MacGillavry (1978) speculate that meiosis may take place in the gamont prior to the formation of the micronuclei and the gametes. This scenario reduces the haploid stage of the life-cycle to only the gametes themselves.

Gamonts are also considered to be uninucleate, whereas agamonts and schizonts are multinucleate, being either homokaryotic or heterokaryotic (somatic and generative nuclei; e.g. Figs 25–27) (Leutenegger, 1977a,b). However, the "uninucleate" gamonts may possess many nuclei during gamogamy (Röttger et al., 1989). Karyokinesis occurs prior to cytokinesis (Winter, 1907). The nucleus divides into smaller nuclei, and hence the daughter nuclei are not produced by a mitotic cycle (Röttger et al., 1986). All of the generations seem to possess large nuclei (Röttger et al., 1989) which are usually somatic (McEnery and Lee, 1981). The generative nuclei are smaller and more numerous (Müller-Merz and Lee, 1976; McEnery and Lee, 1981). The variations in nuclear size and number may simply reflect the phase in which the organism is, rather than for identification of schizont or gamont.

Reproduction by a foraminifera tends to end the life of the parental cell even if some residual cytoplasm remains. There is a possibility that only a portion of the parental organism reproduces, and the residual cell material could continue to grow and possibly reproduce again. This cytotomy was observed in our laboratory in *P. planatus*. Kloos (1981) explains how one specimen of *S. orbiculus* constructed reproductive chambers and then normal cyclic chambers, which were deformed, presumably due to dissolution of part of the brood chambers on the release of the offspring, and subsequent repair. This is not proof of cytotomy in *S. orbiculus*, as reproduction could have been aborted. A direct observation is needed. Röttger (1978) reported cytotomy in *H. depressa* where several individuals continued to live and grow after reproduction. One organism reproduced a second time, although the juveniles formed were irregular and failed to develop.

3.5 Apogamic life-cycles

Although successive agamogonic generations have been observed in many species of foraminifera including *Allogromia laticollaris* (Arnold, 1955b; McEnery and Lee, 1976), *Planorbulina mediterranensis* (Le Calvez, 1938), *Heterostegina depressa* (Röttger et al., 1986, 1990b), *Spiroloculina hyalina* (Arnold, 1964), *Neoconorbina asicularis* (Le Calvez, 1938), *Oolina marginata* (Le Calvez, 1947), *Triloculina linneiana* (Schnitker, 1967), *Trochammina* cf. *T. quadriloba* (Salami, 1976), *Discorinopsis aguagoi* (Arnold, 1954), *Ammonia beccari* (Schnitker, 1974) and *Cyclorbiculina compressa* (Lutze and Wefer, 1980), a distinction has to be made between those species (or strains) in which this reproductive phenomenon is part of a regularly occurring metagenic life-cycle and those species or strains in which a sexual phase is rare or absent. In at least one species, *Allogromia laticollaris,* the mode of asexual reproduction is dependent upon diet (Lee et al., 1969). This, in itself, suggests that in nature, the life-cycle options of some species of foraminifera can be quite adaptive to a variety of different natural community contextual relationships. This is an aspect of foraminiferal life-cycles which should be quite interesting to probe in the future.

Because the life-cycles of many species of foraminifera which have successive agamogonic generations are not yet completely known, it is hard to judge the range of truly apogamic species. One can easily appreciate that species which regularly undergo binary fission [*Allogromia laticollaris*, McEnery and Lee, 1976, Fig. 9; *Allogromia* sp. (NF), Lee and Price, 1963, Fig. 30; *Boderia turneri*, Hedley et al., 1968], budding [*Allogromia laticollaris*, McEnery and Lee, 1976, Fig. 9; *Allogromia* sp. (NF), Lee and Pierce, 1963, Fig. 29; *Saccammina alba*, Goldstein, 1988; *Haliphysema tumanowiceii*, Hedley, 1958] or cytotomy [*Allogromia laticollaris*, McEnery and Lee, 1976, Fig. 9; *Allogromia* sp. (NF), Lee and Pierce, 1963, Figs 31, 32; *Spiroloculina hyalina*, Arnold, 1964], could establish natural self-sustaining apogamic populations.

It is interesting to note, and quite instructive, to look at the results of laboratory studies of the life-cycles of two easily cultured species: *Rosalina leei* and *Allogromia laticollaris*. The populations of *R. leei* which Lee and co-workers (1963) studied, had paraclassical gamontogamic life-cycles. Clone cultures from other localities (Plymouth, U.K. and Wellington, New Zealand), were totally apogamic with repeated generations being produced every 3-5 weeks. Each of the five strains of *Allogromia laticollaris* thus far examined have different fragments of a life-cycle pattern (Figs 4, 7-9) (Arnold, 1955b; McEnery and Lee, 1976; Schwab, 1976).

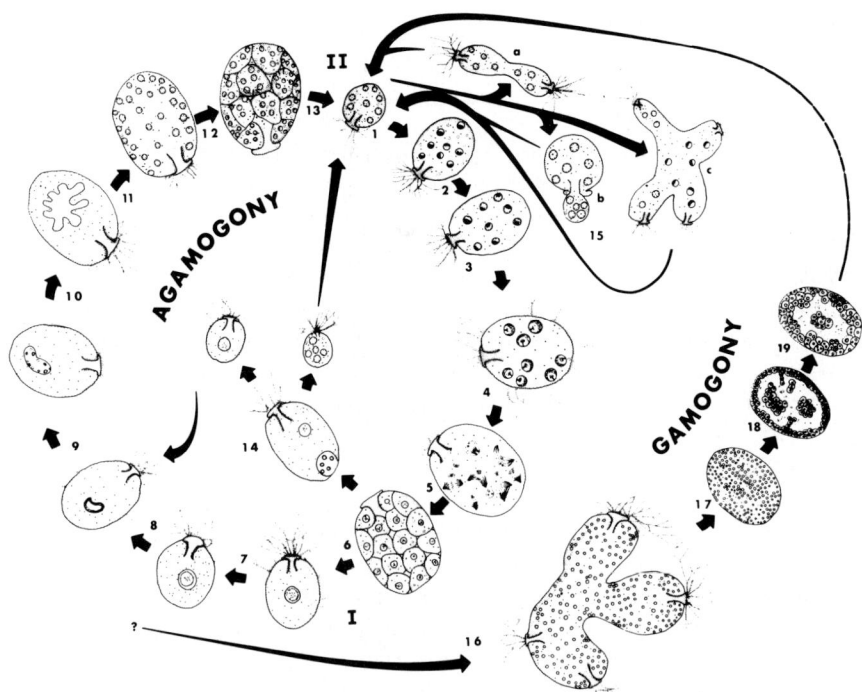

Fig. 9. An illustrated diagram showing the organisms in the life-cycle of *Allogromia laticollaris*. Same strains as in Fig. 8. *Agamont phase*: 1, juvenile agamont II, early G_1; 2, young agamont II, mid-G_1; 3, growing agamont II, late G_1; 4, mature agamont II, chromosomes in "mushroom-like" configuration, RNA accumulated at the periphery; 5, karyokinesis; 6, cytokinesis; 7, young agamont I, S phase; 8, maturing agamont I, G_2 phase; 9, mature agamont I, nucleus differentiating, RNA granules at the periphery of the nucleus; 10, mature agamont I, early "amoeba-form" nucleus; 11, agamont I, "amoeba-form" nucleus; 12, agamont I, post-*Zerfall*; 13, agamont I, shizogony; 14, agamont I, relatively uncommon life-cycle alternate pathway in which budding gives rise to an agamont II; 15, agamont II, relatively uncommon alternate life-cycle pathways including (a) binary fission, (b) budding, (c) cytotomy. *Gamont phase*: 16, giant meiocytotomont; 17, multinucleate gamont prior to the formation of gametes; 18, gamont filled with gametes; 19, gamont with some gametes and zygotes. Reproduced with permission from McEnery and Lee (1976).

Strain SIP had the simplest model (Figs 8, 9), having only a multinucleate asexually reproducing form in its development (McEnery and Lee, 1976). The life-cycle of strain CSH consisted of almost obligatory alternation between a uninucleate diploid agamont I and a multinucleate diploid agamont II. Sexuality was rare (1×10^{-4}) in the CSH strain and was first recognized by the appearance of a giant meiocytotomont (Fig. 40) prior to meiosis, followed by gamogony and the production of gamonts (Fig. 39) and gametes (Figs 36, 38). Specimens of a third strain, TPA, were able to follow both patterns, with a net preference (75%) for a simpler multinucleate agamont II mode (Figs 8 and 9).

In our life-cycle scheme, we have followed the pattern used by McEnery and Lee (1976; Figs 8, 9) for separating the terms agamont I and II from the terms schizont and agamont which have connotations in the life-cycles of other foraminifera (classical, paraclassical, trimorphic, paratrimorphic). Agamonts I and II are probably derivatives of the agamont or schizont which have become reproductively isolated to the extent that life-cycle options are primarily agamogonic. Although the parental–filial relationships were not as fastidiously studied as the above three strains, the strain studied by Arnold (1955b) had five different individuals in the life-cycle, which he recognized as: (1) schizont [mononucleate-$2N$-asexually produced by schizont 5]; (2) schizo-gamont [mononucleate-N-asexually produced]; (3) schizont [multinucleate-$2N$-sexually produced by autogamous fusion of gametes from the schizogamont]; (4) schizont [multinucleate-N-asexually produced by karyokinesis of the schizo-gamont]; (5) schizont [multinucleate-$2N$-asexually produced by schizont 3]. In our composite life-cycle drawing, individuals 3 and 5 are represented by haplogamonts (uninucleate, multinucleate). The other phases could be represented by agamonts I, II, III or by agamont and schizont in our composite cycle but we are not certain of the homology. Strain GF was gametogamic (Schwab, 1976). Although we have only fragmentary knowledge of *Allogromia* sp. (NF) (Figs 28, 33), this species has a recognizable multinucleate haploagamont stage (Lee and Pierce, 1963; Fig. 35).

Taken as a whole, the various strains of *Allogromia laticollaris* seem to follow almost the complete diversity of the life-cycle pathways which occur in foraminifera (Figs 3, 7). This can be viewed in two ways. It could suggest that the diversity in life-cycle options is a basic trait found in the most morphologically primitive foraminifera. It could also be argued that present criteria for recognizing the species *Allogromia laticollaris* are not adequate. Additional criteria might lead to separation into recognizable new species, subspecies or demes which are not dependent upon laborious life-cycle studies for resolution.

4.0 PLANKTONIC FORAMINIFERA

Our knowledge of the reproductive events in planktonic foraminifera is substantially less developed than for benthic species (see also

Chapter 8). It is not feasible to observe planktonic foraminifera *in situ* throughout the life-cycle, because the vastness of the open ocean, and the great depths to which reproductive individuals may sink during the release of gametes, preclude constant monitoring. Moreover, we have not been able to maintain them in continuous culture, as they do not produce viable offspring in the laboratory after the release of gametes. Hence, it is not possible to make detailed observations of physiological and morphological changes during the course of the entire life-cycle. Much of our knowledge comes from combined observations of field data on abundance in space and time combined with laboratory experimental data on the physiology and fine structure of individuals during maturation and gamete release in maintenance cultures.

Only a few species have been studied intensively, either because they occur abundantly in geographical locales where field studies have been done or because they are particularly supple in laboratory culture. The large, thin-shelled species *Hastigerina pelagica* and the more robust-shelled *Globigerinoides sacculifer* have been studied fairly extensively. A general survey of current knowledge is presented with particular emphasis on these two species. More detailed accounts of the historical development of our knowledge, and some current understandings of variations in reproductive events among species, can be found in, for example, Bé and Anderson (1976), Bé *et al.* (1977), Anderson and Bé (1978), Spindler *et al.* (1978, 1979), Hemleben *et al.* (1979, 1989), Spindler and Hemleben (1982) and Anderson (1984).

4.1 Gamete production

It is now clear that biflagellated gametes are released during reproduction in a wide range of planktonic foraminiferan species. The gametes are approximately 2–5 μm in size. Flagellated gametes have been observed in the spinose species *Hastigerina pelagica, Orbulina universa, Globigerinoides conglobatus, G. ruber, G. sacculifer, Globigerina bulloides, Turborotalita humilis* and *Globigerinella aequilateralis*. In non-spinose species, flagellated gametes have been reported in *Globigerinita glutinata, Neogloboquadrina pachyderma, N. dutertrei, Globorotalia inflata* and *G. menardii* (Hemleben *et al.*, 1989). There is no conclusive evidence of gamete fusion and further development, though Ketten and Edmond (1979) have reported the possible attraction and fusion of gametes released by *H. pelagica*. No detailed observations of further development were reported.

4.2 Gametogenesis in planktonic foraminifera

A generalized account of events accompanying gametogenesis in planktonic foraminifera is presented. This is only an outline of events and does not include species-specific variations. At approximately 1–5 days preceding gamete production, the final chamber is added to the

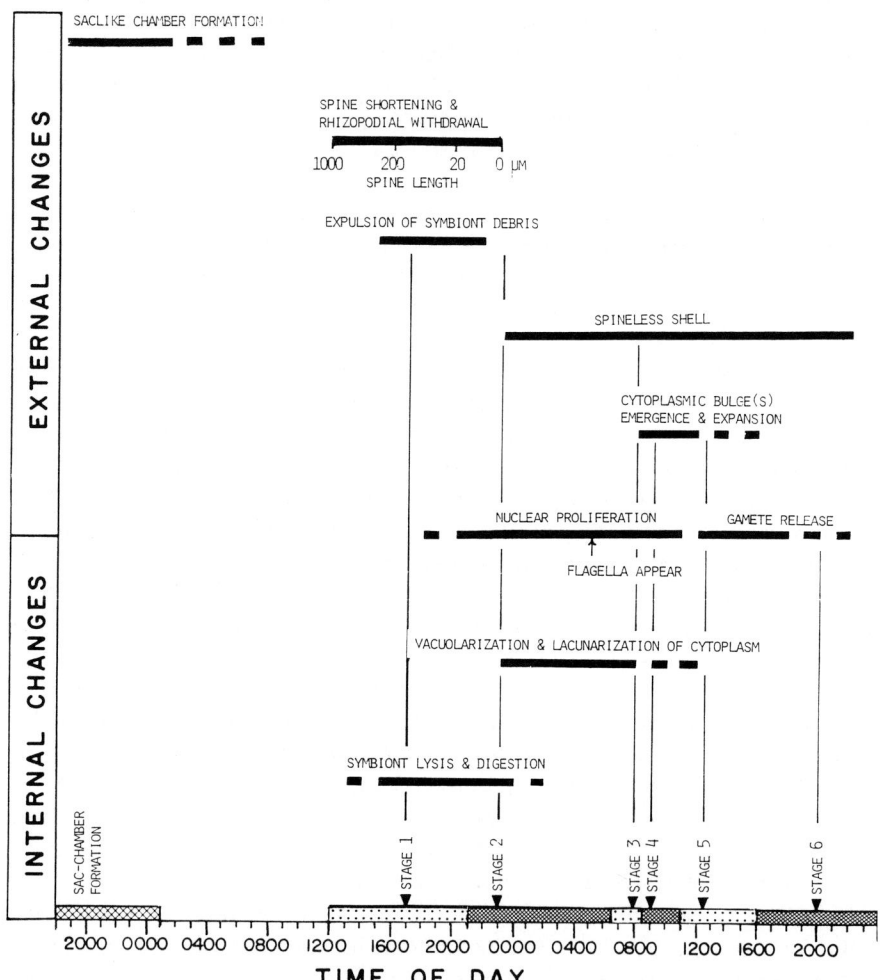

Fig. 10. Diagram of the sequence of internal and external changes occurring during gametogenesis in the planktonic foraminifer *Globigerinoides sacculifer*. Six major stages (horizontal bars with stippled or crossed pattern) mark the key events during the final stages of reproductive maturation, gamete formation and gamete release. Arrows at the baseline indicate peak occurrences of each stage. Reproduced with permission from Bé et al. (1983).

mature shell. Then, 10–24 hours preceding gamete release, the spines are shed, leaving a characteristic alteration of the spine base. Within the same time-frame (4–20 h), nuclear division occurs filling the cytoplasm with myriads of daughter nuclei (Fig. 41). The nuclei are segregated eventually in gametes. At 6–14 h preceding gamete release, the cytoplasm becomes vacuolated and the daughter nuclei become more dispersed. Later, islands of cytoplasm containing one or more nuclei occur within a network of vacuolar lacunae. Flagella develop within the lacunae surrounding the nucleated masses of cytoplasm 7–9 h before release (Fig. 42). Thereafter, the flagellated masses of nucleated cytoplasm gradually become separated into binucleated pregametes (Fig. 43) and individual gametes (Fig. 44), or small clumps of gametes that cohere during release. At 2–6 h before gamete release, a cytoplasmic bulge emerges from the aperture of the shell and numerous gametes can be seen by light optics within this cytoplasmic network. The bulge continues to expand until the myriads of gametes are released with what appears to be an explosive force. At least 10^5 gametes are released from each parent cell (Bé and Anderson, 1976). A more detailed diagrammatic account of events preceding gametogenesis in *Globigerinoides sacculifer* is presented in Fig. 10. A similar sequence of events has been observed in *Hastigerina pelagica*. however, in addition to these major stages, the cytoplasm of *H. pelagic* becomes increasingly reddish orange as reproductive maturity advances. The eruptive release of gametes can be so forceful that regions of the shell are fractured, perhaps favoured in part by the resorption of septa within the shell during the early stages of gametogenesis. Moreover, very large vacuolated bodies, sometimes bearing flagella, are also released with the gametes in *H. pelagica* (Fig. 45). These spherical bodies may be jettisoned waste vacuoles expelled with the gametes (Spindler *et al.*, 1978). Annulate lamellae (see Chapter 2) are also observed in the cytoplasm and appear to provide membranes for the formation of gamete nuclei (Spindler and Hemleben, 1982).

4.3 Reproductive cycles

The complex organization of the shell, elaborate physiological relationships with symbionts, and the complexity of rhizopodial activity of planktonic foraminifera suggest that these are highly evolved protists (e.g. Anderson, 1984). Their reproductive process also provides evidence of highly specialized adaptations. For example, reproductive individuals sink from surface water to a depth where the environment may enhance the survival of the gametes and the earliest stages of the progeny. The release of gametes simultaneously by many individuals of the same species may increase the probability of syngamy (fusion of gametes). Among the planktonic foraminifera, the reproductive cycle in *Hastigerina pelagica* is best understood and the environmental events accompanying or triggering gametogenesis are well documented. *H. pelagica* reproduces on a lunar cycle (Spindler *et al.*, 1979; Hemleben *et al.*, 1989). At 3–7 days following a full moon, numerous individuals collected from the natural environment commence gametogenesis when cultured in the laboratory. During this period, the number of mature individuals in the surface water declines dramatically. However, several days later, juvenile individuals begin to appear once again in plankton samples. This pattern has been observed repeatedly during an intensive sampling period of March–September 1977. In addition to the lunar reproductive cycle, a diurnal cycle of gamete release is superimposed on the longer 29-day cycle. The gametes are typically released during the afternoon and late evening between noon and 20:00 h. There is good evidence that the shedding of spines and sinking of mature organisms may carry them to great depths in the ocean before the gametes are released. There is some evidence that they may settle to the level of the thermocline or deep chlorophyll maximum where potential prey for the young may be more abundant. However, additional research is required to fully determine the depth where gametes are released and where the young develop. The lunar periodic synchronization of gamete production, and the sinking of the reproductive cell into deeper water, may ensure that a large number of gametes are shed simultaneously into the water column to increase the probability of fertilization, and provide a more appropriate environment for survival of the gametes and offspring. The simultaneous release of gametes is consistent with a hypothesis that the planktonic foraminifera are dioecious, i.e. only gametes from different parents undergo syngamy. Other species (e.g. *Globigerinoides ruber, G. sacculifer* and *Globigerinella aequilateralis*) may also reproduce on a lunar cycle (Almogi-Labin, 1981; Reiss and Hottinger, 1984; Bijma, 1986) or semi-lunar cycle (*G. ruber* and *G. aequilateralis*: Hemleben *et al.*, 1989) based on the periodic abundances of mature and juvenile individuals in surface-water plankton samples. However, these inferential conclusions require much further investigation.

4.4 Ontogeny

The growth of planktonic foraminifera from zygote to mature forms has not been documented by direct observation, but has been inferred by examining the shells and cytoplasmic fine structure of individuals of increasingly younger stages in a series beginning with mature stages and progressing backwards (e.g. Brummer et al., 1986, 1987; Hemleben et al., 1989). This same strategy has also been applied to radiolaria (Anderson and Bennett, 1985; Swanberg and Bjørklund, 1987), where direct observation of ontogeny is also not possible, because radiolaria, as with planktonic foraminifera, have not produced successive generations in laboratory cultures. A less accurate method of deducing planktonic foraminiferan shell development involves examining the successive whorls of chambers embedded within the mature spiral (e.g. Desai and Banner, 1985; Sverdlove and Bé, 1985; Brummer et al., 1986). The deposition of calcite, however, on pre-existing chambers during subsequent addition of new chambers, obscures the details of the earlier deposited chambers, and biases morphological reconstruction of the earliest stages.

Using the sequential method of analysing increasingly younger shells of a given species (i.e. from more mature to less mature stages) based on plankton sample specimens, Brummer et al. (1987) proposed a five-stage concept of planktonic foraminiferan development. A general outline is given here. A more detailed discussion is presented in Hemleben et al. (1989).

4.5 Five-stage concept of planktonic foraminiferal ontogenesis

1. *Prolocular stage.* This stage begins when the zygote (not directly observed at present) develops into a spherical first chamber called the proloculus (Fig. 11A). The proloculus is characterized by a larger diameter than the second chamber (deuteroconch) and probably lacks pores, based on its appearance in the two-chambered stage. At the two-chambered stage, the proloculus may possess a few slender spines.

2. *Juvenile stage.* A cytoplasmic growth progresses beyond the prolocular stage, additional chambers are added in a spiral commencing with the deuteroconch, and continuing with several additional chambers of uniform morphology and only slightly increasing size. the spines are thin and sparse, and the pores are few and situated along the sutures of the shell where the chambers join. Prey is largely microphytoplankton.

3. *Neanic stage.* This stage is marked by an overall change in organization and morphology of chambers. The chambers become more inflated in size, and gradually shift position and orientation with subsequent chamber addition until the aperture assumes a position characteristic of the adult form (Fig. 11D). Pores occur over the entire

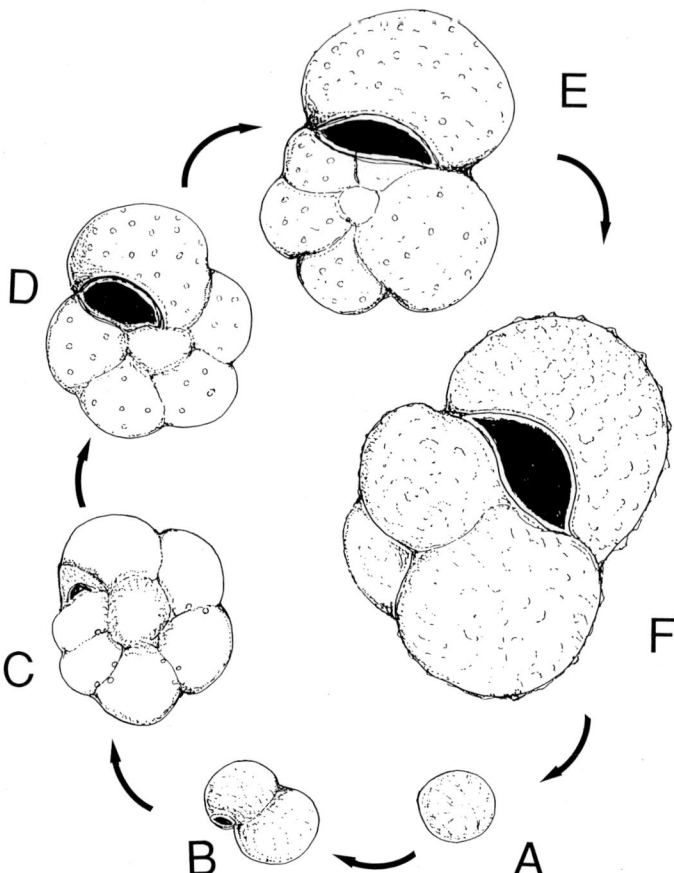

Fig. 11. A generalized diagram of the shell ontogeny of a planktonic foraminiferan, commencing with the proloculus (A) produced by the zygote (not shown), followed by addition of the second chamber or deuteroconch (B), spiral addition of chambers with nearly equal size during the juvenile stage (C), increasing enlargement of later chambers added during the neanic stage (D) where aperture reorientation occurs away from the spiral position, and followed by mature stages (E and F) with much larger chambers and more heavily calcified and ornamented chamber walls. During the juvenile stage (C), pores occur sparsely along the sutures, whereas at later stages the pores are more uniformly dispersed over the surface of the chamber wall (D and E). During reproduction, sufficient additional calcite (known as gametogenic calcite) may be added to partially or completely obliterate the pores (F). In spinose species, spines are absent or very sparse in the deuteroconch stage, but appear and become very dense in the later stages, but are shed at the inception of gametogenesis.

surface of the test and the surface texture becomes more rugose as is characteristic of adult stages. This is fundamentally a transitional stage from juvenile to adult. There is a shift towards larger prey, including zooplankton in carnivorous and omnivorous species.

4. *Adult stage.* Advanced features appear, including deviation from the more spherical form of earlier chambers. Secondary apertures appear in the final chamber. Adult chamber spiral patterns emerge in some species, and species-specific final chambers are added such as the sac chamber in *Globigerinoides sacculifer* and the large multi-apertured, spherical chamber of *Orbulina universa*. Feeding is either omnivorous or carnivorous depending on the species.

5. *Terminal stage.* This stage is characterized by changes related to onset of reproduction, including marked chamber alterations such as shedding of spines, partial wall thickening by deposition of a veneer of calcite (e.g. gametogenic thickening) and alterations of the chamber septa by resorption in some species. The final step is characterized by fully mature non-spinose, gametogenic shells with coarsely cancellate to smooth wall texture. All of the cytoplasm is transformed during gamete production and only the empty test remains.

A generalized diagram of this conceptual plan is presented in Fig. 11.

This is an idealized summary of events during ontogenesis, and substantial differences can occur across species and in relation to environmentally induced (ecophenotypic) alterations during ontogenesis. More detailed analyses of ontogenetic development of several species have been given by Brummer *et al.* (1986, 1987) and Hemleben *et al.* (1989).

5.0 AREAS FOR FURTHER RESEARCH

Additional laboratory research is needed to examine the physiology of reproduction, especially how the organisms are triggered to release gametes. The issue is of particular interest for those species with a lunar cycle. There is some evidence that the process is not based on an endogenous rhythm, as individuals collected from the natural environment within 2 weeks preceding a full moon reproduce at the expected time following a full moon, whereas those collected 13 days or more before the full moon seldom reproduce at the expected period. This suggests that there is an environmental, rather than an endogenous, trigger that needs to be more fully explored. The host–symbiont interaction in the reproductive cycle is not fully understood, though there is some evidence that the symbionts may serve as a food source to provide energy required for gamete production. Large numbers of the symbionts are digested during the onset of gametogenesis. Moreover, we do not know how or when the young planktonic foraminifera acquire symbionts. The gametes are too small to hold symbionts, and we have never observed them being associated with the released gametes. What physiological mechanism mediates host recognition of the symbiont species to be acquired during ontogeny of the foraminiferan? Further research is needed to determine the chromosome number of planktonic foraminifera and to clarify morphological and cytological changes during the life-cycle. It is assumed that planktonic foraminifera do not have alternation of generations as occurs in some benthic species, but this requires further elucidation. We do not know when meiosis occurs, nor the exact sequence of events whereby daughter nuclei are produced during gamete production, including the fine details of the role of annulate lamellae in nuclear membrane formation. We have only generalized information on these significant cellular biological issues. Other major questions of physiological ecology, related to reproductive cycles, include clarification of the physiological characteristics of planktonic foraminifera during ontogeny and how these characteristics influence geographical location and vertical distribution in the water column. We have very little information on the physiological ecology of juvenile planktonic foraminifera or on the conditions that promote growth (including calcite deposition) and reproductive maturation.

Figs 12–18. All except Fig. 16 are light micrographs; Fig. 16 is a scanning electron micrograph. Figure 12 was taken in dark field transmission. Figures 14 and 18 are light micrographs. Figures 13, 15 and 17 were taken with the aid of phase contrast. The nuclei in Figs 13, 15, 17 and 18 were stained by Rafalko's modification of the Fuelgen technique. Figures 12–15 and 17 are reproduced with permission from Lee et al. (1963). Figures 16 and 18 are reproduced from Pawlowski and Lee (in prep.) and are of a new species of *Rotaliella* from Elat on the Red Sea. **Fig. 12.** An agamont shortly after release of gamonts (× 45). **Fig. 13.** Process of cytokinesis during gamogony in an agamont of *Rosalina leei*. Note the multinucleate blocks of cytoplasm in the upper left and uninucleate gamont proloculi in the lower part of the photograph (× 300). **Fig. 14.** Meiosis in an agamont. Photographed with the aid of a deep green filter to enhance contrast (× 300). **Fig. 15.** Meiosis in an agamont. Photographed with the aid of phase contrast and a green filter (× 300). **Fig. 16.** Individual gamonts just prior to their release (× 600). **Fig. 17.** *Rotaliella* sp. in the epiphytic community which develops on the surface of *Enteromorpha*. Note the diatoms and bacteria in the community (scale bar = 100 μm). **Fig. 18.** Young gamonts at time of release (× 750).

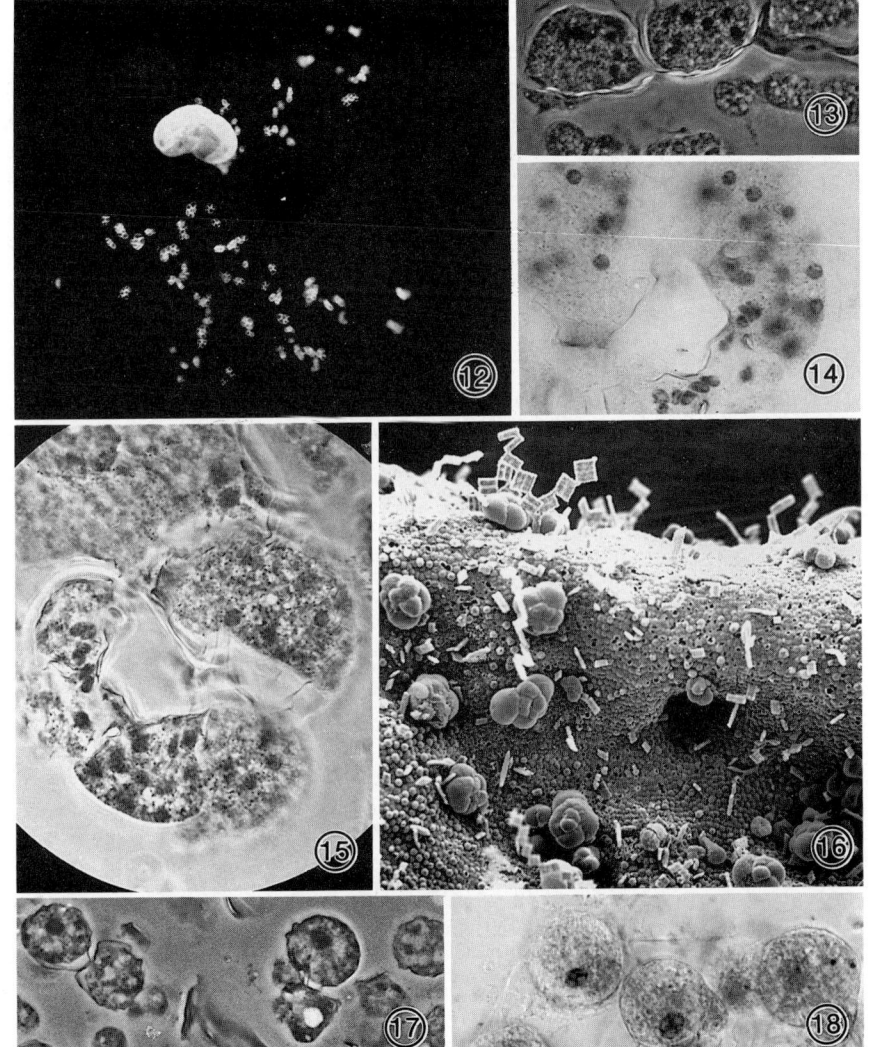

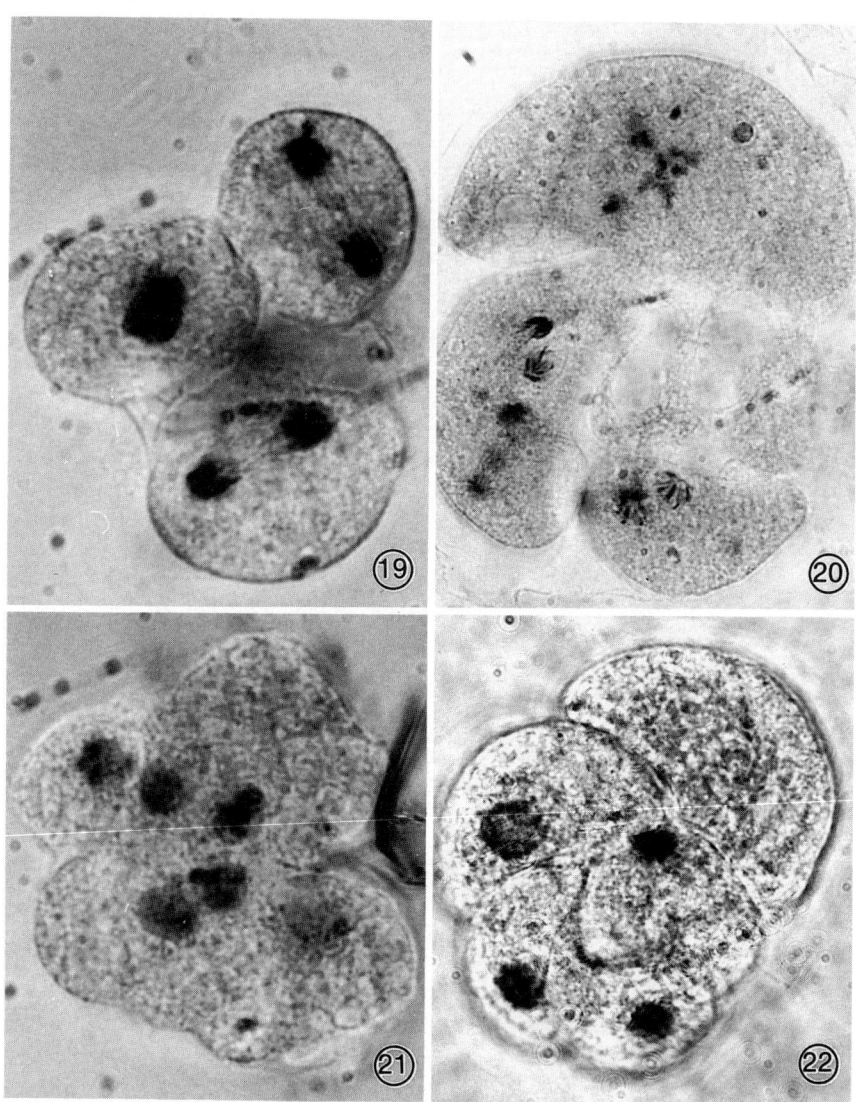

Figs 19–22. Light micrographs of the new species of *Rosalina* illustrated in Fig. 16. The nuclei of the figures were stained with the Fuelgen technique and photographed with the aid of a deep green filter to enhance contrast. Magnification is approximately ×900. **Fig. 19.** First post-zygotic karyokinesis of new foramed agamonts. **Fig. 20.** Meiosis. Remnants of the chromosomes of the somatic nucleus in larger chamber at top of photograph. Optical sections of meiotic generative nuclei at various stages in lower and left chamber. **Fig. 21.** Gamontogamy?/autogamy? **Fig. 22.** Mature agamont.

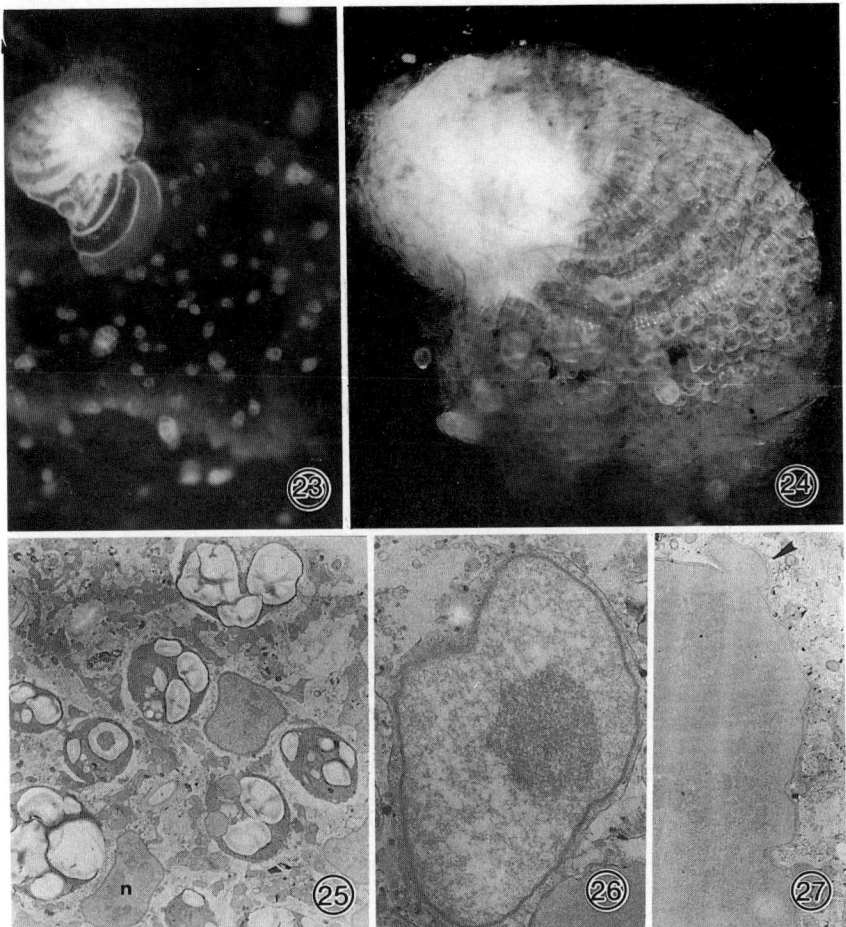

Figs 23–27. Figures 23 and 24 are of *Peneroplis planatus* (Faber and Lee, in press) taken in dark field supplemented with epi-illumination. Figure 25 is a TEM section of *Archaias angulatus* and Figs 26 and 27 TEM sections of *Sorites marginalis*. Reproduced with permission from Müller-Merz and Lee (1976). **Figs 23, 24.** Release of megalospheric juveniles (proloculi ~60 μm) from a schizont (Fig. 23 ×26; Fig. 24 ×75). **Fig. 25** Low-power survey of a chamberlet of an agamont with many symbionts (*Chlamydomonas hedleyi*) and two somatic nuclei (×4050). **Fig. 26.** Small (generative) nucleus with prominent nucleus (×15 000). **Fig. 27.** Portion of a somatic nucleus from the same specimen as Fig. 26. Arrow points to finger-like blebs at the surface of nucleus (×5100).

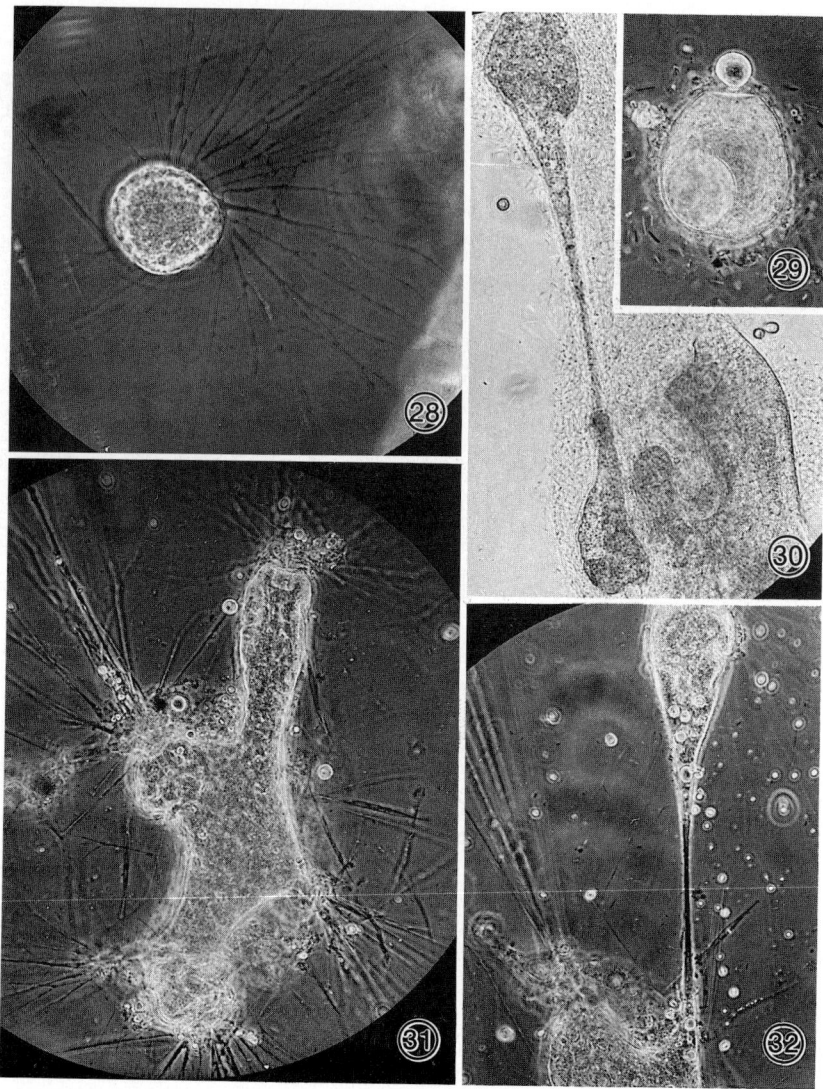

Figs 28–32. *Allogromia* sp. (strain NF). Figures 28, 29, 31, and 32 were taken with the aid of phase contrast; Fig. 30 is a transmission light micrograph. Reproduced with permission from Lee and Pierce (1963). **Fig. 28.** Typical multinucleate individual with six small nuclei in view in this optical section (× 75). **Fig. 29.** A monoral individual undergoing exogenous and endogenous budding. Note all pseudopodia have been withdrawn by this individual (× 262). **Fig. 30.** A bioral organism, in a heavily bacterized culture, in the process of binary fission (× 150). **Fig. 31.** A pentoral individual; note long neck on aperture at the top of photograph (× 150). **Fig. 32.** Late stage in cytotomy of the same individual (Fig. 31) leading to the separation of the long-necked filial cell from the parental cell (× 165).

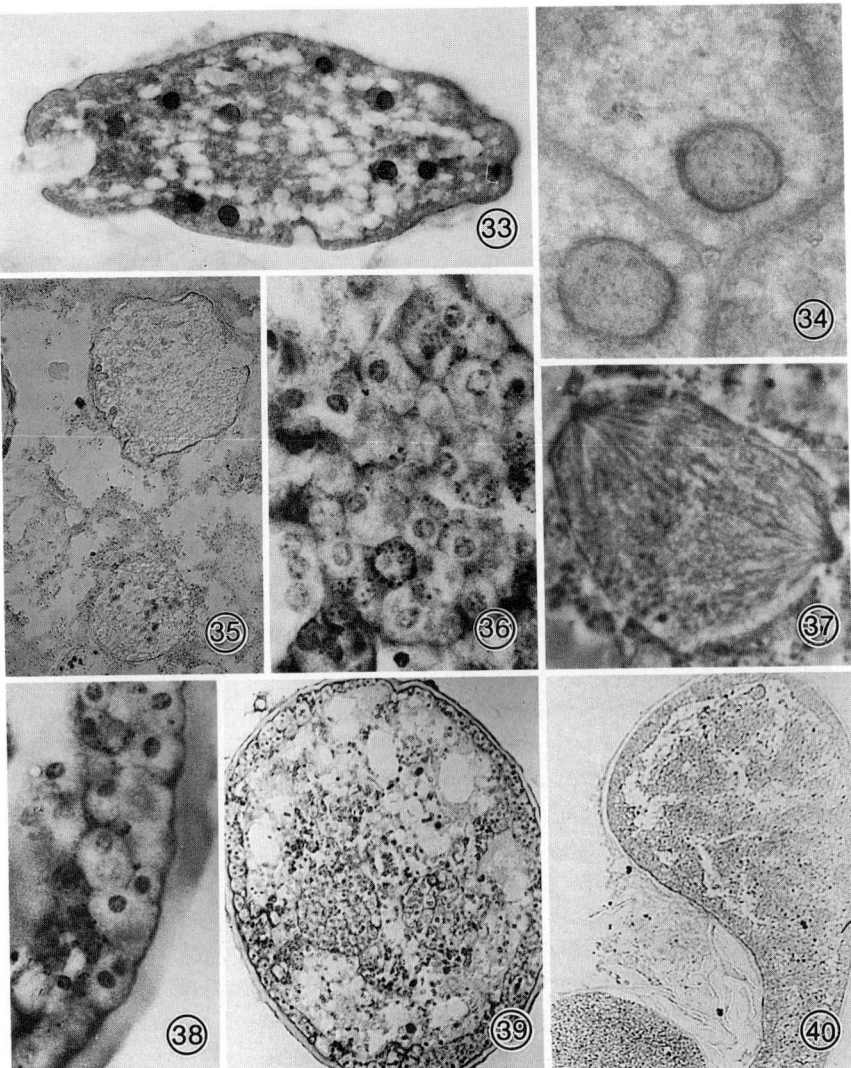

Figs 33–40. All except Fig. 33 are Fuelgen-stained preparations photographed with the aid of a deep green filter; Fig. 33, Heidenhain's azan technique. Figures 33 and 35 are of *Allogromia* sp. (strain NF) and are reproduced with permission from Lee and Pierce (1963). Figures 34, 36–40 are of *Allogromia laticollaris* and are reproduced with permission from McEnery and Lee (1976). **Fig. 33.** A trioral multinucleated organism (only two apertures are visible in this optical section: × 375). **Fig. 34.** Newly formed agamonts I within parental test (× 375). **Fig. 35.** Two multinucleate individuals from the same culture. The intensity of the stain in the nuclei was measured in a microspectrophotometer. The stain in the nuclei of the organism at the bottom of the figure averaged twice the intensity of the nuclei of the organism at the top. The bottom organism was presumed diploid; the top organism was haploid (× 37.5). **Figs 36, 38.** Enlargements of various portions of the gamonts in Fig. 39 showing gametes (average nuclear diameter is 4 μm). **Fig. 37.** Nucleus of an agamont 2 of the CSH strain in anaphase (11 μm). **Fig. 39.** Gamont with cytokinesis nearly completed. Note uninucleate gametes at periphery and in the two central whorls (original size 216 μm). **Fig. 40.** Section of one end of a meiocytotomont. The whole organism was ~1500 μm long by 525 μm in greatest diameter.

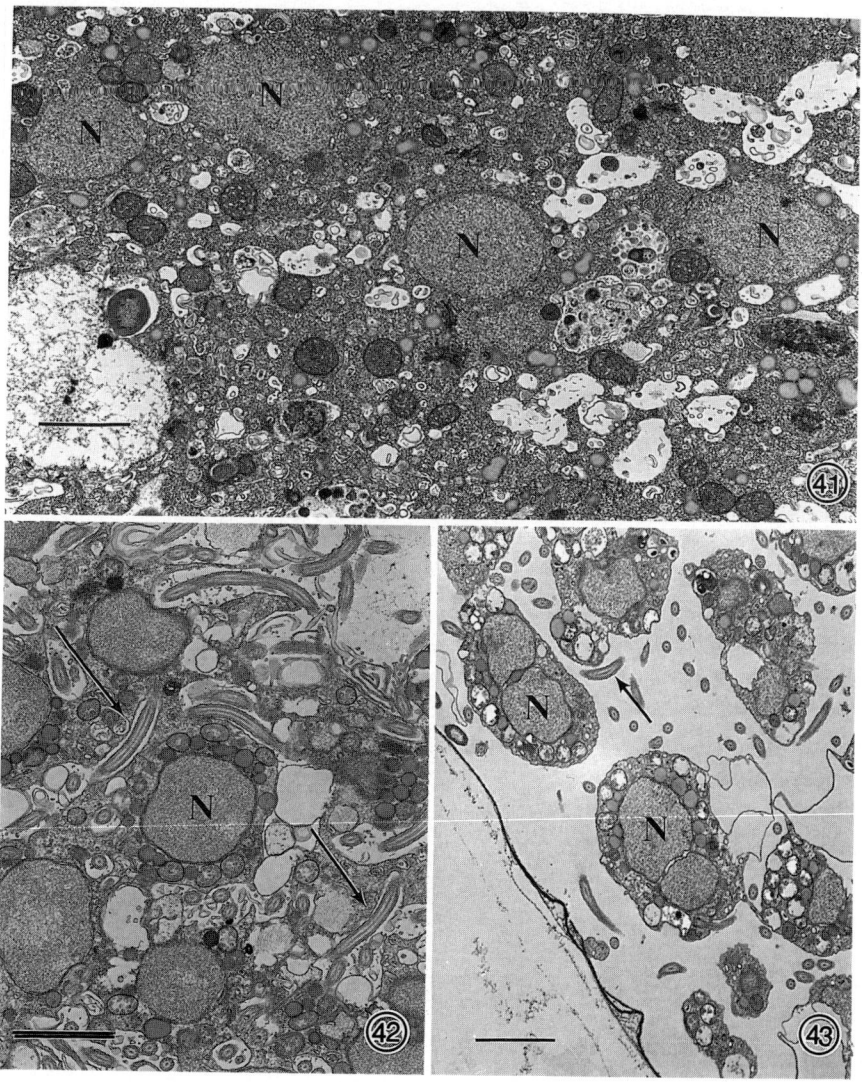

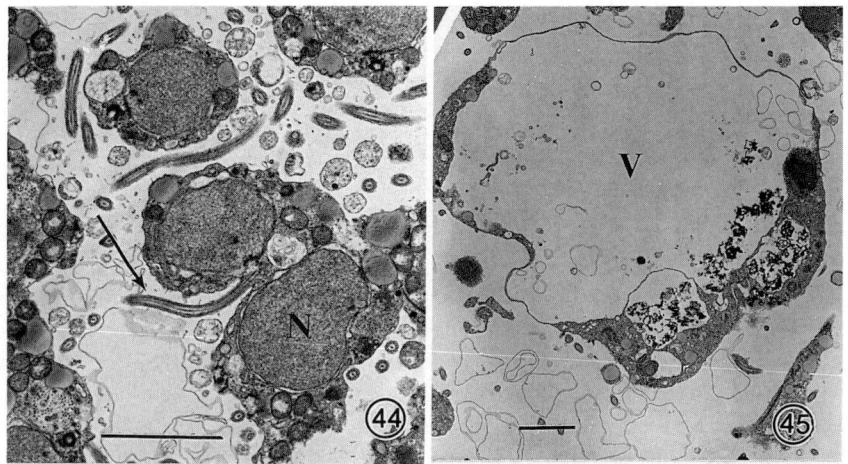

Fig. 44. An individual gamete (above) and a binucleated pregamete (below) with flagella (arrow) prior to release in *G. ruber*. **Fig. 45.** A vacuolated (V) spherical body expelled by *H. pelagica* during gamete release (scale bars = 2 μm).

Figs 41–43. Transmission electron micrographs of cytoplasmic events during gametogenesis in planktonic foraminifera. Proliferation of nuclei (N) within the cytoplasm of a reproductive planktonic foraminiferan (Fig. 41) precedes gradual segregation of the nuclei into interconnected nucleated islands of cytoplasm (*G. ruber*). Numerous flagella (arrows) protrude into the surrounding cisternae (Fig. 42). At a much later stage (Fig. 43), binucleated pregametes become segregated from the main mass of cytoplasm in preparation for release (*Pulleniatina obliquiloculata*).

Clearly, the survival rate of the earliest stages of life are critical in determining the standing stock of mature individuals. Greater attention to major factors that regulate the survival of offspring is warranted, including possible disease organisms, predators and environmental factors leading to mortality. Until we can devise satisfactory methods of observing syngamy, either in the laboratory or by special observational techniques in the natural environment, it is not possible to determine presently what proportion of the many gametes released undergo syngamy and how many zygotes are produced. There must be a large attrition in numbers of individuals compared to the number of gametes released. If the hundreds of thousands of gametes produced by each mature planktonic foraminifer produced zygotes, and thence mature offspring, the number of planktonic foraminifera would exceed the available space. Further research is required to determine what factors limit population growth and how these vary with climate, season, geographical location, and physico-chemical and biotic environmental variables. As a corollary, little is known about the differences in survival strategies across species, especially in relation to the presence of absence of symbionts, mode of nutrition and variation in niche. Particular attention to habitat depth in the water column and its variation with season and geographical locale is warranted to more fully characterize the factors promoting reproductive success and survival of offspring. In sum, many of the most significant questions concerning factors regulating life-cycle events, species abundance and the productivity of planktonic foraminifera remain to be investigated.

Although we know more about the life-cycles of benthic foraminifera than we do about the planktonic forms, in truth we still know very little. The only cytologically completely known life-cycles are in those species which have been cultured in the laboratory for many continuous generations. Cytological details for most other species remain fragmentary. One of the big obstacles is the high cost and consequent unavailability of microspectrophotometers to most foraminiferal researchers. If these devices were more common, we would have more information on the ploidy of critical life-cycle stages. A second major impediment is the relative scarcity of species which have been cultured in the lab. The challenge is to do so.

REFERENCES

Almogi-Labin, A. (1981). Population dynamics of planktonic foraminifera and pteropoda, Gulf of Aqaba, Red Sea. *Paleontol.* 87:481–511.

Anderson, O.R. (1984). Cellular specialization and reproduction in planktonic foraminifera and radiolaria. In *Marine Plankton Life Cycle Strategies* (eds K.A. Steidinger and L.M. Walker), pp. 35–66. Chemical Rubber Co. Press, Boca Raton, Florida.
Anderson, O.R. and Bé, A.W.H. (1978). Recent advances in foraminiferal fine structure research. In *Foraminifera* (eds R.H. Hedley and C.G. Adams), Vol. 1, pp. 122–202. Academic Press, London.
Anderson, O.R. and Bennett, P. (1985). A conceptual and quantitative analysis of skeletal morphogenesis in living species of solitary radiolaria *Euchitonia elegans* and *Spongaster tetras*. *Mar. Micropaleontol.* **9**:441–454.
Angell, R.W. (1971). Observations on gametogenesis in the foraminifer *Myotheca*. *J. Foram. Res.* **1**:39–42.
Arnold, Z.M. (1954). *Discorinopsis aguayi* (Bermudez) and *Discorinopsis vadescens* Cushman and Brönnimann: A study of variation in cultures of living foraminifera. *Contr. Cushman Fdn Foram. Res.* **5**:4–13.
Arnold, Z.M. (1955a). An unusual feature of Miliolid reproduction. *Contr. Cushman Fdn Foram. Res.* **6**:94–96.
Arnold, Z.M. (1955b). Life history and cytology of the foraminifera *Allogromia laticollaris*. *Univ. Calif. Pub. Zool.* **61**:167–252.
Arnold, Z.M. (1964). *Biological Observations on the Foraminifer Spiroloculina hyalina* Schultze. University of California Press, Berkeley.
Arnold, Z.M. (1967). Biological observations on the foraminifer *Calcituba polymorpha*. *Roboz. Arch. Protitenk.* **110**:280–304.
Arnold, Z.M. (1982). *Psammophaga simplora* n. gen., n. sp., a polygenic Californian saccamminid. *J. Foram. Res.* **12**:72–78.
Arnold, Z.M. (1984). The gamontic karyology of the saccamminid foraminifer *Psammophaga simplora*. *J. Foram. Res.* **14**:171–186.
Bé, A.W.H. and Anderson, O.R. (1976). Gametogenesis in planktonic foraminifera. *Science* **12**:890–892.
Bé, A.W.H., Hemleben, Ch., Anderson, O.R., Spindler, M., Hacunda, J. and Tuntivate-Choy, S. (1977). Laboratory and field observations of living planktonic foraminifera. *Micropaleontol.* **23**:155–179.
Bé, A.W.H., Anderson, O.R., Faber, W.W., Jr and Caron, D.A. (1983). Sequence of morphological and cytoplasmic changes during gametogenesis in the planktonic foraminifer *Globigerinoides sacculifer* (Brady). *Micropaleontol.* **29**:310–325.
Bijma, J. (1986). Observations on the life history and carbon cycling of planktonic foraminifera, Gulf of Eilat/Aqaba. Masters thesis, University of Groningen, The Netherlands.
Boltovskoy, E. and Wright, R. (1976). *Recent Foraminifera*. W. Junk, The Hague.
Brummer, G.J.A., Hemleben, Ch. and Spindler, M. (1986). Planktonic foraminiferal ontogeny and new perspectives for micropaleontology. *Nature* **319**:50–52.
Brummer, G.J.A., Hemleben, Ch. and Spindler, M. (1987). Ontogeny of extant globigerinid planktonic foraminifera: a concept exemplified by *Globigerinoides sacculifer* (Brady) and *G. ruber* (d'Orbigny). *Mar. Micropaleontol.* **12**:357–381.
Cesana, D. (1972). Ultrastructure des gamètes chez un Foraminifère: *Iridia lucida* LeCalvez. *Cashiers de Recherches de l'Academie des Sciences* **274**:1044–1047.
Cesana, D. (1975). Les stades initiaux de la gamogonie chez *Iridia lucida* LeCalvez (foraminifera Lagynidae). *Cashiers de Recherches de l'Academie des Sciences* **281**:262–266.

Cesana, D. (1978). La mitose gamogonique chez *Iridia lucida* (Foraminifera, Lagynidae). *Ann. Sci. Natur.* **20**:287–320.
Corliss, J.O. and Lom, J. (1985). An annotated glossary of protozoological terms. In *An Illustrated Guide to the Protozoa* (eds J.J. Lee, S.H. Hutner and E.C. Bovee), pp. 576–602. Society of Protozoologists, Lawrence, Kansas.
Desai, D. and Banner, F.T. (1985). The ontogeny of, and relationship between middle Miocene and Quaternary *Orbulina* (Foraminifera). *J. Micropaleontol.* **4**:81–91.
Erskian, M.G. and Lipps, J.H. (1987). Population dynamics of the foraminiferan *Glabratella ornatissima* (Cushman) in Northern California. *J. Foram. Res.* **17**:240–256.
Faber, W.W., Jr and Lee, J.J. (1991). Feeding and growth of *Peneroplis planatus* (Fichell and Moll) Montfort. *Symbiosis* **10**:63–82.
Fermont, W.J.J. (1977). Biometrical investigation of the genus *Operculina* in recent sediments of the Gulf of Elat, Red Sea. *Utrecht Micropaleontol. Bull.* **15**:111–147.
Føyn, B. (1936). Foraminiferenstudien I. Der Lebenszyklus von *Discorbina vilardeboana* d'Orbigny. *Bergens Mus. Arbok. Naturv. Rek.* **2**:1–22.
Føyn, B. (1937). Über die Kernverthältnisse der Foraminifere *Myxotheca arenilega* Schaudinn. *Archiv. Protistk.* **87**:272–295.
Goldstein, S.T. (1988). On the life cycle of *Saccammina alba* Hedley, 1962. *J. Foram. Res.* **18**:311–325.
Goldstein, S.T. and Barker, W.W. (1990). Gametogenesis in the monothalamous agglutinated foraminifer *Cribrothalammina alba*. *J. Protozool.* **37**:20–27.
Grell, K.G. (1954). Der Generationswechsel der polythalamen Foraminifere *Rotaliella heterocaryotica*. *Archiv der Protistenk.* **100**:211–235.
Grell, K.G. (1957a). Untersuchungen über die Fortpflanzung und sexualität der Foraminiferen I. *Rotaliella roscoffensis*. *Arch. Protistk.* **102**:147–164.
Grell, K.G. (1957b). Sexuelle Differenzierung bei den Gamonten der Foraminifere *Patellina corrugata*. *Zeit. Natur.* **12b**:415.
Grell, K.G. (1958a). Untersuchungen über die Fortpflanzung und sexualität der Foraminiferen II. *Rubratella intermedia*. *Arch. Protistk.* **102**:291–308.
Grell, K.G. (1958b). Untersuchungen über die Fortpflanzung und sexualität der Foraminiferen III. *Glabratella sulcata*. *Arch. Protistk.* **102**:449–472.
Grell, K.G. (1958c). Studien zum Differenzierungsproblem an Foraminiferen. *Naturwissen.* **2**:25–32.
Grell, K.G. (1959). Untersuchungen über die Fortpflanzung und sexualität der Foraminiferen IV. *Patellina corrugata*. *Arch. Protistk.* **104**:211–235.
Grell, K.G. (1960). Nachweis der sexuellen Differenzierung bei *Patellina corrugata* durch Teilbild analyse einer Films. *Zeit. Natur.* **15b**:270.
Grell, K.G. (1967). Sexual reproduction in protozoa. In *Research in Protozoology* (ed. T.T. Chen), pp. 149–213. Pergamon Press, Oxford.
Grell, K.G. (1973). *Protozoology*. Springer-Verlag, Berlin.
Grell, K.G. (1979). Cytogenetic systems and evolution in Foraminifera. *J. Foram. Res.* **9**:1–13.
Grell, K.G. (1988). The life cycle of the monothalamous foraminifer *Heterotheca lobata*, n. gen., n. sp. Karl Grell, 1988. *J. Foram. Res.* **18**:54–74.
Hallock, P.M. (1985). Why are larger foraminifera large? *Paleobiology* **11**:195–208.
Hallock, P.M., Cottey, T.L., Forward, L.B. and Halas, J. (1986). Population biology and sediment production of *Archaias angulatus* (Foraminiferida) in Largo Sound, Florida. *J. Foram. Res.* **16**:1–8.

Hedley, R.H. (1958). A contribution to the biology and cytology of *Haliphysema* (Foraminifera). *Proc. Zool. Soc. Lond.* **130**:569–576.
Hedley, R.H. (1962). the significance of an "inner chitinous lining" in Saccamminid organisation with special reference to a new species of *Saccamminia* (Foraminifera) New Zealand. *N.Z. J. Sci.* **5**:375–389.
Hedley, R.H. (1964). *The Biology of Foraminifera*. In *Internat. Rev. Gen. Exper. Zool.* I. [eds W.J. Felts and R.J. Harrison] pp 1–45. Academic Press, London.
Hedley, R.H. and Wakefield, J.St.J. (1967). Clone culture studies of a new rosalinid foraminifer from Plymouth, England and Wellington, New Zealand. *J. Mar. Biol. Assoc. U.K.* **47**:121–128.
Hedley, R.H., Pardy, D.M. and Wakefield, J.St.J. (1968). Reproduction in *Boderia turneri* (Foraminifera). *J. Nat. Hist.* **2**:147–151.
Hemleben, Ch., Bé, A.W.H., Spindler, M. and Anderson, O.R. (1979). Dissolution effects induced by shell resorption during gametogenesis in *Hastigerina pelagica* (d'Orbigny). *J. Foram. Res.* **9**:118–124.
Hemleben, Ch., Spindler, M. and Anderson, O.R. (1989). *Modern Planktonic Foraminifera*. Springer-Verlag, New York.
Heron-Allen, E. (1915). Contributions to the study of bionomics and reproductive processes in the Foraminifera. *Phil. Trans. R. Soc. Lond.* **B206**:227–279.
Hofker, J. (1930a). *The Foraminifera of the Siboga Expedition, pt II, Families Astrorhizidae, Rhizamminidae, Reophaxidae, Anomalinidae, Peneroplidae*, in *Saboga Expeditie IV*a, pp. 79–170. E.J. Brill, Leiden.
Hofker, J. (1930b). Der Generationswechsel von *Rotalia beccarii* var. *flevensis*, nor. var. *Z. Zell. Mikrosk. Anat.* **10**:756–768.
Hofker, J. (1976). Further studies in Caribbean Foraminifera. Studies on the Fauna of Curaçao and other Caribbean Islands 49.
Jeeps, M.W. (1942). Studies on *Polystomella* Lamarck (Foraminifera). *J. Mar. Biol. Ass. U.K.* **25**:612–665.
Ketten, D.R. and Edmond, J.M. (1979). Gametogenesis and calcification of planktonic foraminifera. *Nature* **278**:546–548.
Kloos, D.P. (1981). Growth and embryogenesis of the foraminifer *Sorites orbiculus*. *Paleontol. Proc.* **B84**:145–159.
Kloos, D.P. (1984). Parents and broods of *Sorites orbiculus* (Forskål), a biometric analysis. *J. Foram. Res.* **14**:277–281.
Kloos, D..P. and MacGillavry, H.J. (1978). Reproduction and life cycle of *Sorites orbiculus* (Forskål) Foraminifera. *Geologie en Mijnbouw* **57**:221–225.
Krüger, R. (1990). Untersuchungen zur Biologie der Großforaminifere *Calcarina gaudichaudii*, d'Orbigny, 1840. Diplomarbeit Universität Kiel, 65 pp.
Lacroix, E. (1941). Les *Orbitolites* de Gulfe d'Akaba. *Bull. Inst. Océanogr.* 794.
Le Calvez, J. (1938). Recherche sur les foraminifères. *Arch. Zool. exp. gén.* **80**:163–333.
Le Calvez, J. (1947). *Entosolenia marginata*, foraminifère apogamique ectoparasite d'un autre foraminifère *Discorbis mediterranensis*. *Cahiers de Recherches de l'Acadèmie des Sciences* **224**:1448–1480.
Le Calvez, J. (1950). Recherches sur les Foraminifères II. Place de meiose et sexualité. *Arch. Zool. Exp. Gén.* **87**:211–243.
Le Calvez, J. (1952). Le couple, *Discorbis patelliformis* (Brady)-*erecta* (Sidebottom) et les *Discorbis plastogamiques*. *Arch. Zool. Exp. Gén.* **89**:56–62.
Lee, J.J. and Pierce, S. (1963). Growth and physiology of foraminifera in the laboratory: part 4—monoalgal culture of an allogromiid with notes on its morphology. *J. Protozool.* **10**:404–411.

Lee, J.J. and McEnery, M.E. (1970). Autogamy in *Allogromia laticollaris* (Foraminifera). *J. Protozool.* **17**:184–195.
Lee, J.J., Freudenthal, H.D., Muller, W.A., Kossoy, V., Pierce, S. and Grossman, R. (1963). Growth and physiology of foraminifera in the laboratory. 3. Initial studies of *Rosalina floridana* (Cushman). *Micropaleontol.* **9**:449–466.
Lee, J.J., Muller, W.A., Stone, R.J., McEnery, M.E. and Zucker, W. (1969). Standing crop of Foraminifera in sublittoral epiphytic communities of a Long Island salt marsh. *Mar. Biol.* **4**:44–61.
Leutenegger, S. (1977a). Ultrastructure de foraminifères perforés et imperforés ainsi que de leurs symbiontes. *Cahiers de Micropaleontol.* **3**:1–52.
Leutenegger, S. (1977b). Reproductive cycles of larger Foraminifera and depth distribution of generations. *Utrecht Micropaleontol. Bull.* **15**:27–34.
Lipps, J.H. and Erskian, M.G. (1969). Plastogamy in foraminifera: *Glabratella ornatisima* (Cushman). *J. Protozool.* **16**:422–425.
Lister, J.J. (1895). Contributions to the life history of the foraminifera. *Phil. Trans. R. Soc. Lond.* **B186**:401–453.
Loeblich, A.R., Jr and Tappan, H. (1964). Sarcodina chiefly "Thecamoebians" and Foraminiferida. In *Treatise on Invertebrate Paleontology*, Part C, *Protista 2*. Geological Society of America/University of Kansas Press, Lawrence.
Lutze, G.F. and Wefer, G. (1980). Habitat and sexual reproduction of *Cyclorbiculina compressa* (d'Orbigny), Soritidae. *J. Foram. Res.* **10**:251–260.
Margulis, L., Corliss, J.O., Melkonian, M. and Chapman, D.J. (1990). *Handbook of Protoctista*. Jones and Bartlett, Boston, Mass.
Matera, N.J. and Lee, J.J. (1972). Environmental factors affecting the standing crop of foraminifera in sublittoral and psammonlittoral communities of a Long Island salt marsh. *J. Mar. Biol.* **14**:89–103.
McEnery, M.E. and Lee, J.J. (1976). *Allogromia laticollaris*: A foraminiferan with an unusual apogamic metagenic life cycle. *J. Protozool.* **23**:94–108.
McEnery, M.E. and Lee, J.J. (1981). Cytological and fine structural studies of three species of symbiont-bearing larger foraminifera from the Red Sea. *Micropaleontol.* **27**:71–83.
Meobius, K. (1880). Foraminifera von Mauritius. In *Beiträge zur Meeresfauna der Insel Mauritius und der Seychellen* (eds K. Moebius, F. Richter and E. Martens), pp. 65–112. Gutman, Berlin.
Muller, P.H. (1974). Sediment production and population biology of the benthic foraminifer *Amphistegina madagascariensis*. *Limnol. Oceanogr.* **19**:802–809.
Müller-Merz, E. and Lee, J.J. (1976). Symbiosis in the larger foraminiferan *Sorites marginalis* (with notes on Archaias spp.). *J. Protozool.* **23**:390–396.
Munier-Chalmas, E. (1880). Dimorphisme in Nummulites et Assilina. *Bull. Soc. Geol. France Ser. 3* **8**:300.
Munier-Chalmas, E. and Schlumberger, C. (1883). Nouvelles observations sue le dimorphisme des Foraminifères. *Acad. Sci. Paris Comtes Rendus* **96**:862–866.
Myers, E.H. (1935). Culture methods for the marine foraminifera of the littoral zone. *Trans. Am. Microsc. Soc.* **54**:264–267.
Myers, E.H. (1936). The life-cycle of *Spirillina vivipara* Ehrenberg, with notes on morphogenesis, systematics and distribution of the foraminifera. *J. R. Micr. Soc.* **56**:120–146.
Myers, E.H. (1940). Observations on the origin and fate of flagellated gametes in multiple tests of *Discorbis* (Foraminifera). *J. Mar. Biol. Assoc. U.K.* **24**:201–226.

Myers, E.H. (1943a). Life activities of foraminifera in relation to marine ecology. *Proc. Am. Phil. Soc.* **86**:439–459.
Myers, E.H. (1943b). Biology, ecology and morphogenesis of a pelagic foraminifer. *Stanford Univ. Biol. Sci.* **9**:5–30.
Nyholm, K.G. (1956). On the life cycle and cytology of the foraminifera *Nemogullmia longevariabilis*. *Zool. Bidr. Uppsala* **31**:483–493.
Nyholm, K.G. (1961). Morphogenesis and biology of the foraminifer *Cibicides lobatulus*. *Zool. Bidr. Uppsala* **33**:157–192.
Pawlowski, J. and Lee, J.J. (in prep.). Life cycle of a new species of *Rotaliella* from the Red Sea.
Reiss, Z. and Hottinger, L. (1984). *The Gulf of Aqaba: Ecological Micropaleontology*. Ecological Studies Vol. 50. Springer-Verlag, Berlin.
Rhumbler, L. (1909). Die Foraminiferen (Thalamophoren) der Plankton Expedition, 1. Teil. In *Ergebnisse der Plankton-Expedition der Humbolt-Stiftung* (ed. V. Hensen), Vol. 3, pp. 1–331. Lipsius und Tischer, Kiel.
Ross, C.A. (1972). Biology and ecology of *Marginopora vertebralis* (Foraminiferida), Great Barrier Reef. *J. Protozool.* **19**:181–192.
Röttger, R. (1974). Larger foraminifer: Reproduction and early stages of development in *Heterostegina depressa*. *Mar. Biol.* **26**:5–12.
Röttger, R. (1978). Unusual multiple fission in the gamont of the larger foraminiferan *Heterostegina depressa*. *J. Protozool.* **25**:5–12.
Röttger, R., Spindler, M., Schmaljohann, R., Richwien, M. and Flaudung, M. (1984). Functions of the canal system in the rotaliid foraminifer, *Heterostegina depressa*. *Nature* **309**:789–791.
Röttger, R., Flaudung, M., Schmaljohann, R., Spindler, M. and Zacharis, H. (1986). A new hypothesis: The so-called megalospheric schizont of the larger foraminifer, *Heterostegina depressa* d'Orbigny, 1826, is a separate species. *J. Foram. Res.* **16**:141–149.
Röttger, R., Schmaljohann and Zacharis, H. (1989). Endoreplication of zygotic nuclei in the larger Foraminifer *Heterostegina depressa* (Nummulitidae). *Euro. J. Protist.* **25**:60–66.
Röttger, R., Krüger, R. and Rijk, S. de (1990a). Trimorphism in foraminifera (protozoa) – verification of an old hypothesis. *Euro. J. Protist.* **25**:226–228.
Röttger, R., Krüger, R. and Rijk, S. de (1990b). Larger foraminifera: Variation in outer morphology and proloculus size in *Calcarina gaudichaudii*. *J. Foram. Res.* **20**:170–174.
Sakai, K. and Nishihira, M. (1981). Population study of the benthic foraminifera. In *Proceedings of the 4th International Coral Reef Symposium*, Vol. 2, pp. 763–766, Manila.
Salami, M.B. (1976). Biology of *Trochammina* cf. *T. quadriloba* Höglund (1947), an agglutinating foraminifera. *J. Foram. Res.* **6**:142–153.
Schaudinn, F. (1895a). Über der dimorphismus der Foraminiferen. *Ges. Naturf. Freunde, Berlin, S.-B.* **5**:87–97.
Schaudinn, F. (1895b). Über plastogamie bei Foraminiferen. *Ges. Naturf. Freunde, Berlin. S.-B.* **5**:170–190.
Schlumberger, C. (1875). *Reproduction des Foraminifères*. Ass. Française pour l'Avancement des Sciences, Nantes.
Schnitker, D. (1967). Variation in test morphology of *Triloculina linneiana* d'Orbigny in laboratory cultures. *Contr. Cushman Fdn Foram. Res.* **18**:84–86.
Schnitker, D. (1974). Ecotypic variation in *Ammonia beccarii* (Linné). *J. Foram. Res.* **4**:217–223.

Schultze, M. (1854). Über den organismus der polythalamen Foraminiferen nebst Bemerkungen über die Rhizopoden im allgemeinen, mit 7 tag, Leipzig.

Schwab (1969). Elektroncnmikroskopishe Untersuchungen an der Foraminifere *Myxotheca arenilega* Schaudinn. *Z. Zellforsch.*

Schwab (1976). Gametogenesis in *Allogromia laticollaris*. *J. Foram. Res.* **6**:251–257.

Sliter, W.V. (1965). Laboratory experiments on the life cycle and ecologic controls of *Rosalina globularis* d'Orbigny. *J. Paleontol.* **12**:210–215.

Sliter, W.V. (1970). *Bolivina doniezi* Cushman & Wickenden in clone culture. *Contr. Cushman Fdn Foram. Res.* **21**:87–99.

Spindler, M. and Hemleben, Ch. (1982). Formation and possible function of annulate lamellae in a planktic foraminifer. *J. Ultrastruct. Res.* **81**:341–350.

Spindler, M., Anderson, O.R., Hemleben, Ch. and Bé, A.W.H. (1978). Light and electron microscopic observations of gametogenesis in *Hastigerina pelagica* (Foraminifera). *J. Protozool.* **25**:427–433.

Spindler, M., Hemleben, Ch., Bayer, U., Bé, A.W.H. and Anderson, O.R. (1979). Lunar periodicity of reproduction in the planktonic foraminifer *Hastigerina pelagica*. *Mar. Ecol. Prog. Ser.* **1**:61–64.

Sverdlove, M.S. and Bé, A.W.H. (1985). Taxonomic and ecological significance of embryonic and juvenile planktonic foraminifera. *J. Foram. Res.* **15**:235–242.

Swanberg, N.R.S. and Bjørklund, K.R. (1987). The pre-cephalic development of the skeleton of *Amphimelissa setosa* (Actinopoda: Nassellarida). *Mar. Micropaleontol.* **11**:333–341.

Voronova, M.N. (1976). Polypoidy of somatic nuclei in agamonts of the foraminifer *Cibicides lobatulus*. *Tsitologiya* **18**:509–512.

Voronova, M.N. (1978). Nuclear dualism in gamonts of the foraminifer *Cibicides lobatulus*. *Tsitologiya* **20**:859–867.

Weber, H. (1965). Über die paarung der Gamonten und den Kerndualismus der Foraminifere *Metarotaliella parva* Grell. *Arch. Protistk.* **108**:217–270.

Winter, F.W. (1907). Zur Kenntnis der Thalamophoren. I. Untersuchung über *Peneroplis pertusus* (Forskål). *Archiv. Protistk.* **10**:1–113.

10
Collection, maintenance and culture methods for the study of living foraminifera

O. ROGER ANDERSON, JOHN J. LEE and
WALTER W. FABER JR

1.0	Introduction	335
2.0	Benthic foraminifera	338
	2.1 Collection	338
	2.2 Initial handling, transportation and processing	339
	2.3 Setting up vessels for maintenance of benthic foraminifera in the laboratory	341
3.0	Planktonic foraminifera	346
	3.1 Collection	346
	3.2 Initial handling and processing	348
	3.3 Setting up vessels for maintenance of planktonic foraminifera in the laboratory	349
	3.4 Feeding and maintenance of planktonic foraminifera	352
	References	354

1.0 INTRODUCTION

Much of our knowledge of the life-cycles and other aspects of the biology of foraminifera is restricted to less than 1% of the modern living species. In part, this is due to the small numbers of species which have been successfully maintained or cultivated in the laboratory. We suspect that these small numbers are more likely due to the lack of sustained efforts by investigators rather than to the tasks themselves. Such a diverse range of littoral, sublittoral, shallow benthic (<40 m) and planktonic foraminifera has been maintained in the laboratory (Table 1) that no researcher should be discouraged from the attempt. Even though few have accepted the challenge, the key principles of

Table 1 Species of foraminifera which have been maintained or cultured in the laboratory

Foraminiferal species	Food source(s)	Reference(s)
Allogromia laticollaris	*Nitzschia aculuris*, *Cylindrotheca closterium*, *Phaeodactylum tricornutum*, *Chlorococcum* sp., *Nannochloris* sp. and mixtures of the above	Arnold (1954), Lee et al. (1969)
	Chlorella	Schwab (1976)
Allogromia sp. (NF)	Monoxenic with an unidentified pseudomonad	Lee and Pierce (1963)
Ammonia beccarii	*Nitzschia* sp., *Chlamydomonas* sp.	Bradshaw (1955)
	Mixture of 8 species of small pennate diatoms and 3 species of cyanobacteria	Lee et al. (1961)
Amphisorus hemprichii	Mixture of *Nitzschia subcommunis*, *Chlorella* sp. (AT) and *Cylindrotheca closterium*	Lee et al. (in press)
Amphistegina lobifera	*Entomoneis densistriata*, *Nitzschia ovalis*, *Navicula* sp., *Nitzschia subtilis* and *Cylindrotheca closterium*	Lee et al. (in press)
Boderia turneri	*Dunaliella praemolecta*	Hedley (1964)
Bolivina doniezi	*Nitzschia angularis*	Sliter (1970)
Calcituba polymorpha	Naviculoid diatoms and cyanobacteria (vague)	Arnold (1967)
Cibicides lobatulus	*Chlamydomonas* sp.	Nyholm (1961)
Discorbinopsis aguayoi	Pennate diatoms and cyanobacteria (vague)	Arnold (1954)
Elphidium crispum	Diatoms (vague)	Jepps (1942)
	Phaeodactylum tricornutum	Murray (1963)
	Cocconeis placentula, *Amphora* sp., *Entomoneis densistriata* and *Nitzschia subcommunis*	Lee and Lee (1990)
Elphidium incertum	*Navicula* sp., *Amphora* sp.	Lee and Lee (1990)
Elphidium translucens	*Navicula* sp., *Cylindrotheca closterium*, *Nitzschia acicularis* and *Amphora tennerima*	Lee and Lee (1990)

Table 1 continued

Foraminiferal species	Food source(s)	Reference(s)
Glabratella sulcata	Pennate diatoms (vague)	Grell (1958)
Globigerinoides sacculifer	Calanoid copepods, Artemia salina nauplii	Anderson et al. (1979)
Globorotalia truncatulinoides	Emiliana huxleyi	Anderson et al. (1979)
Haliphysema tumanowiczii	Diatoms (vague)	Hedley (1958)
Hastergerina pelagica	Artemia salina nauplii	Anderson et al. (1979)
Iridia lucida	Diatoms (vague)	Le Calvez (1938)
Metarotaliella parva	Diatoms (vague) (later SEM of old preparations was shown to be Entomoneis densistriata)	Weber (1965), Lee (1983)
Myxotheca arenilega	Algal zoospores	Føyn (1936)
	Chlorella	Grell (1973)
Orbulina universa	Artemia salina nauplii, Calanoid copepods	Spindler et al. (1984)
Patellina corrugata	Diatoms (vague)	Grell (1959)
Peneroplis planatus	Dunaliella salina Cocconeis placentula, Nitzschia sp., Amphora sp., Navicula sp. and Chlorella sp.	Faber and Lee (1991)
Rosalina leei	Amphora coffeaeformis, Nitzschia acicularis, Cylindrotheca closterium, Nitzschia sp., Dunaliella parva, Dunaliella quartolecta and Dunaliella praemolecta	Hedley and Wakefield (1967), Muller (1975)
Rosalina vailardeboana	Algal zoospores	Føyn (1936)
Rotaliella heterocaryotica	Dunaliella sp.	Grell (1954)
Rotaliella roscoffensis	Diatoms (vague)	Grell (1957)
Saccamina alba	Diatoms (vague)	Hedley (1963)
Spirillina vivipara	Diatoms (vague)	Myers (1936)
Spiroloculina hyalina	Amphora sp. (Halamphora group)	Muller (1975)

maintenance and culture were summarized more than a half century ago (e.g. Myers, 1935, 1937; Galtsoff, 1937; Jepps, 1942) and reiterated almost every other decade since then [e.g. Arnold, 1954, 1974 (includes historical review); Murray, 1973; Lee, 1974; Lee and Muller, 1974; Hemleben et al., 1989]. In addition to updating some of the techniques, it is our goal in this review to encourage others to join us in this worthwhile enterprise.

We make a distinction in this review between maintenance and culture, which to some may seem artificial. However, we feel the distinction is a real one and its recognition will lead to eventual progress in the field. Maintenance is the ability to keep foraminifera alive in the laboratory for sustained periods (days, weeks, months, years). The foraminifera usually feed and grow during this time. Occasionally, they reproduce. Culture is one step beyond. Cultured foraminifera go through continuous growth and reproduction and many generations are produced in the laboratory. When we culture foraminifera in the laboratory, we have gained sufficient knowledge of their physiological, ecological and nutritional needs to be able to sustain them indefinitely.

2.0 BENTHIC FORAMINIFERA

2.1 Collection

The choice of a collection site is a critical starting place. Benthic foraminifera are notoriously patchy in their distribution in the field at distance scales (m^2–100 m^2) meaningful to the collector (Buzas, 1965; Lynts, 1966; Matera and Lee, 1972). In general, broad expanses of coarse quartz sand beaches are poor collecting sites because these are built by vigorous wave or current action (Arnold, 1974).

Living foraminifera are more likely to be found in more protected habitats with less rapidly flowing waters and which are most likely to be sites of deposition of fine sand to silt (Arnold, 1974). Tide pools and rocky shores in calm and/or protected waters are also good collecting sites.

Although not immediately obvious from the perspective of a scientist used to observing the annual production of foraminifera in the sediments, the distribution of living foraminifera in littoral zones can be strongly correlated with the distribution of particular species of macrophytes and affected by seasonal factors. For example, mid-July to mid-August is the easiest time of the year to collect *Elphidium incertum, E. clavatum, E. translucens* and *Quinqueloculina* spp. from epiphytic communities in the northeastern U.S. salt marshes (Matera and Lee, 1972; also discussed in Lee, 1974). The easiest benthic organisms to collect are those which are epiphytic on seaweeds or seagrasses because they are so easily handled. Many foraminiferan species are only weakly attached to the seaweeds or seagrasses on which they are living, and therefore care must be taken to gently collect and transfer substrates to harvesting vessels. Plastic bags with ties carried along in standard net sacks are most convenient when collecting by SCUBA. With experience, a diver can recognize foraminifera (0.5 mm and larger) on natural substrates. In the littoral zone, the substrates are collected in plastic buckets. In either case, one exercises care not to place too much substrate into the harvesting vessels, otherwise it could become rapidly anaerobic under crowded conditions.

If the foraminifera of interest are found on the surfaces of small rocks, one can gently collect these. Some species are most abundant on the surface or in the upper layers of sediment. In littoral and SCUBA depth waters, they are most easily collected by using plastic kitchen utensils (cooking spoons, bent spatulas and pancake flippers). A paint scraper with a curved draw-knife blade is a convenient tool for scraping foraminifera off of the surfaces of larger rocks. A plastic bulb-baster used in the kitchen for basting meat is a useful tool to suck up the scrapings and transfer them to the harvesting container. As did Murray (1963) and Jepps (1942) before us, living benthic foraminifera are also easily collected from slightly deeper or more hazardous habitats like Drake's Island in Plymouth Harbor, by epibenthic dredges (Lee and Lee, 1990). Several custom-designed and -built collection tools are described and illustrated in Arnold's (1974) review.

2.2 Initial handling, transportation and processing

As soon as possible after collection, the samples should be processed. In shallow waters, it is reasonable to process a half bucket of seaweeds or seagrasses as soon as it is collected. The bucket is three-quarters filled with sea water, and the seaweeds and seagrasses are rubbed against each other to dislodge the foraminifera. After each mass of algae or grass is rubbed in the bucket, it is discarded, until only the sea water and the epiphytes remain. The bucket is then decanted through sieves into another bucket. For salt-marsh foraminifera, we have routinely used no. 4 (4.76 mm), no. 18 (1.0 mm) and no. 35 (0.5 mm) sieves to remove small clams, microarthropods, annelids and coarser sand grains. When we have processed collections of larger foraminifera, an ordinary plastic kitchen collander has been a most useful sieve. Some custom-built sieving devices are described by Arnold (1974). The bucket is allowed to stand in the shade for approximately 10–15 min to let the foraminifera settle to the bottom. The upper $\frac{7}{8}$ of the contents of the bucket are decanted and fresh sea water is added to the bucket, which is then placed aside, to let the foraminifera settle again to the bottom. After the bucket is decanted again, a sample can be examined with the aid of a hand lens or a field dissecting microscope. This is particularly important when one is collecting in an unfamiliar location.

A successful collection, containing foraminifera at the bottom of a bucket, should be transferred to wide-mouthed, screw-capped polyethylene bottles or rigid plastic bags. One should not place too much sample in one container. As a general rule, a container should be filled with samples 20% by volume and then overlain with approximately three times their volume with sea water, leaving an air space of approximately 20% at the top. Samples in the field are immediately transferred to an insulated food chiller or submerged under water at the collection site to avoid killing many of the foraminifera by the combined effects of elevated temperature, rapid bacterial growth and anoxia.

As soon as possible on returning to the laboratory, the sediment containing foraminifera is spread out in wide-mouthed, covered containers which are of a size and depth so that they can be easily manipulated under a dissecting microscope. We have used 18 × 12 × 2.5 inch pyrex baking dishes (Corning cat. no. 3170), stacking culture dishes (155 × 55 mm, Wheaton cat. no. 41684) and deep Petri dishes (Corning cat. no. 3160-152, 325 or 3140). It is important to keep the dishes covered, otherwise evaporation causes the salinity to rise rapidly.

We use thin paraffin seals (Parafilm) to join baking and stacking culture dishes. It is very useful at this stage to place ordinary microscope slides into the sediment, letting them stand or lean vertically against the sides of the dishes (Arnold, 1974). We have used plastic racks designed to hold slides for staining for this purpose (Peel-a-way Sci. Div., W. Glen Wunderly Co., South El Monte, Calif.). Arnold (1974) described some custom-built racks.

The containers are left for a number of hours or overnight in a location where they will not be disturbed. During this time, most of the living foraminifera crawl to the surface of the sediment, vertically up the walls of the container or up the microscope slides. Foraminifera are most easily examined and recovered from the surfaces of the microscope slides. The slides are gently removed from the sample containers and transferred to a Petri dish with a layer of sea water. The foraminifera are brushed from one side of the slide with the aid of an artist's brush (#3–5 sable). With the brushed side down, the other surface of the slide is easily examined under a dissection microscope. Many of the foraminifera have very easily observed reticulopodial webs in this situation (relatively low food abundance). Allogromiids are more easily recognized on slides which are viewed by epi-illumination, or dark-field, than they are in transillumination. Foraminifera on the walls of the sample containers can be removed with the aid of a sable brush or an easily manipulated plastic transfer pipette (e.g. Fisher cat. no. 13-711-5A) and transferred to a Petri dish with sea water where they can be examined more closely. Foraminifera on the surface of the sediment are most easily observed in a dissection microscope using epi-illumination. They can be removed using Pasteur or transfer pipettes.

2.3 Setting up vessels for maintenance of benthic foraminifera in the laboratory

Many different methods have been used to successfully maintain foraminifera in the laboratory. They range from placing the foraminifera in circulating marine aquaria to very carefully managed gnotobiotic systems. Before one begins culture or maintenance efforts, some thought has to be given to the aims of the research programme. Crude, or agnotobiotic, maintenance is relatively unsophisticated, easy to do, less time-consuming, but notoriously undependable.

2.3.1 Circulating marine aquaria and other flow-through systems

Foraminifera can thrive in both circulating and recirculating marine aquaria. The rate of water flow should be gentle and the system should be moderately illuminated, by means of natural or artificial light for at least 8 h/day. The system can be aerated but not so vigorously as to cause turbulence. Temperature should be regulated to be at or slightly below the temperature of the sea where the organisms were collected. Many foraminifera are easily observed grazing and reproducing on the aquarium walls. The longevity of particular species of foraminifera in marine aquaria is quite variable. One of us (J.J.L.) has kept particular species of foraminifera (e.g. *Ammonia beccarii, Rosalina leei* and *Bolivina vaughni*) alive and reproducing for several years in recirculating aquaria. For many reasons (e.g. recruitment of new organisms coming in with the sea water, predation, changing community structure), they disappear between sample periods. With recirculating systems, the problem of recruitment of new organisms which might upset the stability of the populations within the aquarium is diminished, but one must guard against shifts in pH and salinity. These are easily measured and adjusted (daily, if practical) parameters. Glass covers are important to retard evaporation and slow salinity shifts. As pointed out by Arnold (1974), the course gravel or sand (2–5 mm) needed for proper filtration of many recirculating systems increases the difficulty of examining, sampling and harvesting the foraminifera.

After some time, the walls of illuminated circulating aquaria become overgrown with *Aufwuchs*, which must be scraped off in order to see the foraminifera. Sometimes the foraminifera do well in aquaria with *Aufwuchs*, often they disappear from the community, being replaced by ciliates and naked amoebae. In our experience, it is better to scrape the sides of the aquaria at regular intervals. The scrapings are sucked up with a meat-baster and they can be agitated and sieved to separate the foraminifera before discarding the rest of the material.

Arnold (1966, 1967) described a closed-circulation system of small volume containers which he built to maintain foraminifera in the laboratory. To have one available for use, a person would have to be as skilled a machinist as Arnold is, or have access to a machine shop which could build one. We found some commercial chemostats (e.g. New Brunswick Scientific Co., model C32) quite adaptable for the purpose (Lee *et al.*, in press). In fact, the chemostat design which provides for the continuous flow of fresh sterile media into the foraminiferal culture (reaction vessel) was superior for the maintenance of several species of larger foraminifera (*Amphistegina* spp., *Amphisorus hemprichii* and *Marginopora kudakajimensis*) than the recirculating systems we tried. These species of larger foraminifera continuously withdraw small amounts (μg at/l) of NO_3^1 and PO_4^{3-} from their medium but cannot tolerate much higher levels (Lee *et al.*, in press).

2.3.2 Small batch type maintenance/culture systems

Many different successful approaches have been used to maintain and culture foraminifera in the laboratory. These can be divided as follows:

1. Fortuitous/agnotobiotic:
    - no intervention,
    - enriched and/or controlled,
    - fed natural mixtures.

2. Fortuitous/partially gnotobiotic:
   - fed live food,
   - fed heat-killed food,
   - food organisms pre-inoculated or encouraged to grow with foraminifera.
3. Inductive methods:
   - tracer feeding,
   - isolation of food organisms from the habitat,
   - setting up partially gnotobiotic/synxenic cultures.

2.3.2.1 *Fortuitous/agnotobiotic.* These are the simplest to set up and maintain. They are always worth the effort. As soon as possible after collection, the foraminifera are picked up with brushes (5–0 sable brushes) and transferred to filtered or sterile sea water in stacking dishes, deep Petri dishes, tissue culture flasks, etc. It is usually prudent to eliminate micro-crustaceans and other micro-herbivores at this stage, because they can compete with the foraminifera for food and because their activities often interfere with the pseudopodia of the foraminifera. The dishes are then placed in moderate natural or artificial light ($\sim 30 \,\mu E/m^2$), in an incubator or room which is approximately the temperature of the habitat where the foraminifera were collected. In the simplest type of fortuitous set up, the media should be changed weekly if evaporation from the dishes is not a problem. Thin paraffin film can be used to retard evaporation when dish lids do not fit tightly. We use an Abbe-refractometer (Bausch and Lomb, Abbe-3L) because it is quick and easy to check the salinity of a drop of sea water from a vessel. If there are no blooms of ciliates, particularly hypotrichs, the foraminifera may do well. Sometimes Føyns Erdschreiber has been used successfully to encourage microalgal growth in the dishes where the foraminifera are being grown or maintained (see Lee *et al.*, 1970, 1975, for formulation and preparation). Erdschreiber sea water enriched with nitrate, phosphate and soil extract encourages the growth of about 50% of the microalgae present in natural marine littoral communities (Lee *et al.*, 1975). To a degree this is a good idea, but, as noted by Arnold (1954, 1974), algal overgrowth can inhibit foraminiferal growth and reproduction. One way to avoid this problem is to transfer the foraminifera every month to fresh vessels.

Bacterial blooms stimulated by the addition of nutrients can be controlled by the addition of antibiotics or antibiotic mixtures (Arnold, 1967; LaGarde, 1967; Lee *et al.*, 1970; Lee and Muller, 1974). Instead of enriching the medium in which the foraminifera are being maintained, some workers have added natural mixtures of microalgae or detritus to foraminifera in unenriched sea water (e.g. Röttger *et al.*, 1980).

2.3.2.2 *Fortuitous/partially gnotobiotic.* Almost all of the earlier students of foraminifera (e.g. Lister, 1895; Jepps, 1926; Sandon, 1932; Myers, 1935, 1936, 1942, 1943) made qualitative microscopic observations as they watched them feed. Diatoms are easily recognized and were the most frequently reported food along with small flagellates, macrophyte gametes, unicellular and filamentous algae, bacteria, faecal pellets, micrometazoa, etc. It is reasonable, therefore, to set up maintenance/culture vessels with potential food organisms which have worked well in the past in maintaining foraminifera (Table 1) or of the types which you observe the foraminifera to consume.

There are four approaches to setting up this type of partially gnotobiotic maintenance/culture vessel: (1) feeding live potential food to foraminifera in a sea water medium; (2) feeding heat-killed potential food to foraminifera in sea water media; (3) inoculation of foraminifera onto lawns of pre-incubated potential food organisms; or (4) encouraging the food organisms to grow along with the foraminifera. Whichever of these approaches is used, it is prudent to brush the foraminifera carefully to eliminate extraneous organisms before inoculating them into sterilized sea water and vessels. If one can work out the concentrations one needs to use to achieve a good balance between consumers and consumed, feeding live food is a good method to maintain/culture foraminifera. Algal overgrowth can be prevented by regulating the amount of light either by shortening the photoperiod, lengthening the distance between the vessels and the light source, or by throwing a layer or two of black nylon window screen over the vessels. In our experience (Lee et al., 1966, 1969, in press; Muller and Lee, 1969), mixtures of food species promote more vigorous growth and reproduction of foraminifera than do single species.

Both Grell (1954) and Bradshaw (1957) used heat-killed algae as food. The algae can be killed by boiling for 3 min or by being placed in a water bath for 30 min at 50°C. Foraminifera readily consume such algae. After a few hours, it begins to decay and bacteria begin to grow in response to excess dead food. Under such conditions, it is reasonable to cut back on food quantity or change the sea water daily.

We have grown a number of species of foraminifera by inoculating them on previously grown lawns of *Dunaliella salina, Cylindrotheca closterium* and *Amphora* sp. (*Halamphora* cluster). The lawns are grown for 5–7 days in Erdschreiber medium which is gently decanted and replaced with sterile sea water just before the foraminifera are introduced into the medium. If vessels with lawns are illuminated by dark field, the "feeding trails" of foraminifera are readily observed.

2.3.2.3 *Inductive methods.* Inductive methods require more time, experience and skills, but they offer greater potential for success than do fortuitous methods. Very briefly, samples of the communities in which the foraminifera are growing are (1) aseptically inoculated into liquid or solidified versions of a series of differential media (e.g. Lee *et al.*, 1975) and (2) oxidized and prepared for examination in the light microscope (Lee *et al.*, 1975) or the SEM (Lee *et al.*, 1980a,b,c, 1989). A study of the microscopic preparations very rapidly gives one a good picture of the diatom population structure of the community in which the foraminifera are feeding. After incubation in the light for approximately 10 days, the diatom and chlorophyte colonies formed on the agar plates are usually ready for picking. Because representatives of different diatom genera form different types of colonies on agar (Lee *et al.*, 1975), it is possible to selectively look for particular colony types most likely formed by desired species in the community. Working under a dissecting microscope, we use a sharpened, alcohol-sterilized and flamed microspatula (Fisher cat. no. 21-401-10) to cut, lift and transfer isolated colonies to the same liquid media.

After the picked colonies have grown in liquid media (7–10 days), an aseptic sample of the culture can be examined by phase microscopy to see if the culture is axenic. If not, it can be restreaked on plates of the same medium. If the cultures are axenic, they are inoculated into fresh medium and incubated for 5 days. Sterile radionuclide label (1–10 μCi/ml; ^{32}P or ^{14}C) is then aseptically added to the medium (Lee *et al.*, 1988). The culture is incubated for an additional 3 days after which it is transferred to a conical bottom plastic centrifuge tube. The algae are gently centrifuged out of the labelling medium in a clinical centrifuge (~ 1000 g). The cells are resuspended in fresh sea water and again gently centrifuged. The cells are washed until no label is left in the medium (usually 5–8 washes) and then resuspended in a small known volume of sea water. Aliquots are taken to assay radioactivity and to enumerate the algal population. The rest of the algal cells are inoculated into small Petri plates or 9-hole spot plates containing the foraminifera.

The foraminifera have been inoculated 4–8 h earlier and have easily observable pseudopodial networks. The tracer-labelled cells are gently pipetted over the foraminifera and their reticulopodia. After incubation with the labelled food for 3–6 h, the foraminifera can be removed from the labelling medium and transferred to 9-hole spot plates with fresh sea water. The foraminifera are brushed free of adhering algae and washed five times until they are free of external label. The organisms are then returned to a Petri plate with fresh unlabelled sea water and unlabelled algae. While the radiation transferred to the foraminifera in the ingested food could be measured at this time (e.g. Lee et al., 1966), we now recognize that very little of some species of algae that is ingested [e.g. *Chlorella* sp. (strain AT)] is assimilated. Much label in these organisms is egested after 24 h (Lee et al., 1988). Therefore, it would seem better to examine the tracer-labelled foraminifera after 24 h in a "cold chase". It is not surprising to find very different algal dietary preferences in foraminifera from the same community (e.g. Muller, 1975; Lee et al., 1988).

Once the tracer-feeding information is at hand, partially gnotobiotic/synxenic maintenance cultures can be set up using mixtures of the preferred food organisms. The techniques and choices for the set up are the same as one would use for fortuitous/partially gnotobiotic cultures except that one knows in advance that the foraminifera will consume and assimilate the organisms in the vessel.

3.0 PLANKTONIC FORAMINIFERA

3.1 Collection

Planktonic foraminifera occur exclusively in open ocean habitats, usually locations of great depth known as blue water environments. It is preferable, when possible, to collect in open ocean locations using a research vessel as a base (see Chapter 8). Planktonic foraminifera are rarely found near land, except where currents bring open ocean water near to shore as occurs on some islands such as Bermuda, Curaçao and Barbados. When working from island locations, it is important to determine the minimum distance from the shore necessary to obtain healthy individuals. Usually, a distance of one to several miles is adequate depending on the "island shadow effect". The quality of the ocean water used for the collection and culturing of planktonic foraminifera must be carefully monitored to ensure that it is of the highest purity. Contaminants from the surface of the boat, or from chemicals introduced into containers used for collecting and holding the organisms, can cause mortality. The water samples that are to be used for culturing should be taken at the sampling site. We use a plastic bellows hand pump (e.g. Fisher cat. no. 01-007-1) fitted with a plastic hose to draw water up from 2 m or more below the boat. This ensures purity as much as possible. The water is collected in non-toxic plastic bottles. We have found that collapsible, plastic water bags used for carrying potable water on camping trips, etc., are very convenient for collecting and storing open ocean water. The water should be refrigerated to prevent growth of microbiota, if it is not used the same day of collection. In general, we prefer not to use water if it is stored for more than several days to a week.

3.1.1 Collection site

The geographical and seasonal distribution of planktonic foraminifera in the North Atlantic has been described by Bé (1959, 1960, 1977). Warm-water, spinose species tend to occur year round in low-latitude locations, especially in equatorial regions. Cold-water species (some are non-spinose) occur at great depths during the summer months in temperate latitudes or can be collected during the winter months (e.g. December to March near Bermuda) from surface water. Some species (e.g. *Hastigerina pelagica*) reproduce at monthly intervals on a lunar cycle, and are not abundant in surface waters for several days during the period of a full moon.

Two methods of collection have successfully been used in our work: (1) capture in plankton nets deployed from a gently drifting boat or (2) hand collection during SCUBA diving. Small glass jars are used to enclose the foraminiferan as it passes by the diver. A more detailed account is presented in Hemleben *et al.* (1989). Only a general survey of methods is given here.

3.1.2 Net collection

Many species of planktonic foraminifera are sufficiently robust to withstand collection using plankton nets. The nets should be deployed from a gently drifting boat to prevent excessive turbulence and clogging of material in the net and cod end. We use nets with a 76-μm mesh to collect smaller species (non-spinose species and juvenile forms of larger species) or a 202-μm mesh for larger species such as *Hastigerina pelagica* (Anderson *et al.*, 1979). The net should be deployed from a gently drifting boat for about 3–5 min depending on the abundance of biota in the water. A trial deployment will indicate how abundant the species are and what sampling time is optimum to obtain a sufficiently rich sample of planktonic foraminifera without excessive aggregation. After collection, the sample should immediately be diluted with, and suspended in, at least a gallon of sea water to reduce the probability of excessive agglomeration. The diluted samples should be placed in an insulated chest for transport to the laboratory.

3.1.3 Collection by SCUBA divers

Where possible, collection by SCUBA divers is preferred to minimize collection trauma (Bé et al., 1977; Hemleben et al., 1989). Only certified divers should do this work and adherence to standard safety rules should be carefully maintained to ensure the well-being of the divers. Samples are usually collected at a depth of 3–5 m. Two nylon mesh bags are suspended from the side of the boat tethered to a line containing a weight at one end. One bag contains the collecting jars filled with sea water and the other is used to hold the jars containing the samples. We use 130-ml glass canning jars with unlined white plastic lids. If sunlight is sufficiently intense, the SCUBA divers can usually see the larger species as they drift past, owing to the sparkling quality of their peripheral cytoplasm. The divers position themselves within the shadow of the hull of the boat and visualize the planktonic foraminifera against the dark background of the bottom of the boat. One individual foraminiferan is collected per jar to minimize aggregation. The collecting jar is opened and rinsed with ambient sea water, when it is removed from the mesh bag. This is done by inserting one's index finger into the jar and creating a vortex. During collection, the lid of the jar is held at an angle near the lip. When a planktonic foraminiferan passes nearby, the lid is displaced slightly creating a current that swirls the organism into the jar. The lid is immediately secured tightly to the jar to prevent water loss during transport to the laboratory. It is possible for a diver to collect 50–80 individuals during a 30- to 60-min period depending on the abundance of organisms. An experienced diver can also identify species by noting the configuration of the spines, organization of the mass of cytoplasm and distribution of symbionts. When necessary, a plankton net can be deployed as a drogue to slow the drift of the boat and aid the divers in maintaining their position relative to the hull. The filled sample jars should be placed in an insulated storage chest to maintain a relatively constant temperature during the return trip to the laboratory.

3.2 Initial handling and processing

3.2.1 Net-collected samples

Upon return to the laboratory, the planktonic foraminifera should be isolated from the plankton sample as soon as possible. A high-quality dissecting microscope is required. We dispense approximately 100 ml of the diluted plankton catch into a pyrex culture dish (90 mm diameter × 40 mm deep). A Pasteur pipette, fitted with a rubber bulb, is used to remove individuals with the least adhering debris while observed with a dissecting microscope. If the diameter of the pipette is too small, it can be fractured at a point higher on the tip to yield a larger diameter opening. Glass needles produced by heating Pasteur pipettes in a flame and pulling them to a fine tip can be used to gently tease away adhering debris from individuals that are not too heavily contaminated. In general, however, only the least agglomerated individuals should be collected. Those that are more heavily contaminated may not recover. Approximately 20 individuals are gently transferred by Pasteur pipette to a second culture dish containing about 200 ml of sea water brought from the collection site. After several hours or overnight, the healthiest individuals will purge the cytoplasm of debris by expulsion during cytoplasmic streaming. Each of the more robust organisms can be transferred to individual culture vessels and identified to species level using an inverted microscope. It is important to separate the planktonic foraminifera as soon as possible after they have shed debris to prevent aggregation and cannibalism. Some species such as *Globigerinoides ruber* must be illuminated to encourage recovery and regrowth of spines.

3.2.2 SCUBA-collected samples

Upon return to the laboratory, the lids of the collection jars should be removed as soon as possible to permit gas exchange with the surrounding air. Gently rock the jar to ensure that the planktonic foraminiferan is near the bottom before removing the lid, else it may adhere to the lid or be lost with the small volume of water that typically spills from the jar when the lid is removed. Because SCUBA-collected individuals usually have a robust layer of cytoplasm surrounding the spines and shell, it may be necessary to use the broad end of a Pasteur pipette to transfer them to culture vessels. We shorten the tapered end of the pipette to produce a nib and attach a rubber bulb. The organism is drawn up into the broad end. It is important to gently deposit the foraminiferan into the receiving dish to prevent damage.

3.3 Setting up vessels for maintenance of planktonic foraminifera in the laboratory

3.3.1 Basic maintenance requirements

Symbiont-bearing planktonic foraminifera require illumination to maintain the health of the symbionts and to ensure optimum longevity of the host. If the symbiont-bearing species are placed in darkness for more than 24 h, physiological changes begin to occur leading towards expulsion of the symbionts and onset of reproduction. Fluorescent illumination using daylight bulbs with a 12-h light/12-h dark cycle is adequate. The cultures should be positioned to produce a light intensity of approximately 150–350 $\mu E/m^2/s$. The symbiont population density, quality of their pigment, and the general health of the foraminiferan should be carefully monitored, and the light intensity adjusted to obtain optimum growth and longevity. The temperature of the water should be kept at or below the temperature of the water at the collection site. For routine maintenance cultures, pyrex culture dishes containing 100–200 ml of sea water recently collected from the sampling site are adequate. Only one planktonic foraminiferan should be included in each dish, otherwise, they eventually become entangled leading to cannibalism. The dish is covered with a non-metallic closure. We have found that a square piece of non-toxic plastic is satisfactory and permits the dishes to be stacked in rows in front of the fluorescent fixtures (Fig. 1). It is essential to have an efficient and briskly circulating air conditioning system to maintain the cultures at constant temperature. If the dishes are placed sufficiently far from the fluorescent fixtures, and air is directed towards them from a fan near the source of the air conditioner, this system is adequate for routine observation and maintenance.

3.3.2 Environmentally controlled maintenance systems

For experimental research studies, and more exact observations of the planktonic foraminifera, the temperature and light should be carefully

Fig. 1. Pyrex culture dishes, each containing one planktonic foraminiferan and closed by a square piece of non-toxic plastic, are stacked in front of a bank of fluorescent bulbs. The distance must be carefully controlled, and a brisk source of circulation of cooled air should be provided to prevent heating of the sea water.

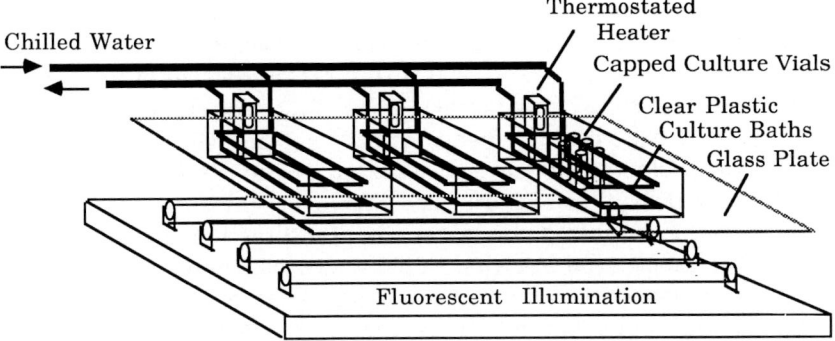

Fig. 2. An environmentally controlled culture system for experimental studies on planktonic foraminifera. Clear plastic or glass temperature baths supported on a plate glass sheet are illuminated from beneath by fluorescent bulbs. A chilled water source circulates through copper coils surrounding the inside periphery of the water bath and a thermostated, circulating heater maintains constant temperature. Closed, glass vials containing one planktonic foraminiferan each are suspended in the water baths. Reproduced with permission from Anderson et al. (1989).

controlled. Optically clear glass vials (30 mm diameter × 90 mm height) containing c. 30 ml of sea water, closed with a plastic cover to prevent evaporation, are immersed within a temperature-controlled bath. A single planktonic foraminiferan is placed in each vial (Fig. 2). The water baths contain a coil of copper tubing connected to a cold water source. A thermostated heater with circulating pump is inserted in each bath to maintain a preset temperature. Immersion heaters are available with a precision of $\pm 0.02°C$. Fluorescent fixtures are placed beneath a sheet of plate glass that supports the clear plastic water baths. This provides an unobstructed source of illumination. To change the intensity of the light, the vials can be surrounded by layers of nylon mesh that serve as neutral density filters to reduce incident light intensity. The individual planktonic foraminifera are observed daily in each vial using an inverted microscope with long working distance objective lenses. It is important to obtain glass vials with flat, optically clear bottoms. A sample of the vials should be requested from the supplier to determine their optical clarity.

For carefully controlled feeding experiments, the sea water can be filtered through a 0.22-μm pore size Millipore filter to remove naturally occurring microbiota. The filter should be rinsed with distilled water to remove possible inhibitory substances before use. If appropriate, however, unfiltered sea water can be used. Freshly collected sea water contains natural sources of microbiota that may serve as a food source.

It is not always possible to determine the species of microbiota in the water sample and it is often difficult to monitor what species proliferate in the culture vials. Hence, it is best to filter the sea water prior to use for carefully controlled nutritional studies. If the planktonic foraminifera are grown in large culture dishes, it may not be necessary to change the sea water during the several weeks that the organisms are being maintained. If small vials are used, however, it is wise to carefully monitor the condition of the water including testing the salinity. We use a refractometer that requires only a small drop of water as a sample. Also, the water should be inspected for algal or other microbiotic blooms using light optics. In general, it is wise to transfer the planktonic foraminifera only when the quality of the water suggests it is necessary, typically once per week. On the whole, the organisms do best when they are least disturbed. If it is necessary to transfer an individual, the broad end of a large-bore Pasteur pipette is used to withdraw the planktonic foraminiferan and gently deposit it in the freshly prepared sea water in a clean vial. The vials should be cleaned with tap water using a test tube brush, rinsed with distilled water, and air-dried. We avoid the use of detergents. If the species tends to adhere to the wall of the pipette, we feed the individuals several hours before transfer. This reduces the adhesiveness of the rhizopodia, and decreases the tendency of the foraminifera to cling to foreign surfaces.

3.4 Feeding and maintenance of planktonic foraminifera

Juvenile spinose species and some non-spinose species of planktonic foraminifera consume algal prey. Food vacuoles contain a variety of small diatoms, dinoflagellates, other thecate algae and small plastid-bearing protista (Anderson et al., 1979). We routinely maintain cultures of a coccolithophorid *Emiliana huxleyi* as potential algal prey. Standard culture media such as sterile F/4 medium (Guillard, 1975) can be used to grow the algae using fluorescent illumination. The cultures need to be transferred at regular intervals of several days if a relatively large inoculum is used. For maintenance, a smaller inoculum (c. 1 drop of inoculum per 30 ml) can be used. When the algae are harvested for feeding, a small aliquot (c. 10 ml) is gently sedimented by centrifugation to separate the algal cells from the growth medium. The pellet is resuspended in sea water to produce a concentration of about 10^3 cells/ml. The suspension is drawn into a pipette and a small drop deposited near the planktonic foraminifera. We usually supply food once every 1 or 2 days. Some small marine diatoms maintained in laboratory culture may also be added.

Mature non-spinose species and spinose species of planktonic foraminifera grow best when supplied with a zooplankton food source. We have found that 1-day-old *Artemia* (brine shrimp) nauplii are accepted as animal prey. Non-spinose species may not be able to capture and hold actively swimming nauplii. The non-spinose species typically adhere to the bottom of the culture dish with an extended halo of rhizopodia resembling benthic species. Perhaps, due to the attachment to the substratum, the rhizopodia do not hold prey as effectively as the rhizopodia of some spinose species. Therefore, we partially immobilize the prey either by gently warming an aliquot of the brine shrimp until they are heat-stunned, or by repeatedly forcing them in and out through the tip of a Pasteur pipette. For the latter method, it is best to suspend the nauplii in a small volume of sea water such as in a watch glass or embryo dish. The weakened nauplii are then pipetted near the planktonic foraminifera where they will be snared and gradually enclosed in feeding rhizopodia. Larger spinose species do not usually require the prey to be disabled.

We maintain a continuous supply of *Artemia* nauplii by adding cysts (labelled *Artemia* eggs by the suppliers) to dishes of sea water at 48-h intervals. A large surface-to-volume ratio is required in the culture dishes and we sometimes use gentle aeration to maintain circulation of the cysts during hatching. It is essential that the cysts have adequate aeration or they will not hatch. Hence, only a small quantity of the cysts, as recommended in the accompanying instructions sent by the supplier, should be deposited in the culture container. It is also wise to check carefully that the cysts are not contaminated with pesticides or other potentially toxic substances that may inhibit growth of the planktonic foraminifera. If the foraminifera consistently reject the nauplii, or if there is high foraminiferan mortality, it is advisable to consider finding a different source for the *Artemia* cysts. The supplier may also be able to verify if the cysts are free of pesticides and other potentially toxic contaminants (such as San Francisco Brand). Nauplii are offered once daily as food. With this feeding regime, the planktonic foraminiferan will typically add one new chamber each day. If algal symbionts are present, the numbers increase forming a tawny golden-green halo within the cytoplasm surrounding the shell. When mature, the spinose planktonic foraminifera shed their spines, digest some of the symbionts, eject the remainder and the cytoplasm becomes milky white. Within a day or two, a massive cloud of biflagellated gametes is released into the water. The total mass of cytoplasm is consumed in this process and only the empty shell remains.

Thus far, it has not been possible to induce the gametes to fuse and produce a second generation. Hence, we are not able to grow planktonic foraminifera in continuous culture in the laboratory. However, small individuals collected from the natural environment can be grown in maintenance cultures until reproductive maturity. During the course of their maturation, they usually add several chambers of increasing size. Their developmental stages can clearly be observed by close inspection with light optics.

Additional detailed information on observation protocols and specific requirements for maintaining a range of planktonic foraminiferan species can be found in Bé et al. (1977) and Hemleben et al. (1989).

REFERENCES

Anderson, O.R., Spindler, M., Bé, A.W.H. and Hemleben, Ch. (1979). Trophic activity of planktonic foraminifera. *J. Mar. Biol. Assoc. U.K.* **59**:791–799.

Arnold, Z.M. (1954). Culture methods in the study of living Foraminifera. *Paleontol.* **28**:404–416.

Arnold, Z.M. (1966). A laboratory system for maintaining small-volume cultures of Foraminifera and other organisms. *Micropaleontol.* **12**:109–118.

Arnold, Z.M. (1967). Utilisation des antibiotiques dans la réalisation des cultures de foraminifères sous faible volume. II. Application a la technique des cultures des foraminifères. *Vie et Milieu* **18**:36–45.

Arnold, Z.M. (1974). Field and laboratory techniques for the study of living foraminifera. In *Foraminifera* (eds R.H. Hedley and C.G. Adams), Vol. 1, pp. 153–206. Academic Press, London.

Bé, A.W.H. (1959). Ecology of recent planktonic foraminifera. Part 1. Areal distribution in the western North Atlantic. *Micropaleontol.* **5**:77–100.

Bé, A.W.H. (1960). Ecology of recent planktonic foraminifera. Part 2. Bathymetric and seasonal distributions in the Sargasso Sea off Bermuda. *Micropaleontol.* **6**:373–392.

Bé, A.W.H. (1977). An ecological, zoogeographic and taxonomic review of recent planktonic foraminifera. In *Oceanic Micropaleontology* (ed. A.T.S. Ramsay), pp. 1–100. Academic Press, London.

Bé, A.W.H., Hemleben, Ch., Anderson, O.R., Spindler, M., Hacunda, J. and Tuntivate-Choy, S. (1977). Laboratory and field observations of living planktonic foraminifera. *Micropaleontol.* **23**:155–179.

Bradshaw, J.S. (1955). Preliminary laboratory experiments on ecology of foraminiferal populations. *Micropaleontol.* **1**:351–358.

Bradshaw, J.S. (1957). Laboratory studies on the rate of growth of the foraminifer '*Streblus beccarii* (Linné) var. *tepida* (Cushman)'. *J. Paleontol.* **31**:1138–1147.

Buzas, M.A. (1965). The distribution and abundance of Foraminifera in Long Island Sound. *Smithson. misc. Cllns* **149**:1–89.

Faber, W.W., Jr and Lee, J.J. (1991). Feeding and growth of *Peneroplis planatus* (Fichel and Moll) Montfort. *Symbiosis* **10**:63–82.

Føyn, B. (1936). Foraminiferenstudien I. Der Lebenszyklus von *Discorbina vilardeboana* d'Orbigny. *Bergens Mus. Arbok. Naturv. Rek.* **2**:1–22.

Galtsoff, P.S. (1937). General methods of collecting, maintaining, and rearing marine invertebrates in the laboratory. In *Culture Methods for Invertebrate Animals* (eds P.S. Galtsoff, F.E. Lutz, P.S. Welch and J.G. Needham), pp. 5–40. Dover, New York.

Grell, K.G. (1954). Der Generationswechsel der Polythalamen Foraminifera *Rotaliella heterocaryotica. Archiv der Protistk.* **100**:211–235.

Grell, K.G. (1957). Untersuchengen über die Fortpflanzung und sexualität der Foraminiferen I. *Rotaliella roscoffensis. Arch. Protistk.* **102**:147–164.

Grell, K.G. (1958). Untersuchengen über die Fortpflanzung und sexualität der Foraminiferen III. *Glabratella sulcata. Arch. Protistk.* **102**:449–472.

Grell, K.G. (1959). Untersuchengen über die Fortpflanzung und sexualität der Foraminiferen IV. *Patellina corrugata. Arch. Protistk.* **104**:211–235.

Grell, K.G. (1973). *Protozoology*, 554 pp. Springer-Verlag, Berlin.

Guillard, R.R.L. (1975). Culture of phytoplankton for feeding marine invertebrates. In *Culture of Marine Invertebrate Animals* (eds W.L. Smith and M.H. Chanley), pp. 29–60. Plenum Press, New York.

Hedley, R.H. (1958). A contribution to the biology and cytology of *Haliphysema* (Foraminifera). *Proc. Zool. Soc. Lond.* **130**:569–576.

Hedley, R.H. (1964). The Biology of Foraminifera. In *Internat. Rev. Gen. Exper. Zool.* I. (Eds W.J. Felts and R.J. Harrison) pp 1–45. Academic Press, London.

Hedley, R.H. and Wakefield, J.St.J. (1967). Clone culture studies of a new rosalinid foraminifer from Plymouth, England and Wellington, New Zealand. *J. Mar. Biol. Assoc. U.K.* **47**:121–128.

Hemleben, Ch., Spindler, M. and Anderson, O.R. (1989). *Modern Planktonic Foraminifera.* Springer-Verlag, New York.

Jepps, M.W. (1926). Contributions to the study of *Gromia oviformis* Dujardin. *Quart. J. Micro. Sci.* **70**:(5), 701–720.

Jepps, M.W. (1942). Studies on *Polystomella* Lamarck (Foraminifera). *J. Mar. Biol. Assoc. U.K.* **25**:612–665.

LaGarde, E. (1967). Utilisation des antibiotiques dans le réalisation des cultures de Foraminifères sous faible volume. I. Étude de l'action des antibiotiques sur les microflores hétérotrophes marines. *Vie et Milieu* **18**:27–35.

Le Calvez, J. (1938). Recherche sur les foraminifères. *Arch. Zool. Exp. Gén.* **80**:163–333.

Lee, J.J. (1974). Towards understanding the niche of foraminifera. In *Foraminifera* (eds R.H. Hedley and G. Adams), Vol. 1, pp. 208–257. Academic Press, London.

Lee, J.J. (1980). Nutrition and physiology of the foraminifera. In *Biochemistry and Physiology of Protozoa* (eds M. Levandowsky and S. Hutner), Vol. 3, pp. 43–66. Academic Press, London.

Lee, J.J. (1983). Perspective on algal endosymbionts in larger foraminifera. *Int. Rev. Cytol. Suppl.* **11**:49–77.

Lee, J.J. and Lee, R.E. (1990). Chloroplast retention in elphidids (Foraminifera). In *Endocytobiology IV* (eds P. Nardon, V. Gianinazzi-Pearson, A.M. Grenier, L. Margulis and D.C. Smith), pp. 215–220. Institut National de la Recherche Agronomique, Paris.

Lee, J.J. and Muller, W.A. (1974). Trophic dynamics and niches of salt marsh foraminifera. *Amer. Zool.* **13**:215–223.

Lee, J.J. and Pierce, S. (1963). Growth and physiology of foraminifera in the laboratory: Part 4: Monoalgal culture of an allogromiid with notes on its morphology. *J. Protozool.* **10**:404–411.

Lee, J.J., Pierce, S., Tentchoff, M. and McLaughlin, J.J.A. (1961). Growth and physiology of foraminifera in the laboratory: part 1 – collection and maintenance. *Micropaleontol.* **7**:461–466.

Lee, J.J., McEnery, M., Pierce, S., Freudenthal, H.D. and Muller, W.A. (1966). Tracer experiments in feeding littoral foraminifera. *J. Protozool.* **13**:659–670.

Lee, J.J., Muller, W.A., Stone, R.J., McEnery, M.E. and Zucker, W. (1969). Standing crop of Foraminifera in sublittoral epiphytic communities of a Long Island salt marsh. *Mar. Biol.* **4**:44–61.

Lee, J.J., Tietjen, J.H., Stone, R.J., Muller, W.A., Rullman, J. and McEnery, M. (1970). The cultivation and physiological ecology of members of salt marsh epiphytic communities. *Helgolander Wiss. Meeres.* **20**:136–156.

Lee, J.J., McEnery, M.E., Kennedy, E.M. and Rubin, H. (1975). A nutritional analysis of a sublittoral epiphytic diatom assemblage from a Long Island salt marsh. *J. Phycol.* **11**:14–49.

Lee, J.J., McEnery, M.E. and Garrison, J.R. (1980a). Experimental studies of larger foraminifera and their symbionts from the Gulf of Elat on the Red Sea. *J. Foram. Res.* **10**:31–47.

Lee, J.J., McEnery, M. and Reimer, C.W. (1980b). The isolation, culture and identification of endosymbiotic diatoms from *Heterostegina depressa* d'Orbigny and *Amphistegina lessonii* d'Orbigny (larger Foraminifera) from Hawaii. *Bot. Mar.* **23**:297–302.

Lee, J.J., Reimer, C.W. and McEnery, M.E. (1980c). The identification of diatoms isolated as endosymbionts from larger foraminifera from the Gulf of Elat (Red Sea) and the description of 2 new species, *Fragilaria shiloi* sp. nov. and *Navicula reissii* sp. nov. *Bot. Mar.* **23**:41–48.

Lee, J.J., Erez, J., ter Kuile, B., Lagziel, A. and Burgos, S. (1988). Feeding rates of two species of larger foraminifera *Amphistegina lobifera* and *Amphisorus hemprichii*, from the Gulf of Eilat (Red Sea). *Symbiosis* **5**:61–102.

Lee, J.J., McEnery, M.E., ter Kuile, B., Erez, J., Röttger, R., Rockwell, R.F., Faber, W.W., Jr and Lagziel, A. (1989). Identification and distribution of endosymbiotic diatoms in larger foraminifera. *Micropalteontol.* **35**:353–366.

Lee, J.J., Sang, K., ter Kuile, B., Strauss, E., Lee, P.J. and Faber, W.W., Jr (in press). Nutritional and related experiments aimed at the laboratory maintenance of three species of symbiont-bearing larger foraminifera. *Mar. Biol.*

Lister, J.J. (1895). Contributions to the life history of the foraminifera. *Phil. Trans. R. Soc.* **B186**:401–453.

Lynts, G.W. (1966). Variation of foraminiferal standing crop over short lateral distances in Buttonwood Sound, Florida Bay. *Limnol. Oceanogr.* **11**:562–566.

Matera, N.J. and Lee, J.J. (1972). Environmental factors affecting the standing crop of foraminifera in sublittoral and psammonlittoral communities of a Long Island salt marsh. *J. Mar. Biol.* **14**:89–103.

Muller, W.A. (1975). Competition for food and other niche-related studies of three species of salt marsh foraminifera. *Mar. Biol.* **31**:339–351.

Muller, W.A. and Lee, J.J. (1969). Apparent indispensibility of bacteria in foraminiferan nutrition. *J. Protozool.* **16**:471–478.

Murray, J.W. (1963). Ecological experiments on Foraminiferida. *J. Mar. Biol. Assoc. U.K.* **43**:621–642.

Murray, J.W. (1973). *Distribution and Ecology of Living Benthic Foraminiferids.* Heinemann, London.

Myers, E.H. (1935). Culture methods for the marine foraminifera of the littoral zone. *Trans. Am. Microsc. Soc.* **54**:264–267.

Myers, E.H. (1936). The life-cycle of *Spirillina vivipara* Ehrenberg, with notes on morphogenesis, systematics and distribution of the foraminifera. *J. R. Micr. Soc.* **56**:120–146.

Myers, E.H. (1937). Culture methods for marine foraminifera of the littoral zone. In *Culture Methods for Invertebrate Animals* (eds P.S. Galtsoff, F.E. Lutz, P.S. Welch and J.G. Needham), pp. 93–96. Dover, New York.

Myers, E.H. (1942). Biology of the foraminifera and their significance in paleoecology. *N.Y. Acad. Sci. Trans. Ser. 2* **50**:191–194.

Myers, E.H. (1943). Life activities of foraminifera in relation to marine ecology. *Proc. Am. Phil. Soc.* **86**:439–459.

Nyholm, K.G. (1961). Morphogenesis and biology of the foraminifer *Cibicides lobatulus*. *Zool. Bidr. Uppsala* **33**:157–192.

Röttger, R., Irwan, A., Schmajohan, R. and Franzisket, L. (1980). Growth of the symbiont-bearing foraminifera *Amphistegina lessonii* d'Orbigny and *Heterostegina depressa* d'Orbigny (Protozoa). In *Endocytobiology* (eds W. Schwemmler and H.E.A. Schenk), Vol. 1, pp. 125–132. Walter de Gruyter, Berlin.

Sandon, H. (1932). The food of protozoa. *Publ. Fac. Sci. Egypt. Univ.* **1**:1–187.

Schwab (1976). Gametogenesis in *Allogromia laticollaris*. *J. Foram. Res.* **6**:251–257.

Sliter, W.V. (1970). *Bolivina doniezi* Cushman & Wickenden in clone culture. *Contr. Cushman Fdn Foram. Res.* **21**:87–99.

Spindler, M., Hemleben, Ch., Salomons, J.B. and Smit, L.P. (1984). Feeding behavior of some planktonic foraminifers in laboratory cultures. *J. Foram. Res.* **14**:237–249.

Weber, H. (1965). Über die paarung der Gamonten und den Kerndualismus der Foraminifere *Metarotaliella parva* Grell. *Arch. Protistk.* **108**:217–270.

Index

Abbe refractometer, 343
Abiotic factors, 232, 277–278
Abyssal, 240–242
Abyssotherma pacifica, 32
Acervulina sp., 240
Achnanthes spp., 183, 188
  A. haukiana, 201
  A. maceneryae, 185, 188, 193
Acidosomes, 22
Acid phosphatase, 172
Actin, 29
  filaments, 123–125
Adercotryma sp., 238–239, 240
Agamont, 290, 291
Agamont II, 290, 291
Agamont III, 290, 291
Agamogony, 291
Agglutinated tests, 31, 35
Alabaminella spp., 227
Algae, 29–31
Algal symbiont-bearing foraminifera, calcium/carbon cycling, 73–89
Algal symbionts, 29–31
Algal symbiosis, 44–45
  mathematical models, 49
Allogromia spp., 11, 96, 98, 99, 101, 104–7, 110, 113, 114, 115, 116, 117, 118, 119, 120, 121, 122, 123, 125, 126, 127–129, 130, 132, 133, 135, 137, 193, 232, 310, 324, 325, 336
  A. laticollaris, 93, 94, 95, 96, 100, 107, 109, 110, 111, 112, 121, 125, 138, 289, 290, 291, 293, 294, 298, 302, 303, 310–312, 325, 336
Alpha (α) index, 227
Ammobaculites spp., 226, 234, 236–237
Ammonia spp., 224, 233, 234, 236, 237, 238
  A. beccarii, 310, 336, 341
Ammonium chloride, 27
Ammotium spp., 236

Amoeba proteus, 116, 123
Amphidinium sp., 181, 182, 184, 185
Amphidinium carterae, 181
Amphidinium rhychocephalum, 181
Amphisorus spp., 10, 44, 49, 95, 96, 97, 112, 125, 177
  A. hemprichii, 14, 75, 77, 78, 81, 82, 83, 85, 86, 114, 118, 159, 165, 172, 175, 177, 181, 182, 184, 185, 186, 187, 193, 196, 197, 199, 202, 203, 204, 205, 206, 306, 307, 336, 342
Amphistegina spp., 44, 46, 49, 60, 66, 96, 112, 162, 165, 166, 168, 175, 176, 177, 192, 342
  A. gibbosa, 47, 168, 196
  A. lessonii, 47, 66, 75, 162, 168, 178, 179, 183, 185, 187, 188, 193, 196, 198, 206, 306
  A. lobifera, 47, 66, 75, 76, 77, 78, 81, 82, 83, 84, 85, 86, 159, 166, 168, 169, 183, 187, 188, 193, 196, 197, 198, 199, 202, 203, 205, 206, 306, 336
  A. papillosa, 47
Amphora spp., 183, 185, 188, 193, 196, 344
  A. bigibba, 202
  A. coffeaeformis, 201
  A. erezii, 188
  A. roettgerii, 183, 187, 188
  A. tenerrima, 187, 188, 190, 191, 206, 212
Androsina lucasi, 162, 165, 168, 169, 175
Anlagen, 75
Annulate lamellae, 25–26
Antarctic bottom water, 249
Apogamic life cycle, 310–312
Archaias sp., 47, 247
  A. angulatus, 82, 86, 162, 175, 189, 190, 192, 196, 198, 199, 201, 202, 205, 308, 323

360 Index

Arenoparrella spp., 233
Artemia spp., 353
  A. salina, 200
Assembly regulators, 128
Astrammina spp., 96, 104, 112, 114, 118, 120
  A. rara, 100, 102, 107, 123, 127, 144
ATP, 24, 76, 115, 131
ATPase, 130
Aufwuchs spp., 342
Autogamy, 302–305
Autonomous motility, 110–112
Axial microtubule movements, 131–135
'Axopods', 93

Bacteria, 49, 50, 116
Baculogypsina sp., 33
  B. sphaerulata, 163, 166, 172, 173, 192
Batch maintenance, 342–346
  fortuitous/agnotobiotic, 343
  fortuitous/partially gnotobiotic, 344
  inductive methods, 345–346
Bathyal, 240–242
Behaviour, 193–195
Bending, *see* Reticulopod movements
Benthic foraminifera, 8–29
  abiotic factors, 232
  abyssal, 240–242
  bathyal, 240–242
  biomass, 225
  biotic factors, 232–233
  carrying capacity, 230
  collection and handling, 338–341
  dead assemblages, 243
  density, 225
  distribution, 223–224
  diversity, 227–229
  dominance, 227–229
  ecology, 221–253
  ecophenotypes, 244
  environmental characteristics, 233–243
  estuaries, 233–235
  feeding strategies, 246–247
  growth, 245–246
  laboratory maintenance of, 341–346
  lagoons, 233–235
  larger foraminifera, 242–243
  life length, 248
  mangals, 233
  marshes, 233
  morphogroups, 244–245
  ontogeny, 245–246
  palaeoecological studies, 248–249
  patchiness, 225–226
  population dynamics, 229–232
  processing of, 339–341
  production, 231
  sampling, 223
  seasonality, 227
  shelf seas, 235–240
  stable isotopes, 246
  substrate, 224–225
  tests, 243–246
Bicarbonate, 79
  splitting theory, 74–75
Bigenerium sp., 240
Billiella adamsi, 260
Binary fission, 291
Biomass, 225
Biomineralizing process, 1
Biotic factors, 232–233, 278
Boderia turneri, 298, 310, 336
Bolivina spp., 25, 238, 239, 240
  B. argentea, 25
  B. doniezi, 299, 303, 336
  B. vaughni, 341
Bolliella adamsi, 260
Borelis schlumbergerii, 86, 165, 166, 183, 185, 188
Brizalina, 238–240
Brood chamber, 291–306
'Bubble chambers', 93
Buccella spp., 44, 240
Budding, 291
Bulimina spp., 238–240
Buliminella spp., 44, 240
  B. tenuata, 24, 25
Bulk streaming, 112–113
Buoyancy, 26–27, 42, 273–274

Calcarina spp., 58, 163
  C. calcar, 163, 166, 168, 169, 183, 185, 188, 192
  C. gaudichaudii, 52, 56, 58, 59, 306, 308
  C. hispida, 57, 163

Index   361

Calcarinids, 58–59, 158
Calcification, 46, 73–89, 198–199
　Amphisorus hemprichii, 81–82
　Amphistegina lobifera, 78–81
　bicarbonate splitting theory, 74–75
　carbon budget, 82–86
　energy theory of, 75–76
　organic matrix theory of, 75
　poison removal theory, 76–77
　schemes for, 77–82
　theories of, 74–77
Calcite compensation depth (CCD), 242, 249
Calcite needles, 35
Calcite shells, 31, 35
Calcium, 27–28, 31, 35, 44, 46, 76
Calcium-ATP, 76
Calcituba polymorpha, 299, 336
Canal systems, 52–59
　calcarinid, 58–59
　excretion, 55–56
　nummulitids, 52
Cancris spp., 238–240, 249
Candeina nitida, 164, 267
Carbon, 198–199
Carbon budgets, 82–86
　Amphisorus hemprichii, 83
　Amphistegina lobifera, 82–83
　other species, 83–86
Carbon cycling, 73–89
Carbon dioxide, 46, 62, 73–76, 78, 79, 80
Carbon flow, 203–207
Carbonic anhydrase, 80–82
Carrying capacity, 231
Cassidulina sp., 224, 238–240
Cassidulinoides cornuta, 25
Catalase, 22
Cellular organelles, 15–29
　annulate lamellae, 25
　cytoskeletal structures, 28–29
　digestive vacuoles, 20–22
　endoplasmic reticulum, 18–19
　fibrillar bodies, 26–28
　Golgi body, 19–20
　lysosomes, 21–22
　microbodies, 22–23
　mitochondria, 24–25
　nucleus, 14, 15, 18
　peroxisomes, 22–23
　ribosomes, 18–19
　sexual reproduction and, 25–26

Cell wall, 3
Chaos carolinensis, 123
Chlamydomonas hedleyi, 110, 160, 175–178, 179, 189, 201, 205, 323
Chlamydomonas provasolii, 162, 175–179, 190, 204, 205
Chlorella, 177, 202, 346
　C. marina, 177
　C. nana, 177
　C. saccharophila, 177
　C. salina, 177
　C. spaercki, 177
　C. stigmatophora, 177
Chlorophytes, 175–177
Chloroplast husbandry, sepucstration, retention, 207–212
Chrysophytes, 182–189
Cibicides spp., 236, 240, 249
Classification, 7–8
Cladophora socialis, 196
CLIMAP, 278, 279
Clumped distribution, 225–226
Coated vesicles, 117
Cocconeis spp., 170, 183, 187
　C. andersonii, 184, 185, 188
　C. placentula, 196, 201, 202, 212
Coiling mode, 42–43
Coiling ratios, 269–270
Collection, 338–339, 346–348
　net, 347
　SCUBA, 348
　sites of, 347
Compartmentalization, of cytoplasm, 9–18
Contractile proteins, 28–29
Cornuspira sp., 240
Cribrostomoides sp., 238–240
Crosslinkers, 128
Crosslinks, microtubular, 133
Crushing, resistance to, 64
Crystalline inclusions, 22
Culture, see maintenance
Cyclammina sp., 47
Cyclopeus carpenteri, 163, 192
Cyclorbiculina compressa, 86, 162, 175, 190, 192, 199, 203, 204, 205, 306, 308, 310
Cylindrotheca spp., 344
Cytochalasin D, 135, 137
Cytology, 7–40

Cytoplasm, 9–14, 48, 56–57
  and cytoskeleton, 28–29
  diffuse, 9
  sexual reproduction, 25–26
  transitional, 9
  zonal, 9
  gelation, 2
  fibrils, 107–110
  and cytoplasmic transport, 107–109
  dynamics, 109–110
  streaming, 9, 12, 21–22, 28–29, 102–106, 194
  reticulopod contents, 139–141
  transport, 107–109
Cytoskeletal structures, 28–29
Cytotomy, 291

DCMU, 75–76, 80, 81, 185, 187, 194, 199
Dead assemblages, 243
Defecation vacuoles, 116
Density, 26–27
  benthic foraminifera, 225
Depth distribution, of planktonic foraminifera, 264–265
Detritovore, 234
Diatoms, 182–189
Diet, *see* Nutrition
Difflugia spp., 123
Diffuse cytoplasm, 9
Digestion, 14, 116
Digestive vacuoles, 20–22
Dinoflagellates, 177–182
Disassembly regulators, 128
Discorbis spp., 240, 249, 297
  D. bertheoti, 297
  D. vilardeboanus, 297, 299
Discorinopsis agvagoi, 310, 336
Distribution, of planktonic foraminifera, 259–262, 264–265
Disturbance, of planktonic foraminifera, 276–277
Diversity, 227–229
DNA, 15, 18, 145
Dominance, 227–229
Dormancy, 62
Dunaliella spp., 344
  D. salina, 201
Dynein, 128, 130, 134

Ecology, 2
  and benthic foraminifera, 221–253
  and planktonic foraminifera, 255–284
Ecophenotypes, 244
Eggerella sp., 238–240
Elastic processes, 57–58
Elliptical vesicles, 117
Elphidella hanni, 63
Elphidium spp., 95, 112, 116, 172, 173, 210, 224, 227, 233, 234, 236–238, 244
  E. clavatum, 339
  E. crispum, 207, 212, 286, 293–295, 336
  E. excavatum, 207, 209
  E. insertum, 207, 212, 295, 336, 339
  E. translucens, 207, 209, 212, 295, 336, 339
  E. williamsoni, 207, 209, 210
Emiliana huxleyi, 200, 352
Endoplasmic reticulum, 18–19
  RER, 18–19
  SER, 18–19
Endosymbionts, 3
Endosymbiotic algae, 162–164, 174–189
  chlorophytes, 162, 175–177
  chrysophytes, 163, 164, 182–189
  diatoms, 162, 163, 182–189
  dinoflagellates, 162, 163, 177–182
  rhodophytes, 162, 182
Energy, 48, 131, 134
  calcification, 75–76
  requirements, 24–25
Entomoneis spp., 163, 188, 202, 212
Environmental types, 233–243
  abyssal, 240–241
  bathyal, 240–242
  estuaries, 233–235
  lagoons, 233–235
  mangals, 233
  marshes, 233
  shelf seas, 235–240
Epistominella spp., 227, 236, 237, 241, 242
Eponides spp., 227
Estuaries, 233–235
Environmentally controlled maintenance systems, 350–352
Enzymes, 22

Ethoxyzolamide, 80
Excretion, 55–56
Extension, *see* Reticulopod movements
Extracellular surface transport, 102–106
Extrashell cytoplasm, 9–14

F-actin, 124, 125
Fasciolites spp., 249
Feeding strategies, 42, 246–247
Feeding systems, 352–354
Fibrillar bodies, 26–28
Fibrils, *see* Cytoplasmic fibrils
Filopods, 108–110, 119
　bending, 136–137
　extension/withdrawal, 136
Flagellated reproductive cells, 26
Flow-through systems, 341–342
Fontbotia spp., 236, 237, 241
Food, 14, 16
Foraminifera –
　benthic, *see under* Benthic foraminifera
　benthic vs. planktonic, 8–29
　concept of, 7–8
　planktonic, *see under* Planktonic foraminifera
　see also under specific aspects of
Form-function relations, 41–72
Fossils, 3
Fragilariz shiloi, 78, 162, 178, 179, 183, 185–191, 193, 206–209
Fursenkoina spp., 226

Gamete production, planktonic foraminifera, 313–316
Gametes, 56, 289, 314–315
Gametogamy, 289, 294–297
Gamogony, 291
Gamont, 290
Gamont II, 290
Gamontogamy, 290, 297–302
Gaudryina sp., 240
Gavelinopsis spp., 238–240
Generative nuclei, 291
Genetics, 3, 15, 18
Glabratella mediterranensis, 297, 302, 310

Glabratella ornatissima, 297
Glabratella patelliformis, 297, 301
Glabratella sulcata, 293, 297, 301, 337
Globigerina bulloides, 93, 163, 256, 257, 260, 264, 266, 272, 278
Globigerina cristata, 163
Globigerinella aequilateralis, 163, 180–182, 197, 200, 260, 266, 267, 313, 316
Globigerinella falconensis, 200
Globigerinita glutinata, 183, 200, 260, 264, 266, 267, 275, 313
Globigerinita juvenilis, 267
Globigerinoides spp., 10
　G. conglobatus, 179, 193, 200, 260, 264, 267, 313
　G. ruber, 35, 163, 177, 193, 198, 200, 256, 257, 260, 263, 264, 266, 267, 272–276, 313, 316, 326, 327
　G. sacculifer, 12, 33, 35, 163, 179, 193, 194, 197, 200, 201, 256–260, 263, 264, 266, 267, 268, 270, 271, 273, 274, 275, 276,, 278, 299, 313–316, 319, 337
Globobulina pacifica, 25
Globoquadrina dutertrei, 200
Globorotalia hirsuta, 200, 256, 257, 260, 265, 266, 268, 273, 275
　G. inflata, 164, 200, 256–257, 260, 266, 275, 313
　G. menardii, 164, 200, 256, 257, 260, 266, 267, 275, 313
　G. theyeri, 256, 257, 260, 264, 266
　G. truncatulinoides, 200, 265, 266, 268, 272, 273, 275, 278, 337
　G. tumida, 256–257, 260
Glycocalyx, 114
Glyoxisomes, 22–23
Golgi bodies, 19–20
　contents of, 20
'Grip and tug' model, 143–144
Growth, 245–246
Gymnodinium spp., 200
　G. beii, 163, 181, 189

Habitats, see environmental types
Halamphora spp., 196
Halimeda tuna, 80
Haliphysema tumanowiceii, 298, 310, 337

Halophila stipulacea, 192, 201
Hanzawaia sp., 224
Hastigerina pelagica, 10, 25, 26, 33, 35, 37, 116, 158, 164, 179, 200, 260, 264, 272, 274, 275, 276, 293, 313, 315, 316, 337, 347
Haynesina spp., 104, 105, 227, 233, 234, 236, 237
  H. germanica, 207–209, 210, 212
Helical filaments, 121–123
Herbivorous feeding, 163, 247
Heterocycline tuberculata, 55, 192
Heterostegina spp., 44, 49, 52, 174, 177, 200
  H. depressa, 52, 53, 54, 55, 57, 58, 59, 83, 86, 163, 165, 183, 188, 191, 192, 195, 206, 209, 305, 306–309, 310
Heterokaryotic nuclei, 14, 292
Heterotheca lobata, 298
Hill-Whittingham equation, 78, 81
Host-symbiont associations, 1
Hyaline walls, 61–62
Hydrodynamics, and shape, 46–47
Hydrogenosomes, 24
Hydrogen peroxide, 22
Hydroxyl ion, 79
Hypotheses, on form-function, 41–72

Imperforate tests, 62
Inner organic lining, 65
Internal morphology and structural complexity, 47–49
Intracellular organelle streaming, 102–106
Intrashell cytoplasm, 9–14
Intrathalamous movements, 112–113
Iridia spp., 107, 114
  I. lucida, 297, 298, 337
Islandiella sp., 240
Isotope tracing, 1
Isotopes, stable, 269–273

Jadammina spp., 233, 236, 237

Karyokinesis, 292
Kinesin, 314
Kummerform chamber, 273

Lagoons, 233–235
Lamellipods, 95, 103, 119
Larger foraminifera, 42–59, 74–85, 242–243
Lateral microtubule movements, 135–136
Lepidodeuterammina sp., 224
Lepidotrochammina spp., 224
Life cycles, 3, 285–334
  apogamic, 293, 294, 310–312
  autogamy, 289, 302–305
  classical, 284–289, 292
  gametogamy, 294–297
  gamic, 293
  gamontogamy, 297–302
  metagenic, 293
  paraclassical, 293–297
  paratrimorphic, 287–289, 293, 305–309
  programme, 292–294
  terminology for, 289–292
  trimorphic, 288, 293, 305–309
Life history strategies, 45
Life length, 248, 276
Light, 195–198
Lipoproteins, 117
Locomotion, 95–98, 143–144
Loxostomum pseudobeyrichi, 25
Lysosomes, 20–22, 116

Macrogametes, 26
Magnesium calcium carbonate, 35
Maintenance, of collections, 341–346, 349–352
Mangals, 233
Marginal cords, 174
Marginopora spp., 49, 249
  M. kudakajimensis, 162, 172, 173, 177, 179, 182, 342
  M. vertebralis, 162, 165, 177, 198, 306, 308
Marine aquaria, 341–342
Marshes, 233
Mastogloea spp., 186, 187, 201
Mechanochemical transducers, 127
Mechanoenzymes, 127, 140
Meiocytotomont, 238, 239, 290
Melonis spp., 242
Membrane-bound organelles, 115–118

Index 365

Membrane domain transport, 139–141
Membrane surface markers, 105–106
Metarotaliella parva, 210, 287, 297, 301, 337
Michaelis-menten kinetics, 79
Microbodies, 22–23
Microfilaments, 28–29
Microtubules, 118–121
 and actin, 123–125
 associated proteins, 127–128
 axial, 131–135
 crosslinks, 133
 density, 118
 lateral movements, 135–136
 sliding, 128–130
 zipping and unzipping, 135–136
Miliammina spp., 233, 234, 236–237
Miliola spp., 107
Mineralized tests, 31, 35
 agglutinated, 31, 35
 calcite needles, 35
 calcite shells, 35
Mitochondria, 24–25
Morphogroups, 244–245
Morphology, 8–9, 41–42, 47–49, 60–63, 244, 245
 and symbionts, 167–174
Motility, 91–155
 autonomous, 110–112
 cytoplasmic fibrils, 107–110
 extracellular surface transport, 102–107
 mechanism of, 130–144
 network morphogenesis, 100–101
 network withdrawal, 101–102, 141–143
 organizing vesicles, 117
 pseudopodal tension, 143–144
 reticulopod, 130–139
 reticulopod movements, 99–100
 streaming, 102–107, 139–141
mRNA, 18
Myosin, 29
Myxotheca arenilega, 112, 297, 298, 337

Navicula spp., 183, 188, 202
 N. hanseniana, 188, 206
 N. menisculus, 212

N. muscatinei, 183, 188, 212
N. reissii, 188, 190, 191
Nemogullmia longevariabilis, 298
Neoconorbina asicularis, 310, 313
 N. Orbicularis, 299
Neogloboquadrina spp., 256, 257, 260, 262, 264–266, 269, 272, 275, 278, 313
 N. dutertrei, 164, 260, 264, 266, 267, 272, 275, 313
 N. hexagona, 260
Net collection, 347–349
Network morphogenesis, 100–101, 136–138
Network withdrawal, 101–102, 141–143
Nexin links, 136
Nitrate, 191
Nitzschia spp., 183
 N. acicularis, 201
 N. frustulum, 183, 185, 187, 188, 190, 193, 206, 209
 N. frustulum var *symbiotica*, 162, 163, 183–185, 187, 188, 193, 206, 208, 209
 N. laevis, 183–185, 187, 188, 190, 191, 193, 206
 N. panduriformis, 179, 183, 185–188, 190, 191, 193, 206
 N. subcommunis, 212
 N. valdestriata, 184, 185, 187, 188, 190, 191, 192, 206
Nonion depressulus, 230, 231, 248, 249
Nonionella stella, 25, 240
Nonionoides sp., 238, 240
Nubercularia sp., 224, 240
Nucleoli, 15
Nucleus, 14, 15, 18
 generative, 14, *see also* heterokaryotic nuclei
 somatic, 14, *see also* heterokaryotic nuclei
Nummulites cumingii, 163
Nummulitids, 52, 55
Nutrients, absorption of, 49–51
 regeneration, 49
Nutrition, 14, 20–22, 45, 190, 199–203, 246–247, 274–275
 planktonic foraminifera, 352–354
Nuttallides sp., 236–237, 249

366 Index

Ontogenesis, 316–319
　stages, 317–319
Ontogeny, 245–246
Oolina marginata, 299, 310
Operculina sp., 33, 47, 174
　O. ammonoides, 55, 86, 163, 166, 183, 185, 188, 192
Orbitolites sp., 249
Orbulina universa, 27, 163, 179, 181, 189, 193, 195, 198, 201, 256, 257, 260, 264, 266, 267, 270–273, 275, 313, 319, 337
Organelle streaming, 102–106
Organelles, *see* Cellular organelles
Organic carbon, 45, 49
Organic debris, 26
Organic layers, 34–35
Organic matrix theory, 75
Organismic-environmental interactions, 2

Paracrystals, 121–123
Pararotalia sp., 224
Paratrimorphic life cycle, 305–309
Particulate organic carbon, 45
Patchiness, 262–263
　and distribution, 225–226
Patellina corrugata, 25, 240, 288, 297, 299, 301, 302, 337
Peneroplis spp., 182, 234, 236, 237
　P. acicularis, 174, 182
　P. arietina, 162, 165, 206
　P. pertusus, 174, 182, 204–206, 299, 305, 306
　P. planatus, 162, 168, 169, 172, 173, 196, 201, 202, 203, 306, 307, 309, 323, 337
　P. proteus, 162, 175
Perforate vs. imperforate tests, 62
Peroxidases, 22
Peroxisomes, 22–23
pH, 23, 76, 78, 79
Phaeodactylum tricornutum, 191, 201
Phagolysosome, 22, 116
Photocomphensation depth, 192, 195
Phototaxis, 193
Photosynthesis, 78–79
　see also Light

Photosystem II, 75
Poison removal theory, 76, 77
Placopsilina sp., 240
Planar surfaces, 96
Planktonic foraminifera, 8–29, 255–284
　abiotic factors, 277–278
　biogeographic faunal provinces, 261–262
　biotic factors, 278
　buoyancy, 273–274
　coiling ratios, 269–270
　collection of, 346–348
　depth distribution, 264–265
　distribution of, 259–262
　feeding, 274–275
　gamete production, 313
　gametogenesis, 314–315
　life length, 276, 312–319
　maintenance, 349–352
　morphology, 273
　oligotrophic conditions, 275–276
　ontogeny, 267–268, 316–319
　patchiness, 262–264
　processing, 348–349
　reproductive cycles, 315–316
　sampling, 258–259
　sediment trap studies, 265–267
　stable isotopes, 270–273
　symbiosis, 275–276
　test aspects, 267–273
Planorbulina mediterranensis, 238–240, 287, 293, 297, 310
Plasma membrane, 113–115
Plasmotomy, 292
Poison removal theory, 76–77
Polystomella, see Elphidium
Population dynamics, 229–232
Pores, 35
Porosity, 270
Porphyridium spp., 182
　P. purpureum, 162, 174, 178, 179, 182, 203–205
Pressure flow, and streaming, 2
Prey, 21, 199–203, 274–275, 336–337, 352–353 (see also Feeding Strategies)
Primary production, 198–199
Primary Organic Membrane (POM), 34

Processing, 339–341, 348–349
  net samples, 348–349
  SCUBA samples, 349
Production, 231
Protokeelia spp., 183
  P. hottingeri, 185, 188, 193
Pseudopod(ia), 28, 52, 107, 112
  branching, 100–101
  functions of, 55
  fusion, 100–101
  locomotion, 95–99
  morphology, 92–95
  movement of, 99–112
  networks, 92–112
  satellites, 112
  tension, 143–144
  ultrastructure, 113–115
Pseudopores, 48
Pseudorotalia gaimardii, 98
Pulleniatina obliquiloculata, 164, 200, 256, 257, 260, 264, 266, 272, 275, 326, 327
Pyrocystis noctiluca, 179
Pyrocystis robusta, 179

Quinqueloculina spp., 224, 234,236–238, 293, 297, 339
  Q. circularis, 287, 293
  Q. impressa, 98

Radial spokes, 136
Regionization, *see* compartmentalization
Reophax sp., 238–240
Reproduction, 25–26, 29, 31, 56, 248, 315–316
  see also Life cycles
Reticulomyxa spp., 92, 99, 102, 112, 115, 117, 123, 127–133, 135, 137, 140, 142
Reticulopod/Rhizopod, 8–9, 14
  actin, 123–125
  components, 113–130
  cytoskeleton, 118–130
  helical filaments, 121–123
  membrane-bound organelles, 115–118
  microtubules, 118–125
  movements, 99–113

network morphogenesis, 100–101
  network withdrawal, 101–102
  substrate adhesion, 137
  paracrystals, 121–123
  plasma membrane, 113–115
  streaming, 1
  ultrastructure, 113–115
Rhodophytes, 182
Ribosomes, 18–19
RNA, 15, 18, 145
Rosalina spp., 224, 232, 240, 337
  R. floridana, 81
  R. globularis, 299
  R. leei, 160, 296, 297, 299, 310, 320, 321, 337, 341
Rotaliella spp., 288, 302, 320–322
  R. heterocaryotica, 287, 288, 293, 302, 337
  R. roscoffensis, 287, 302, 337
Rough endoplasmic reticulum, 18–19
Rubisco enzyme, 78–79
Rubratella intermedia, 287, 290, 297

Saccammina alba, 238–240, 298, 310, 337
Salicornia spp., 233
Saltation, 102–103
Sampling, 223, 258–259
Satellites, *see* Pseudopodial satellites
Schizogomy, 292
Schizont, 292
SCUBA, 339, 348, 349
Seagrasses, 50
Seasonality, 226–227, 263–264
Secretory activity, 26–28
Sediment trap studies, 265–267
Sexual differentiation, 3
Sexual reproduction, and cytoplasm, 25–26, *see also* gametogamy, gamontogamy, gametes
Shape, 42–43, 46–47
Shelf seas, 235–240
Size, 45–46
Skyllocytosis, 99
Smooth endoplasmic reticulum, 18–19
Somatic nuclei, 292

Sorites marginalis, 14, 82, 86, 162, 170–172, 175, 177–179, 181, 190, 198, 199, 201, 202, 203, 234, 236, 237, 323
Sorites orbiculus, 162, 306, 308, 309
Sorites orbitolitoides, 162
Spartina spp., 233
Sphaeroidinella dehiscens, 156, 157, 264, 268
Spinose form, 9–12
Spirillina vivipara, 288, 297, 298, 301, 337
Spiroloculina spp., 10, 112, 232, 236, 237, 246
   S. hyalina, 105, 114, 299, 310, 337
Sticholonche spp., 123
Streaming, *see* Cytoplasmic streaming
Streblus flavensis, 287
Strength, 48–49, 64–65
Stress, 64
Substrate relations, 224–225
Surface transport, 102–106
Surface/volume ratio, 50
Symbiodinium spp., 162, 178, 179, 181, 182, 184, 185, 190
   S. microadriaticum, 179, 181, 182
   S. pilosum, 182
Symbiosis, 29, 31, 44–45, 73–85, 157–220, 275–276
   adaptations, 159–161
   behavioural studies of, 193–195
   calcification, 198–199
   carbon, 203–207
   chloroplast retention/husbandry, 207–212
   hosts, 161–166
   hypotheses, 160–161
   light and, 191–198
   morphology, 167–174
   nutrition, 189, 199–203
   organization of cells and, 167–174
   primary production and, 73–85, 198–199
   regionalization, 176–174
   released metabolites, 205–206

Terminology, for foraminifera, 289–292
   asexual reproduction, 291–292
   life-cycle programme, 292–294
   sexual reproduction, 289–291
Test function theory, 42, 62, 63
Test shape, 46–47
Tetrahymena spp., 130
Textularia spp., 112, 238–240
Thalassia testudinum, 201
Thalassiosira pseudonana, 200
Tinophodella glutinata, 164
Traction forces, 98–99
Transitional cytoplasm, 9
Transport, 102–105
   extracellular surface, 102–106
Tretomphalus bulloides, 297, 299, 313
Trifarina sp., 238–240
Triloculina sp., 224, 234
Triloculina linneiana, 310
Triloculina rotunda, 297
Trimorphic life cycle, 305–309
Trochammina spp., 224, 233, 298, 303, 304, 310
Tubular inclusions, 22
Tubulins, 124, 125, 127
Turborotalita humilis, 163, 313
Turborotalita quinqueloba, 260, 264

Unzipping, lateral microtubules, 135–136
Uvigerina spp., 224, 238–240, 242

Vitellogenins, 117

Wall structure, 48, 60–62
Withdrawal, *see* Network withdrawal *and also* Reticulopod movements

Zipping, lateral microtubules, 135–136
Zonal cytoplasm, 9